THE FOSECO
FOUNDRYMAN'S HANDBOOK

NINTH EDITION

Pergamon Titles of Related Interest

ASHBY & JONES	Engineering Materials 1 and 2
COUDURIER et al.	Fundamentals of Metallurgical Processes, 2nd Edition
GILCHRIST	Extraction Metallurgy, 3rd Edition
HEARN	Mechanics of Materials, 2nd Edition
INTERNATIONAL INSTITUTE OF WELDING	The Physics of Welding, 2nd Edition
KETTUNEN et al.	Strength of Metals and Alloys (ICSMA8)
LANSDOWN	Lubrication Materials to Resist Wear
MORGAN	Tinplate and Modern Canmaking Technology
NIKU-LARI	Advances in Surface Treatments, Volumes 1–5
SHEWARD	High Temperature Brazing in Controlled Atmospheres
WATERHOUSE & NIKU-LARI	Metal Treatments Against Wear, Corrosion, Fretting and Fatigue

Pergamon Related Journals *(Sample Copy Gladly Sent on Request)*

Acta Metallurgica

Canadian Metallurgical Quarterly

Corrosion Science

Engineering Fracture Mechanics

Fatigue & Fracture of Engineering Materials & Structures

International Journal of Impact Engineering

Journal of the Mechanics & Physics of Solids

Metals Forum

The Physics of Metals & Metallography

Scripta Metallurgica

Welding in the World

THE FOSECO
FOUNDRYMAN'S HANDBOOK

FACTS, FIGURES AND FORMULAE

NINTH EDITION

Revised and Edited by

T. A. Burns

FOSECO (F.S.) LIMITED,
Tamworth, Staffordshire, B78 3TL

PERGAMON PRESS

OXFORD · NEW YORK · BEIJING · FRANKFURT
SÃO PAULO · SYDNEY · TOKYO · TORONTO

U.K.	Pergamon Press plc, Headington Hill Hall, Oxford OX3 0BW, England
U.S.A.	Pergamon Press, Inc., Maxwell House, Fairview Park, Elmsford, New York 10523, U.S.A.
PEOPLE'S REPUBLIC OF CHINA	Pergamon Press, Room 4037, Qianmen Hotel, Beijing, People's Republic of China
FEDERAL REPUBLIC OF GERMANY	Pergamon Press GmbH, Hammerweg 6, D-6242 Kronberg, Federal Republic of Germany
BRAZIL	Pergamon Editora Ltda, Rua Eça de Queiros, 346, CEP 04011, Paraiso, São Paulo, Brazil
AUSTRALIA	Pergamon Press Australia Pty Ltd, P.O. Box 544, Potts Point, N.S.W. 2011, Australia
JAPAN	Pergamon Press, 5th Floor, Matsuoka Central Building, 1-7-1 Nishishinjuku, Shinjuku-ku, Tokyo 160, Japan
CANADA	Pergamon Press Canada Ltd, Suite No. 271, 253 College Street, Toronto, Ontario, Canada M5T 1R5

Copyright © 1986 Foseco (F.S.) Limited

All Rights Reserved. No part of this publication may be reproduced, stored in a retrieval system or transmitted in any form or by any means: electronic, electrostatic, magnetic tape, mechanical, photocopying, recording or otherwise, without permission in writing from the copyright holders.

Eighth revised and enlarged edition 1975
Reprinted 1978, 1979, 1982, 1985
Ninth edition 1986
Reprinted 1989

Library of Congress Cataloging in Publication Data

Main entry under title:
Foundryman's handbook.
Ninth ed. of: The Foseco foundryman's handbook.
8th rev. and enl. ed. 1975.
1. Founding—Handbooks, manuals, etc.
I. Foseco (F.S.) Limited.
TS235.F63   1986   671.2′02′02   85-19196

British Library Cataloguing in Publication Data

Foseco (F.S.) Limited
Foundryman's handbook facts, figures and formulae.
—9th ed.
1. Founding—Handbooks, manuals, etc.
I. Title   II. Foseco International. Foseco
foundryman's handbook.
671.2   TS235

ISBN 0 08 032549 1 Hard cover
ISBN 0 08 033448 2 Flexicover

Printed in Great Britain by BPCC Wheatons Ltd, Exeter

FOREWORD

This reference book has been compiled to help all those concerned with making castings by most of the usual routes except possibly investment casting which uses a rather specialised mould material. Otherwise all the basic metallurgical considerations with regard to metal treatment will still apply. It is hoped the information provided will prove accurate and useful.

Very many important advances in all branches of foundrywork have been made over recent years and this new edition has been enlarged to include as many of the new techniques as possible. Some sections have been amalgamated; for example, those connected with all aspects of the feeding of castings and sand bonding materials. New areas include metal filtration and an extension to the specifications and grades of cast iron now commercially available. Where possible, suggestions and recommendations are made of a practical nature without undue emphasis on the underlying theory. A major effort has been made to standardise on metric or SI units but not all conversions are possible nor are all specifications available in these units.

Suggestions for inclusions, alterations, corrections, etc., in possible future editions would be welcomed, as would constructive criticism or comment. All submissions will be given very careful consideration.

CONTENTS

SECTION I Tables and General Data

The International System of Units (SI) and their relationship to United Kingdom and other units	2
Table of physical properties of metals	10
Corrodibility of some common metals and alloys	12
Conversion tables	18
Approximate weights of materials in pounds per cubic foot	45
Solders and fusible alloys	46
Weight of casting from weight of pattern	47
Shrinkage and contraction of casting alloys	48
Air consumption of various pneumatic tools	50
Hardness conversion chart	51
Seger cones	52
Usual thickness of chills for chilled rolls	53
Method for finding the weight of molten metal in pouring ladles	53
Gating terms and methods	55
Lining systems for pouring ladles and hand shanks	57
Other lining methods	58
Botting practice for cupola operators	59
Botting practice	60
Metal filtration	62

SECTION II Sands and Sand Bonding Systems, Alternative Moulding Systems

Sand test data which have proved satisfactory for various castings in different casting alloys	70
Green sand facing mixtures for iron castings using naturally bonded sands	70
Synthetic sand mixtures for iron castings	71
Sand additives for iron castings	71
Example of routine sand control tests	72
Art castings	74
Moulding sand mixtures for copper-base alloys	74
Moulding and core sand mixtures for steel castings	76
Rammed refractory linings	78
Comparison of standard sieves	79
Comparisons between sieve analysis and index numbers	80
Resin-bonded sand	82
How to calculate sand fineness number	86
Silicate-bonded sand	87

The self-setting processes	100
Alternative moulding systems—the Replicast processes	108

SECTION III Mould and Core Coating

Introduction	114
Water-based dressings	116
Spirit-based dressings	120
Proprietary coating types	123

SECTION IV Light Casting Alloys

Aluminium casting alloys—LM series	126
Aluminium casting alloys—Aerospace series	136
Standard melting and fluxing procedures for aluminium alloys	139
Aluminium silicon alloys	140
Standard test bar for aluminium alloys sand cast	159
Aluminium–magnesium casting alloy BS 1490—LM5	160
Aluminium–magnesium casting alloy BS 1490—LM10	167
Aluminium–zinc–magnesium casting alloy. Frontier 40E alloy	176
Melting, fluxing, grain refining etc., as applied to magnesium base alloys	180
Melting and recovery of aluminium swarf and small scrap	190
Grain refinement of aluminium alloys with titanium and boron	192
Modern methods of fluxing molten aluminium	193

SECTION V Non-Ferrous Casting Alloys

The order of alloying	200
British Standard specifications for the principal non-ferrous casting alloys	201
The importance of furnace atmosphere when melting copper-based alloys	210
Melting, fluxing and degassing procedure (brasses, bronzes and gunmetals)	212
Melting and fluxing procedure for high conductivity and commercial copper castings	215
Commercial copper castings	220
Moulding procedure for high conductivity and commercial copper castings	221
Copper–zinc alloys—brasses	223
Copper alloys—bronzes and gunmetals	230
Casting of cored and solid bronze sticks	238

Test bars suitable for use with bronzes and gunmetals	241
The recovery of non-ferrous swarfs, scraps etc.	242
Pickling of brass and gunmetal castings to improve appearance	243
Recommended standard melting and fluxing procedure for non-ferrous metals and alloys	244

SECTION VI   Iron Castings

Composition of typical cupola charge materials	256
Influence of normal constituents in cast iron	260
Cupola operation data	263
Influence of some alloying elements used in the production of cast iron upon its structure	265
Cupola charge calculations	267
Curtailment of sulphur pick-up during cupola melting of iron charges	269
Chemical composition of unalloyed irons suitable for different classes of castings	272
British Standard specifications for cast iron	274
Relationship between carbon equivalent and tensile strength of grey cast iron	280
Grey cast iron—classification of graphite flake size and shape	281
Malleable cast iron	286
Nodular cast iron	295
Compacted graphite (C.G.) cast iron	305
Scrap diagnosis—its cause and cure	310
Removal of sand from iron castings by pickling	314
Cast irons for resisting heat	315
Cast irons for resisting corrosion	321
Cast irons for resisting wear	326
Inoculated grey cast iron	329
The wedge chill test	334

SECTION VII   Die-castings

Gravity die or permanent mould castings	340
Pressure die-casting	348

SECTION VIII   Steel Casting Specifications

American specifications for steel castings	356
British specifications for steel castings	375

SECTION IX Application of Insulating and Exothermic Risers to Castings

Introduction	378
Important notes for cast irons—breaker cores	384
How to make FEEDEX exothermic sleeves	388
Ramming FEEDEX shapes in sand moulds	389
Practical guide to the determination of feeder sleeve dimensions	390
New technology in feeding aids	393
Application of feeding aids. Steel castings	395
Application of feeding aids. Iron castings	408

SECTION X Principal FOSECO Products

Products for aluminium and magnesium alloys	418
Products for copper and nickel base alloys	419
Products for ferrous metals	421
Products for the white metals	423
Miscellaneous products	423
FOSECO publications	428
Index	431

SECTION I
TABLES AND GENERAL DATA

The tables appearing on pages 18–26, 29–33 and 39–41 are reproduced by kind permission of ROLLS ROYCE LIMITED, from their publication TSD 201 ©.

THE INTERNATIONAL SYSTEM OF UNITS (SI) AND THEIR RELATIONSHIP TO UNITED KINGDOM AND OTHER UNITS

The International System of Units, officially abbreviated SI, is a modernised version of the metric system established by international agreement. It is built upon a foundation of six base-units and their definitions which appear on the following pages. All other SI units are derived from these base-units. Multiples and sub-multiples are expressed in a decimal system.

Common equivalents and conversions to the more usual imperial weights and measures are given in the individual base-unit and also in a conversion table.

    Acknowledgements: Ministry of Technology.
                                        National Physical Laboratory.
                                        United States Department of Commerce.
                                        National Bureau of Standards.

LENGTH   METRE—m

The metre is defined as 1 650 763.73 wavelengths in vacuum of a specified orange-red line of the spectrum of krypton-86 atom.

The SI unit of area is the *square metre* (m^2). Land is often measured by the hectare (10 000 m^2 or approximately 2.5 acres).

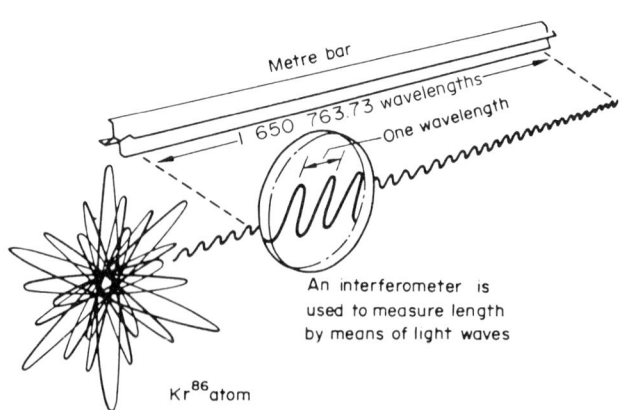

Fig. 1.1.

The SI unit of volume is the *cubic metre* (m³). Fluid volume is often measured by the litre (0.001 m³).

TIME SECOND—s

The second is defined as the duration of 9 192 631 770 cycles of radiation associated with a specified transition of the caesium-133 atom. It is realised by tuning an oscillator to the resonance frequency of the caesium atoms as they pass through a system of magnets and a resonant cavity into a detector.

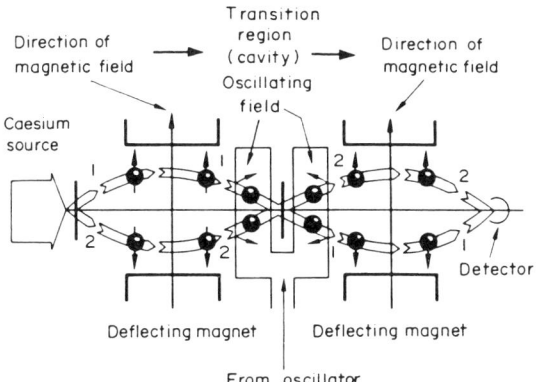

Fig. 1.2. A schematic of an atomic beam clock. The trajectories are drawn for those atoms whose magnetic moments are reversed in the transition region.

The number of periods or cycles per second is called frequency. The SI unit for frequency is the *hertz* (Hz). One hertz equals 1 c/s.

Standard frequencies and correct time are broadcast from GPO station Rugby on frequencies 16, 60 kHz, 2.5, 5, 10 MHz, and the BBC Droitwich transmitter 200 kHz. The standard radio broadcast band extends from 525 to 1605 kHz (medium frequency broadcasting band).

Dividing distance by time gives speed. The SI unit for speed is the *metre per second* (m/s), approximately 3 ft/s.

Rate of change in speed is called acceleration. The SI unit for acceleration is the *metre per second* squared (m/s²).

MASS KILOGRAM—kg

The standard for the unit of mass, the kilogram, is a cylinder of platinum–iridium alloy kept by the International Bureau of Weights and Measures at Paris. A duplicate in the custody of the National Physical

Laboratory, Teddington, serves as the mass standard for the United Kingdom. This is the only base-unit still defined by a material standard.

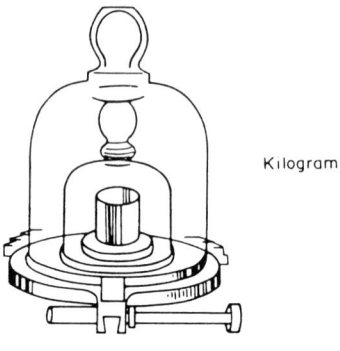

Fig. 1.3.

Closely allied to the concept of mass is that of force. The SI unit of force is the newton (N). A force of 1 N, when applied for 1 s, will give to a 1 kg mass a speed of 1 m/s (an acceleration of 1 m/s^2).

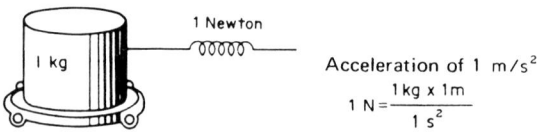

Fig. 1.4.

One N equals approximately two-tenths of a pound-force.

The weight of an object is the force exerted on it by gravity. Gravity gives a mass a downward acceleration of about 9.8 m/s^2.

The SI unit for work and energy of any kind is the *joule* (J).

$$1 \text{ J} = 1 \text{ N} \times 1 \text{ m}$$

The SI unit for power of any kind is the *watt* (W).

$$1 \text{ W} = \frac{1 \text{ J}}{1 \text{ s}}$$

TEMPERATURE KELVIN—K

The thermodynamic or Kelvin scale of temperature used in SI has its origin or zero point at absolute zero and has a fixed point at the triple point of water defined as 273.16 K. The Celsius* scale is derived from the Kelvin scale. The triple point is defined as 0.01°C on the Celsius scale, which is approximately 32.02°F on the Fahrenheit scale. The relationship of the Kelvin, Celsius and Fahrenheit temperature scales is shown below.

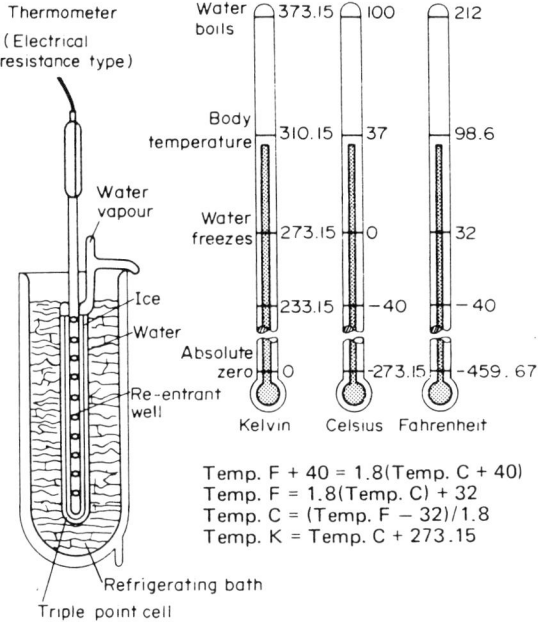

Temp. F + 40 = 1.8(Temp. C + 40)
Temp. F = 1.8(Temp. C) + 32
Temp. C = (Temp. F − 32)/1.8
Temp. K = Temp. C + 273.15

Fig. 1.5.

The triple-point cell, an evacuated glass cylinder filled with pure water, is used to define a known fixed temperature. When the cell is cooled and some ice is formed in the water, the temperature at the interface of solid, liquid, and vapour is 0.01°C. Thermometers to be calibrated are placed in the re-entrannt well.

*The metric practical unit of temperature formerly Centigrade.

LUMINOUS INTENSITY CANDELA—cd

The candela is defined as the luminous intensity of 1/600 000 of a square metre of a radiating cavity at the temperature of freezing platinum under a specified pressure (about 2045 K).

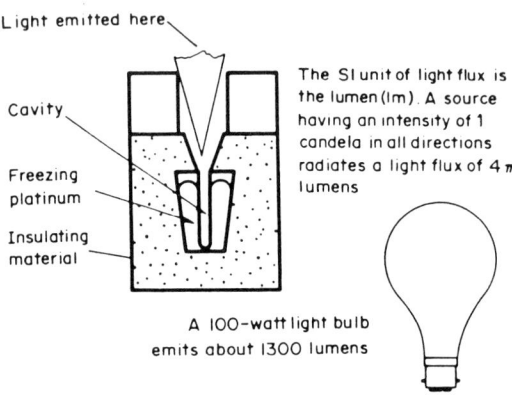

Fig. 1.6.

These Prefixes may be Applied to all SI Units

Multiples and Submultiples	Prefixes	Symbols
$1\,000\,000\,000\,000 = 10^{12}$	tera (ter'a)	T
$1\,000\,000\,000 = 10^{9}$	giga (ji'ga)	G
$1\,000\,000 = 10^{6}$	mega (meg'a)	M*
$1\,000 = 10^{3}$	kilo (kil'o)	k*
$100 = 10^{2}$	hecto (hek'to)	h
$10 = 10$	deca (dek'a)	da
$0.1 = 10^{-1}$	deci (des'i)	d
$0.01 = 10^{-2}$	centi (sen'ti)	c
$0.001 = 10^{-3}$	milli (mil'i)	m*
$0.000\,001 = 10^{-6}$	micro (mi'kro)	μ*
$0.000\,000\,001 = 10^{-9}$	nano (nan'o)	n
$0.000\,000\,000\,001 = 10^{-12}$	pico (pe'ko)	p
$0.000\,000\,000\,000\,001 = 10^{-15}$	femto (fem'to)	f
$0.000\,000\,000\,000\,000\,001 = 10^{-18}$	atto (at'to)	a

*Most commonly used

ELECTRIC CURRENT AMPERE—A

The ampere is defined as the constant current that, when flowing through each of two long, thin parallel wires separated by 1 m in free space, results in a force between the two wires (due to their magnetic fields) of 2×10^{-7} N for each metre of length (simplified version).

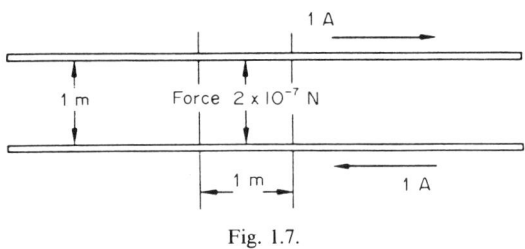

Fig. 1.7.

The SI unit of voltage is the *volt* (V).

$$1 \text{ V} = \frac{1 \text{ W}}{1 \text{ A}}$$

The SI unit of electrical resistance is the *ohm* (Ω).

$$1 \text{ } \Omega = \frac{1 \text{ V}}{1 \text{ A}}$$

Common Equivalents and Conversions

Approximate common equivalent		Conversions accurate to parts per million	
1 inch	= 25 millimetres	inches × 25.4	= millimetres
1 foot	= 0.3 metre	feet × 0.3048	= metres
1 yard	= 0.9 metre	yards × 0.9144	= metres
1 mile	= 1.6 kilometres	miles × 1.60934	= kilometres
1 square inch	= 650 square millimetres	square inches × 645.16	= square millimetres
1 square foot	= 0.09 square metre	square feet × 0.0929030	= square metres
1 square yard	= 0.84 square metre	square yards × 0.836127	= square metres
1 acre	= 4047 square metres	acres × 4046.86	= square metres
1 cubic inch	= 16 cubic centimetres	cubic inches × 16.3871	= cubic centimetres
1 cubic foot	= 0.028 cubic metre	cubic feet × 0.0283168	= cubic metres
1 cubic yard	= 0.8 cubic metre	cubic yards × 0.764555	= cubic metres
1 pint	= 0.568 litre	pints × 0.568261	= litres or cubic decimetres
1 gallon	= 4.5 litres	gallons × 4.54609	= litres
1 ounce	= 28 grams	ounces (avdp) × 28.3495	= grams
1 pound	= 0.45 kilograms	pounds (avdp) × 0.453592	= kilograms
1 cwt	= 51 kilograms	cwt × 50.8023	= kilograms
1 ton	= 1016 kilograms or 1.016 tonnes	tons × 1016.05	= kilograms
		tons × 1.01605	= tonnes (or metric tons)
1 horsepower	= 0.75 kilowatt	horsepower × 0.745700	= kilowatts

1 millimetre	= 0.04 inch	millimetres × 0.039 370 1	= inches
1 metre	= 3.3 feet	metres × 3.280 84	= feet
1 metre	= 1.1 yards	metres × 1.093 61	= yards
1 kilometre	= 0.6 mile	kilometres × 0.621 371	= miles
1 square millimetre	= 0.0015 square inch	square millimetres × 0.001 550	= square inches
1 square metre	= 11 square feet	square metres × 10.763 9	= square feet
1 square metre	= 1.2 square yards	square metres × 1.195 99	= square yards
1 square kilometre	= 247 acres	square kilometres × 247.105	= acres
1 cubic metre	= 35 cubic feet	cubic metres × 35.3147	= cubic feet
1 litre	= 0.22 gallon	litres × 0.219 69	= gallons
1 gram	= 0.035 ounce (avdp)	grams × 0.035 274	= ounces
1 kilogram	= 2.2 pounds (avdp)	kilograms × 2.204 62	= pounds

Table of Physical Properties of Metals

Element	Symbol	Atomic weight	Melting point °C	Boiling point °C	Latent heat of fusion cal/g	Specific heat cal/g/°C
Aluminium	Al	26.97	660	1800	92.4	0.2096
Antimony	Sb	121.76	630	1645	24.3	0.0495
Arsenic	As	74.93	*Volatilises*	*S'blimes* 450	—	0.0758
Barium	Ba	137.37	704	1000 appr.	—	0.068
Beryllium	Be	9.02	1281	1500	31.89	0.425
Bismuth	Bi	209.0	269	1560	13.0	0.0304
Cadmium	Cd	112.41	321	767	14.0	0.0547
Calcium	Ca	40.08	851	1170	78.5	0.145
Carbon	C	12.005	3500	—	—	0.168
Cerium	Ce	140.13	804	1400	—	0.0447
Chromium	Cr	52.01	1830	2260	31.70	0.104
Cobalt	Co	58.94	1490	3467	58.4	0.103
Copper	Cu	63.57	1083	2325	43.0	0.0909
Gallium	Ga	69.74	29.8	2000	19.16	0.079
Gold	Au	197.2	1063	2530	16.1	0.0316
Indium	In	114.8	155	1450	—	0.0570
Iridium	Ir	193.1	2350	4800	—	0.0323
Iron	Fe	55.84	1530	3235	47.9	0.1045
Lead	Pb	207.22	327	1755	5.0	0.0302
Lithium	Li	6.94	186	1400	32.81	0.837
Magnesium	Mg	24.32	659	1107	46.5	0.246
Manganese	Mn	54.93	1242	1900	36.5	0.122
Mercury	Hg	200.61	—38.80	356.7	3.0	0.0333
Molybdenum	Mo	96.0	2622	3550	—	0.0659
Nickel	Ni	58.69	1454.9	3075	73.0	0.103
Niobium	Nb	92.91	2500	3700	—	—
Osmium	Os	190.9	2700	—	—	0.0311
Palladium	Pd	106.7	1549	2800	36.0	0.0592
Phosphorus (Yellow)	P	31.04	44.1	279	5.0	0.189
Platinum	Pt	195.23	1774	4300	27.0	0.0324
Potassium	K	39.10	63.6	758	16.0	0.187
Rhodium	Rh	102.91	1966	2520	—	0.058
Silicon	Si	28.3	1415	2392	120	0.123
Silver	Ag	107.88	960.5	2150	22	0.0556
Sodium	Na	22.997	97.6	877	27.5	0.2829
Strontium	Sr	87.63	771	1366	—	0.0550
Sulphur (rhombic)	S	32.0	115	444.5	9.0	0.0163
Tantalum	Ta	180.89	2996	—	37.0	0.0301
Tellurium	Te	127.6	450	989.8	7.4	0.0525
Thallium	Tl	204	302	1457	—	0.0326
Tin	Sn	118.7	232	2270	14.6	0.0536
Titanium	Ti	47.9	1800	3000	90.0	0.122
Tungsten	W	184.0	3400	4830	40.0	0.034
Uranium	U	238.2	1300	—	—	0.028
Vanadium	V	50.95	1720	—	80.0	0.1153
Zinc	Zn	65.38	419.4	913	26.3	0.0918
Zirconium	Zr	90.6	1857	2900	—	0.0660

TABLES AND GENERAL DATA

Element	Thermal conductivity g-cal/cm sec/°C	Electrical resistivity microhms cm³	Change in volume on melting %	Density g/cm³	Thermal co-efficient linear expansion ×10⁻⁶	Brinell Hardness Number
Aluminium	0.480	2.65	6.6	2.70	23.86	17
Antimony	0.044	39.00	1.4	6.68	11.52	30.0
Arsenic	—	35.00	—	5.73	5.59	—
Barium	—	—	—	3.75	—	—
Beryllium	0.393	18.50	—	1.93	12.3	—
Bismuth	0.0194	115.00	−3.32	9.80	13.46	9.0
Cadmium	0.222	7.50	4.74	8.64	30.69	20
Calcium	—	10.50	—	1.54	25.0	13
Carbon	0.039	—	—	2.30	7.9	—
Cerium	—	—	—	6.92	—	—
Chromium	0.65	13.1	—	7.10	8.1	350
Cobalt	0.1654	9.71	—	8.60	12.36	125
Copper	0.941	1.69	4.05	8.94	16.6	48
Gallium	—	57.1	—	5.95	18.3	—
Gold	0.700	2.42	5.2	19.32	14.43	18.5
Indium	0.057	9.00	—	7.31	33.0	1.0
Iridium	0.141	5.30	—	22.41	7.00	172
Iron	0.176	10.00	5.5	7.86	11.82	66
Lead	0.083	20.80	3.4	11.37	29.24	5.5
Lithium	0.1673	8.40	1.5	0.534	60.0	—
Magnesium	0.376	4.35	4.2	1.74	26.94	25
Manganese	—	5.00	—	7.39	22.8	—
Mercury	0.0148	21.30	3.75	13.56	182.0	—
Molybdenum	0.3489	4.77	—	10.00	5.01	147
Nickel	0.1428	6.40	—	8.90	12.79	80
Osmium	—	—	—	22.48	6.8	—
Palladium	0.17	11.0	—	11.4	11.76	50
Phosphorus (Yellow)	—	10.70	—	1.83	6.24	0.6
Platinum	0.166	10.50	—	21.50	8.99	52
Potassium	0.2365	6.64	2.8	0.862	83.0	0.04
Rhodium	0.2101	5.10	—	12.44	8.5	156
Silicon	0.20	—	—	2.30	7.63	—
Silver	1.006	1.62	4.5	10.50	19.3	25
Sodium	0.3225	4.74	2.5	0.971	72.0	0.10
Strontium	—	25.0	—	2.54	—	—
Sulphuric (rhombic)	0.65	—	—	2.07	70.0	—
Tantalum	0.130	14.6	—	16.60	8.0	40
Tellurium	0.0144	21.0	—	6.25	16.80	—
Thallium	—	—	—	11.88	30.21	—
Tin	0.155	11.30	2.8	7.29	22.34	—
Titanium	—	3.0	—	4.50	7.14	—
Tungsten	0.35	5.48	—	19.30	4.44	—
Uranium	—	—	—	18.7	—	—
Vanadium	—	26.0	—	6.0	—	—
Zinc	0.265	6.1	6.5	7.10	30	35
Zirconium	—	—	—	6.5	—	—

CORRODIBILITY OF SOME COMMON METALS AND ALLOYS

Estimates are relative only: service temperatures are usually the deciding factors in designing equipment.
Originally compiled (1933) by Jerome Strauss; revised (1946) by E. E. Thum using data furnished by several specialists.

Class of material	Carbon	Chromium	Nickel	Silicon	Copper	Other constituents	Sea shore	Industrial	Domestic	Mine	Sea	Saline with H$_2$S	Brackish with MCl	Wet steam
Ingot iron or wrought iron	0.03				0.08		P	P	F	P	F	F	F	FG
Low C steel	0.10				0.08		P	P	F	P	F	F	F	FG
Copper and high tensile steels	0.10				0.25		F	F	F	PF	F	P	F	FG
Hot galvanised iron and steel							G	G	FG	F	FG	FG	FG	FG
Calorised iron and steel							G	G	P	F	P	P	P	
Grey cast iron	2.8 graphite, 0.7 combined carbon						F	F	G	F	F	F	F	F
High silicon iron	0.60			14.25			E	E	E	G	E	E	E	E
Nickel cast iron	3.30		3.50	1.50			FG	FG	F	F	FG	FG	FG	F
Chromium cast iron	2.50	25.00		2.00			G	G	G	G	FG	FG	FG	G
Ni-Cr-Cu cast iron	3.00	2.00	14.00	1.50	6.00		G	G	G	F	G	G	G	G
Nickel steel:														
Low Ni	0.18		3.00				FG	FG	F	F	F	FG	F	F
High Ni	0.30		28.00				G	G	G	FG	G	G	G	G
Chromium steels:														
5% Cr	0.15	5.00					F	F	F	P	F	P	F	F
7% Cr	0.15 max.	7.00					F	F	F	P	F	P	P	G
9% Cr	0.15 max.	9.00		0.75		Mo 0.5	F	F	F	P	F	P	P	FG
12% Cr	0.10 max.	12.00		0.50		Mo 0.5	G	G	G	F	PF	PF	PF	G
17% Cr	0.10 max.	17.00				Mo 1.0	G	G	G	FG	PG	PG	PG	G

TABLES AND GENERAL DATA

17% Cr, 4% Mo	0.10 max.	17.00			Mo 4.0	G	G	G	G	FG	PG	PG	PG	G
27% Cr	0.35 max.	27.00			Ni 0.25	GE	G	E	E	GE	G	FG	FG	E
Silcrome (8% Cr, 3% Si)	0.45	8.25		0.75		F	FG	F	F	F	PF	PF	PF	FG
Cr-Ni steels:				3.50										
8–20	0.20	8.00	20.00			G	G	G	G	FG	PG	F	FG	GE
18–12, 2½% Mo	0.15	18.00	12.00		Mo 2.5	E	E	E	E	E	FE	FE	FE	E
18–8	0.10	18.00	8.00			E	G	E	E	FE	FG	PG	FG	E
18–12	0.10	18.00	12.00			E	G	E	E	FE	PG	PG	PG	E
18–35	0.50 max.	18.00	35.00	1.00		E	E	E	E	G	G	PG	PG	E
25–12	0.25 max.	25.00	12.00			E	E	E	E	G	G	G	G	E
25–20	0.25 max.	25.00	20.00	1.50		E	G	E	E	G	G	E	G	E
Stellite	3.0 max.	30.00			Co 55.0, W 15.0	G	G	G	G	E	G	F	G	E
Hastelloy A			55.00		Mo 20.0, Fe 20.0	G	G	G	G	F	F	G	G	E
Hastelloy B			64.00		Mo 28.0, Fe 6.0	G	E	E	E	G	G	G	G	E
Hastelloy C		14.00	58.00		Mo 17.0, W 5.0, Fe 6.0	E	G	E	E	E	E	E	E	E
Hastelloy D			85.00	10.0		G	G	G	G	P	G	F	F	G
Commercially pure Ni			99.20			E	G	E	E	FG	FE	FE	FE	E
Nickel alloys:														
Monel metal	0.15	15.00	67.50	0.50	Mn 1.50	E	G	E	E	PE	GE	GE	GE	E
Nichrome, 60–15	0.12	14.00	60.00			E	G	E	E	G	FE	FE	FE	E
Inconel, 14% Cr		14.00	80.00			E	G	E	E	G	FE	FE	FE	E
80% Ni, 20% Cr		20.00	80.00			E	G	E	E	G	G	PF	FG	E
Commercially pure Cu						G	G	G	G	FG	G	G	FG	G
Copper alloys:														
Red brass					Zn 15.0	G	G	FG	FG	FG	G	PF	FG	G
Tobin bronze					Zn 39.2, Sn 0.75	G	FG	FG	FG	FG	G	F	FG	G
Phosphor bronze					Sn 5.0	G	FG	G	G	G	PG	FG	G	G
Silicon bronze				3.00	Mn 1.0 or Zn 1.5, Sn 0.5	G	G	G	PG	FG	G	PF	PG	G
Aluminium bronze					Al 9.0, Fe 0.5	G	G	G	G	G	G	G	G	G
Nickel silver		0.50			Zn 5.0	G	G	G	G	FG	GE	FG	FG	G
Admiralty metal		20.00			Zn 29.0, Sn 1.0	G	G	GF	GG	G	FG	PG	G	G
Commercially pure Al-SIC					Al 99.20	GE	EF							

13

Class of material	Nominal composition					Atmospheres		Water				
	Carbon	Chromium	Nickel	Silicon	Copper	Other constituents	Sea shore	Industrial	Domestic	Mine	Sea	Saline with H_2S	Brackish with MCl	Wet steam
Aluminium alloys:														
LM4M (DTD.424A)							PF	PG	P	P	P	P	P	F
LM6M (3L33)							GE	GE	FG	FG	G	G	G	G
LM5M or LM10W (L53)							GE	GE	FG	FG	G	G	G	G
Commercially pure Mg							G	G	F	P	P	P	P	P
Cast Mg alloys	A.S.T.M. A	Z63, A	Z92, A	Z90			F	G	F	P	P	P	P	P
Wrought Mg alloys	A.S.T.M. M	1, AZ3	1X, A	Z61X	AZ80X		G	G	F	P	P		F	P
Commercially pure Sn							G	G	E	G	F	F	G	G
Commercially pure Pb					0.06	Pb 99.90	G	G	PG	G	G		G	G

E—Excellent; almost unlimited service.
G—Good; will give good service.
F—Fair.
P—Poor.
§Below 50°C.
¶May crack under certain conditions.
†GE for fuming H_2SO_4.
‡E for HNO_3 above 80%.
*E for glacial acetic.

TABLES AND GENERAL DATA

CORRODIBILITY OF SOME COMMON METALS AND ALLOYS

Class of material	Oxidation resistance max. safe temp. °F — Oxidising gases	Reducing fuel gas	Sulphur rich gas	Fruit, vegetable juices	Dairy products	Other food products	HCl	H₂SO₄	HNO₃	Acetic	Phosphoric	Alkalies 1 to 20% sol.	Fused	NH₄Cl	MgCl₂	MgSO₄	Hot sulphite liquor	Dye liquor	Sweet	Sour
Ingot iron or wrought iron	1000			P	P	P	P	P	P	P	P	G	G	P	FG	FG	P		F	P
Low C steel		1000		P	P	P	P	P	P	P	P	G	G	P	FG	FG	P		F	P
Copper and high tensile steels				P	P	P	P	P	P	P	P	E	F	P	FG	FG	P	F	F	P
Hot galvanised iron and steel				P	P	P	P	P	P	P	P	F	F	P	FG	FG	P	P	F	P
Calorised iron and steel	1650	1650	1650	P	P	P	P	P	P	P	P	F	G	P	P	P	P		G	FG
Grey cast iron				E	E	E	FG	E	E	G	G	FG	P	G	FG	FG	P		G	F
High silicon iron				F		P	P	P	P	G	P	GE	FG	P	G	G	P		G	F
Nickel cast iron		1200	1750	G	G	G	P	P	G	G	F	E	FG	P	G	G	P	P	G	PF
Chromium cast iron	1750	1750	1750	PF	F	F	P	P	P	P	FG	FG	E	FG	FG	FG	PF		G	G
Ni-Cr-Cu cast iron	1400	1400	1200				P	P	G	FG	FG								G	G
Nickel steel:																				
Low Ni		1150	1150	P	P	P	P	P	P	P	P	GE	G	FG	P	FG	FG	P	FG	PF
High Ni		1250	1200	F	F	FG	F	FG	P	FG	F	E	G	FG	G	G	P		G	G
Chromium steels:																				
5% Cr	1150	1150	1150	P	P	P	P	P	P	P	P	G	G	P	P	P	P	P	G	F
7% Cr	1250	1250	1200	P	P	P	P	P	P	P	P	G	G	P	P	P	P	P		G
9% Cr	1300	1300	1250	P	P	P	P	P	P	P	P	G	G	P	P	F	P	P		G
12% Cr	1350	1350	1300	G	FG	G	P	P	G	G	F	G	P	P	F	G	P	P		G
17% Cr	1550	1550	1500	GE	G	GE	P	P	E	E§	F§	G	P	FG	G	G	P	F	E	GE
17% Cr, 4% Mo				G	G	G	P	P	G	G	P	G	P	FG	G	G	P	F	E	GE
27% Cr	1900	1900	1800	E	G	E	P	P	E	F	G	G	P	E	E	E	GE	F	E	E

CORRODIBILITY OF SOME COMMON METALS AND ALLOYS

Class of material	Oxidising gases	Reducing fuel gas	Sulphur rich gas	Fruit, vegetable juices	Dairy products	Other food products	HCl	H₂SO₄	HNO₃	Acetic	Phosphoric	1 to 20% sol.	Fused	NH₄Cl	MgCl₂	MgSO₄	Hot sulphite liquor	Dye liquor	Sweet	Sour
Silcrome (8% Cr, 3% Si)	1500	1500		F	F	F	P	P	FG	FG	P	G	P	F	F	FG	P	PG	E	GE
Cr-Ni steels:																				
8–20	1650	1650	1100	G	FG	G	P	FG	F	G	FG	E	G	FG	G	G	F		GE	GE
18–12, 2½% Mo	1650	1650	1100	E	E	E	PF	G	E	E	E	E	F	GE	G	G	E		E	E
18–8	2100	2000	1900	GE	E	E	P	PG	E	E	E	E	F	G	GE	G	GE	F	E	E
18–12	2100	2000		E	E	E	P	F	E	G	G	G	F	G	G	G	G	G	E	E
18–35	2100	2000		E	E	E	F	G	G	E	G	E	G	G	G	G	G	E	E	E
25–12	2100	2100		E	E	E	P	P	E	E	E	E	G	G	GE	G	PG	P	E	E
25–20	1400	1800	P		E		P	F	E	E	E	E	FG	G	G	E	GE	P	E	E
Stellite	1400	1800	P	E	E	G	F	F	P	E	E	E	P	E	G	G	G	G	G	E
Hastelloy A	2100	2100	P	G	E	G	G	G	P	E	E	E	P	E	E	E	P	P	E	E
Hastelloy B				G	G	E	E	E	G	E	E	E	P	G	E	E	P	P	E	E
Hastelloy C				E	E	G	G	E	P	E	E	E	P	E	E	E		E	E	E
Hastelloy D	1900	2300	1000	G	G	GE	FG	FG	P	FG	FG	E	E	FG	GE	G	G	E	GE	GE
Commercially pure Ni																				
Nickel alloys:																				
Monel metal	1000	2000	1000	G	FG	FG	FG	G	P	FG	G	E	E	GE	GE	E	P	E	E	G
Nichrome, 60–15	2050	1900	P	E	E	E	F	F	F	G	G	G	G	G	E	G	P		E	E
Inconel, 14% Cr	2000	2100	1500	E	E	E	F	F	G	G	G	E	G	G	E	G	P		E	E
80% Ni, 20% Cr	2000	2100	1500		P	E	FG	FG	P	FG	FG	E	F	FG	G	G	G	FG	FG	PF
Commercially pure Cu				FG		FG	PG	FG	P	FG	FG	G		FG	G	G	G	FG	FG	

TABLES AND GENERAL DATA

Copper alloys:																			
Red brass				P	P	FG	PF	FG	P	P	FG	FG	G	F	FG	G	G	FG	F
Tobin bronze				FG	P	FG	P	F	P	F	F	FG	PF	F	F	FG	F	G	F
Phosphor bronze				PF	P	F	P	FG	P	FG	FG	G	F	FG	G	G	FG	G	F
Silicon bronze				F	P	F	F	FG	P	G	G	FG	F	FG	G	G	FG	G	F
Aluminium bronze				F	PF	F	F	G	P	FG	FG	FG	PF	G	G	G	FG	G	FG
Nickel silver				FG	P	G	G	FG	P	FG	FG	E	F	FG	G	G	FG	G	FG
Admiralty metal				FG	G	FG	P	F	P	FG	FG	FG	F	G	G	F	FG	E	F
Commercially pure Al-SIC	800	800	800	G	G	F			P	G	P	P	P	FG	G	G	P		E
Aluminium alloys:																			
LM4M	P	P	P	U	U	U	P	P	P	P	P	P	P	PF	PF	F	P	F	F
LM6M	300	300	300	FG	G	G	P	F	P	G	P	P	P	G	G	F	FG	G	G
LM5M or 10W	800	800	800	FG	G	G	P	F	G	FG	P	P	P	P	P	F	FG	G	G
Commercially pure Mg		400	400	P	G	PE	P	P	G	P	P	E	P	P	P	P	PE		
Cast Mg alloys		400	400	P	P	PE	P	P	P	P	P	E		P	P	P	PE		
Wrought Mg alloys		400	400	P	P	PE	P	P	P	P	P	E		P	P	P	PE		
Commercially pure Sn				G	G	G	P	P	P	P	F	F		P	P	P		E	
Commercially pure Pb				P	P	P	F	E	P	E	P	G	P	FG	FG	G	F	G	G

E—Excellent; almost unlimited service. P—Poor. †GE for fuming H_2SO_4.
G—Good; will give good service. §Below 50°C. ‡E for HNO_3 above 80%.
F—Fair. U—Unsuitable. ¶May crack under certain conditions. *E for glacial acetic.

THE FOUNDRYMAN'S HANDBOOK

CONVERSION TABLES

Centimetres—Inches

cm		in	cm		in
2.54	1	0.3937	129.54	51	20.0787
5.08	2	0.7874	132.08	52	20.4724
7.62	3	1.1811	134.62	53	20.8661
10.16	4	1.5748	137.16	54	21.2598
12.70	5	1.9685	139.70	55	21.6535
15.24	6	2.3622	142.24	56	22.0472
17.78	7	2.7559	144.78	57	22.4409
20.32	8	3.1496	147.32	58	22.8346
22.86	9	3.5433	149.86	59	23.2283
25.40	10	3.9370	152.40	60	23.6220
27.94	11	4.3307	154.94	61	24.0157
30.48	12	4.7244	157.48	62	24.4094
33.02	13	5.1181	160.02	63	24.8031
35.56	14	5.5118	162.56	64	25.1969
38.10	15	5.9055	165.10	65	25.5906
40.64	16	6.2992	167.64	66	25.9843
43.18	17	6.6929	170.18	67	26.3780
45.72	18	7.0866	172.72	68	26.7717
48.26	19	7.4803	175.26	69	27.1654
50.80	20	7.8740	177.80	70	27.5591
53.34	21	8.2677	180.34	71	27.9528
55.88	22	8.6614	182.88	72	28.3465
58.42	23	9.0551	185.42	73	28.7402
60.96	24	9.4488	187.96	74	29.1339
63.50	25	9.8425	190.50	75	29.5276
66.04	26	10.2362	193.04	76	29.9213
68.58	27	10.6299	195.58	77	30.3150
71.12	28	11.0236	198.12	78	30.7087
73.66	29	11.4173	200.66	79	31.1024
76.20	30	11.8110	203.20	80	31.4961
78.74	31	12.2047	205.74	81	31.8898
81.28	32	12.5984	208.28	82	32.2835
83.82	33	12.9921	210.82	83	32.6772
86.36	34	13.3858	213.36	84	33.0709
88.90	35	13.7795	215.90	85	33.4646
91.44	36	14.1732	218.44	86	33.8583
93.98	37	14.5669	220.98	87	34.2520
96.52	38	14.9606	223.52	88	34.6457
99.06	39	15.3543	226.06	89	35.0394
102.60	40	15.7480	228.60	90	35.4331
104.14	41	16.1417	231.14	91	35.8268
106.68	42	16.5354	233.68	92	36.2205
109.22	43	16.9291	236.22	93	36.6142
111.76	44	17.3228	238.76	94	37.0079
114.30	45	17.7165	241.30	95	37.4016
116.84	46	18.1102	243.84	96	37.7953
119.38	47	18.5039	246.38	97	38.1890
121.92	48	18.8976	248.92	98	38.5827
124.46	49	19.2913	251.46	99	38.9764
127.00	50	19.6850	254.00	100	39.3701

To use this type of table, take the unit which is to be converted from centre column and read conversion direct from left- or right-hand column as the case may be. For example, in above table 10 in is equal to 25.40 cm and 10 cm is equal to 3.9370 in.

METRES—FEET

m		ft	m		ft
0.3048	1	3.28084	15.5448	51	167.323
0.6096	2	6.562	15.8496	52	170.604
0.9144	3	9.843	16.1544	53	173.884
1.2192	4	13.123	16.4592	54	177.165
1.5240	5	16.404	16.7640	55	180.446
1.8288	6	19.685	17.0688	56	183.727
2.1336	7	22.966	17.3736	57	187.008
2.4384	8	26.247	17.6784	58	190.289
2.7432	9	29.528	17.9832	59	193.570
3.0480	10	32.808	18.2880	60	196.850
3.3528	11	36.089	18.5928	61	200.131
3.6576	12	39.370	18.8976	62	203.412
3.9624	13	42.651	19.2024	63	206.693
4.2672	14	45.932	19.5072	64	209.974
4.5720	15	49.213	19.8120	65	213.255
4.8768	16	52.493	20.1168	66	216.535
5.1816	17	55.774	20.4216	67	219.816
5.4864	18	59.055	20.7264	68	223.097
5.7912	19	62.336	21.0312	69	226.378
6.0960	20	65.617	21.3360	70	229.659
6.4008	21	68.898	21.6408	71	232.940
6.7056	22	72.178	21.9456	72	236.220
7.0104	23	75.459	22.2504	73	239.501
7.3152	24	78.740	22.5552	74	242.782
7.6200	25	82.021	22.8600	75	246.063
7.9248	26	85.302	23.1648	76	249.344
8.2296	27	88.583	23.4696	77	252.625
8.5344	28	91.863	23.7744	78	255.906
8.8392	29	95.144	24.0792	79	259.186
9.1440	30	98.425	24.3840	80	262.467
9.4488	31	101.706	24.6888	81	265.748
9.7536	32	104.987	24.9936	82	269.029
10.0584	33	108.268	25.2984	83	272.310
10.3632	34	111.549	25.6032	84	275.591
10.6680	35	114.829	25.9080	85	278.871
10.9728	36	118.110	26.2128	86	282.152
11.2776	37	121.391	26.5176	87	285.433
11.5824	38	124.672	26.8224	88	288.714
11.8872	39	127.953	27.1272	89	291.995
12.1920	40	131.234	27.4320	90	295.276
12.4968	41	134.514	27.7368	91	298.556
12.8016	42	137.795	28.0416	92	301.837
13.1064	43	141.076	28.3464	93	305.118
13.4112	44	144.357	28.6512	94	308.399
13.7160	45	147.638	28.9560	95	311.680
14.0208	46	150.919	29.2608	96	314.961
14.3256	47	154.199	29.5656	97	318.241
14.6304	48	157.480	29.8704	98	321.522
14.9352	49	160.761	30.1752	99	324.803
15.2400	50	164.042	30.4800	100	328.084

Square Centimetres—Square Inches

cm²		in²	cm²		in²
6.452	1	0.155	329.032	51	7.905
12.903	2	0.310	335.483	52	8.060
19.355	3	0.465	341.935	53	8.215
25.806	4	0.620	348.386	54	8.370
32.258	5	0.775	354.838	55	8.525
38.710	6	0.930	361.290	56	8.680
45.161	7	1.085	367.741	57	8.835
51.613	8	1.240	374.193	58	8.990
58.064	9	1.395	380.644	59	9.145
64.516	10	1.550	387.096	60	9.300
70.968	11	1.705	393.548	61	9.455
77.419	12	1.860	399.999	62	9.610
83.871	13	2.015	406.451	63	9.765
90.322	14	2.170	412.902	64	9.920
96.774	15	2.325	419.354	65	10.075
103.226	16	2.480	425.806	66	10.230
109.677	17	2.635	432.257	67	10.385
116.129	18	2.790	438.709	68	10.540
122.580	19	2.945	445.160	69	10.695
129.032	20	3.100	451.612	70	10.850
135.484	21	3.255	458.064	71	11.005
141.935	22	3.410	464.515	72	11.160
148.387	23	3.565	470.967	73	11.315
154.838	24	3.720	477.418	74	11.470
161.290	25	3.875	483.878	75	11.625
167.742	26	4.030	490.322	76	11.780
174.193	27	4.185	496.773	77	11.935
180.645	28	4.340	503.225	78	12.090
187.096	29	4.495	509.676	79	12.245
193.548	30	4.650	516.128	80	12.400
200.000	31	4.805	522.579	81	12.555
206.451	32	4.960	529.031	82	12.710
212.903	33	5.115	535.483	83	12.865
219.354	34	5.270	541.934	84	13.020
225.806	35	5.425	548.386	85	13.175
232.258	36	5.580	554.838	86	13.330
238.709	37	5.735	561.289	87	13.485
245.161	38	5.890	567.741	88	13.640
251.612	39	6.045	574.192	89	13.795
258.064	40	6.200	580.644	90	13.950
264.516	41	6.355	587.096	91	14.105
270.967	42	6.510	593.547	92	14.260
277.419	43	6.665	599.999	93	14.415
283.870	44	6.820	606.450	94	14.570
290.322	45	6.975	612.902	95	14.725
296.774	46	7.130	619.354	96	14.880
303.225	47	7.285	625.805	97	15.035
309.677	48	7.440	632.257	98	15.190
316.128	49	7.595	638.708	99	15.345
322.580	50	7.750	645.160	100	15.500

Cubic Centimetres—Cubic Inches

cm³		in³	cm³		in³
16.387064	1	0.061024	835.740	51	3.1122
32.774	2	0.1220	852.127	52	3.1732
49.161	3	0.1831	868.514	53	3.2343
65.548	4	0.2441	884.901	54	3.2953
81.935	5	0.3051	901.288	55	3.3563
98.322	6	0.3661	917.675	56	3.4173
114.709	7	0.4272	934.062	57	3.4784
131.097	8	0.4882	950.449	58	3.5394
147.484	9	0.5492	966.837	59	3.6004
163.871	10	0.6102	983.224	60	3.6614
180.258	11	0.6713	999.611	61	3.7225
196.645	12	0.7323	1015.998	62	3.7835
213.032	13	0.7933	1032.385	63	3.8445
229.419	14	0.8543	1048.772	64	3.9055
245.806	15	0.9154	1065.159	65	3.9666
262.193	16	0.9764	1081.546	66	4.0276
278.580	17	1.0374	1097.933	67	4.0886
294.967	18	1.0984	1114.320	68	4.1496
311.354	19	1.1595	1130.707	69	4.2107
327.741	20	1.2205	1147.094	70	4.2717
344.128	21	1.2815	1163.481	71	4.3327
360.515	22	1.3425	1179.868	72	4.3937
376.902	23	1.4036	1196.255	73	4.4548
393.290	24	1.4646	1212.642	74	4.5158
409.677	25	1.5256	1229.295	75	4.5768
426.064	26	1.5866	1245.417	76	4.6378
442.451	27	1.6476	1261.804	77	4.6988
458.838	28	1.7087	1278.191	78	4.7599
475.225	29	1.7697	1294.578	79	4.8209
491.612	30	1.8307	1310.965	80	4.8819
507.999	31	1.8917	1327.352	81	4.9429
524.386	32	1.9528	1343.739	82	5.0040
540.773	33	2.0138	1360.126	83	5.0650
557.160	34	2.0748	1376.513	84	5.1260
573.547	35	2.1358	1392.900	85	5.1870
589.934	36	2.1969	1409.288	86	5.2481
606.321	37	2.2579	1425.675	87	5.3091
622.708	38	2.3189	1442.062	88	5.3701
639.096	39	2.3794	1458.449	89	5.4311
655.483	40	2.4410	1474.836	90	5.4922
671.876	41	2.5020	1491.223	91	5.5532
688.257	42	2.5630	1507.610	92	5.6142
704.644	43	2.6240	1523.997	93	5.6752
721.031	44	2.6851	1540.384	94	5.7363
737.418	45	2.7461	1556.771	95	5.7973
753.805	46	2.8071	1573.158	96	5.8583
770.192	47	2.8681	1589.545	97	5.9193
786.579	48	2.9292	1605.932	98	5.9804
802.966	49	2.9902	1622.319	99	6.0414
819.353	50	3.0512	1638.706	100	6.1024

Grams—Ounces (Avoirdupois)

g		oz	g		oz
28.3495	1	0.03527	1445.8245	51	1.79877
56.6990	2	0.07054	1474.1740	52	1.83404
85.0485	3	0.10581	1502.5235	53	1.86931
113.3980	4	0.14108	1530.8730	54	1.90458
141.7475	5	0.17635	1559.2225	55	1.93985
170.0970	6	0.21162	1587.5720	56	1.97512
198.4465	7	0.24689	1615.9215	57	2.01039
226.7960	8	0.28216	1644.2710	58	2.04566
255.1455	9	0.31743	1672.6205	59	2.08093
283.4950	10	0.35270	1700.9700	60	2.11620
311.8445	11	0.38797	1729.3195	61	2.15147
340.1940	12	0.42324	1757.6690	62	2.18674
368.5435	13	0.45851	1786.0185	63	2.22201
396.8930	14	0.49378	1814.3680	64	2.25728
425.2425	15	0.52905	1842.7175	65	2.29255
453.5920	16	0.56432	1871.0670	66	2.32782
481.9415	17	0.59959	1899.4165	67	2.36309
510.2910	18	0.63486	1927.7660	68	2.39836
538.6405	19	0.67013	1956.1155	69	2.43363
566.9900	20	0.70540	1984.4650	70	2.46890
595.3395	21	0.74067	2012.8145	71	2.50417
623.6890	22	0.77594	2041.1640	72	2.53944
652.0385	23	0.81121	2069.5135	73	2.57471
680.3880	24	0.84648	2097.8630	74	2.60998
708.7375	25	0.88175	2126.2125	75	2.64525
737.0870	26	0.91702	2154.5620	76	2.68052
765.4365	27	0.95229	2182.9115	77	2.71579
793.7860	28	0.98756	2211.2610	78	2.75106
822.1355	29	1.02283	2239.6105	79	2.78633
850.4850	30	1.05810	2267.9600	80	2.82160
878.8345	31	1.09337	2296.3095	81	2.85687
907.1840	32	1.12864	2324.6590	82	2.89214
935.5335	33	1.16391	2353.0085	83	2.92741
963.8830	34	1.19918	2381.3580	84	2.96268
992.2325	35	1.23445	2409.7075	85	2.99795
1020.5820	36	1.26972	2438.0570	86	3.03322
1048.9315	37	1.30499	2466.4065	87	3.06849
1077.2810	38	1.34026	2494.7560	88	3.10376
1105.6305	39	1.37553	2523.1055	89	3.13903
1133.9800	40	1.41080	2551.4550	90	3.17430
1162.3295	41	1.44607	2579.8045	91	3.20957
1190.6790	42	1.48134	2608.1540	92	3.24484
1219.0285	43	1.51661	2636.5035	93	3.28011
1247.3780	44	1.55188	2664.8530	94	3.31538
1275.7275	45	1.58715	2693.2025	95	3.35065
1304.0770	46	1.62242	2721.5520	96	3.38592
1332.4265	47	1.65769	2749.9015	97	3.42119
1360.7760	48	1.69296	2778.2510	98	3.45646
1389.1255	49	1.72823	2806.6005	99	3.49173
1417.4750	50	1.76350	2834.9500	100	3.52700

KILOGRAMS—POUNDS

kg		lb	kg		lb
0.45359243	1	2.20462	23.133	51	112.436
0.907	2	4.409	23.587	52	114.640
1.361	3	6.614	24.040	53	116.845
1.814	4	8.818	24.494	54	119.050
2.268	5	11.023	24.948	55	121.254
2.722	6	13.228	25.401	56	123.459
3.175	7	15.432	25.855	57	125.663
3.629	8	17.637	26.308	58	127.868
4.082	9	19.842	26.762	59	130.073
4.536	10	22.046	27.216	60	132.277
4.990	11	24.251	27.669	61	134.482
5.443	12	26.455	28.123	62	136.687
5.897	13	28.660	28.576	63	138.891
6.350	14	30.865	29.030	64	141.096
6.804	15	33.069	29.484	65	143.300
7.257	16	35.274	29.937	66	145.505
7.711	17	37.479	30.391	67	147.710
8.165	18	39.683	30.844	68	149.914
8.618	19	41.888	31.298	69	152.119
9.072	20	44.092	31.751	70	154.324
9.525	21	46.297	32.205	71	156.528
9.979	22	48.502	32.659	72	158.733
10.433	23	50.706	33.112	73	160.937
10.886	24	52.911	33.566	74	163.142
11.340	25	55.116	34.019	75	165.347
11.793	26	57.320	34.473	76	167.551
12.247	27	59.525	34.927	77	169.756
12.701	28	61.729	35.380	78	171.961
13.154	29	63.934	35.834	79	174.165
13.608	30	66.139	36.287	80	176.370
14.061	31	68.343	36.741	81	178.574
14.515	32	70.548	37.195	82	180.779
14.969	33	72.753	37.648	83	182.984
15.422	34	74.957	38.102	84	185.188
15.876	35	77.162	38.555	85	187.393
16.329	36	79.366	39.009	86	189.598
16.783	37	81.571	39.463	87	191.802
17.237	38	83.776	39.916	88	194.007
17.690	39	85.980	40.370	89	196.211
18.144	40	88.185	40.823	90	198.416
18.597	41	90.390	41.277	91	200.621
19.051	42	92.594	41.731	92	202.825
19.504	43	94.799	42.184	93	205.030
19.958	44	97.003	42.638	94	207.235
20.412	45	99.208	43.091	95	209.439
20.865	46	101.413	43.545	96	211.644
21.319	47	103.617	43.999	97	213.848
21.772	48	105.822	44.452	98	216.053
22.226	49	108.026	44.906	99	218.258
22.680	50	110.231	45.359	100	220.462

LONG TONS (2240 lb)—Short tons (2000 lb)

long tons		short tons	long tons		short tons
0.89286	1	1.12	45.5357	51	57.12
1.7857	2	2.24	46.4286	52	58.24
2.6786	3	3.36	47.3214	53	59.36
3.5714	4	4.48	48.2143	54	60.48
4.4643	5	5.60	49.1071	55	61.60
5.3571	6	6.72	50.0000	56	62.72
6.2500	7	7.84	50.8929	57	63.84
7.1429	8	8.96	51.7857	58	64.96
8.0357	9	10.08	52.6786	59	66.08
8.9286	10	11.20	53.5714	60	67.20
9.8214	11	12.32	54.4643	61	68.32
10.7143	12	13.44	55.3571	62	69.44
11.6071	13	14.56	56.2500	63	70.56
12.500	14	15.68	57.1429	64	71.68
13.3929	15	16.80	58.0357	65	72.80
14.2857	16	17.92	58.9286	66	73.92
15.1786	17	19.04	59.8214	67	75.04
16.0714	18	20.16	60.7143	68	76.16
16.9643	19	21.28	61.6071	69	77.28
17.8571	20	22.40	62.5000	70	78.40
18.7500	21	23.52	63.3929	71	79.52
19.6429	22	24.64	64.2857	72	80.64
20.5357	23	25.76	65.1786	73	81.76
21.4286	24	26.88	66.0714	74	82.88
22.3214	25	28.00	66.9643	75	84.00
23.2143	26	29.12	67.8571	76	85.12
24.1071	27	30.24	68.7500	77	86.24
25.0000	28	31.36	69.6429	78	87.36
25.8929	29	32.48	70.5357	79	88.48
26.7857	30	33.60	71.4286	80	89.60
27.6786	31	34.72	72.3214	81	90.72
28.5714	32	35.84	73.2143	82	91.84
29.4643	33	36.96	74.1071	83	92.96
30.3571	34	38.08	75.0000	84	94.08
31.2500	35	39.20	75.8929	85	95.20
32.1429	36	40.32	76.7857	86	96.32
33.0357	37	41.44	77.6786	87	97.44
33.9286	38	42.56	78.5714	88	98.56
34.8214	39	43.68	79.4643	89	99.68
35.7143	40	44.80	80.3571	90	100.80
36.6071	41	45.92	81.2500	91	101.92
37.5000	42	47.04	82.1429	92	103.04
38.3929	43	48.16	83.0357	93	104.16
39.2857	44	49.28	83.9286	94	105.28
40.1786	45	50.40	84.8214	95	106.40
41.0714	46	51.52	85.7143	96	107.52
41.9643	47	52.64	86.6071	97	108.64
42.8571	48	53.76	87.5000	98	109.76
43.7500	49	54.88	88.3929	99	110.88
44.6429	50	56.00	89.2857	100	112.00

METRIC TONNES (1000 kg)—TONS (2240 lb)

tonnes		long tons	tonnes		long tons
1.01605	1	0.984206	51.818	51	50.195
2.032	2	1.968	52.834	52	51.179
3.048	3	2.953	53.850	53	52.163
4.064	4	3.937	54.867	54	53.147
5.080	5	4.921	55.883	55	54.131
6.096	6	5.905	56.899	56	55.116
7.112	7	6.889	57.915	57	56.100
8.128	8	7.874	58.931	58	57.084
9.144	9	8.858	59.947	59	58.068
10.160	10	9.842	60.963	60	59.052
11.177	11	10.826	61.979	61	60.037
12.193	12	11.810	62.995	62	61.021
13.209	13	12.795	64.011	63	62.005
14.225	14	13.779	65.027	64	62.989
15.241	15	14.763	66.043	65	63.973
16.257	16	15.747	67.059	66	64.958
17.273	17	16.732	68.075	67	65.942
18.289	18	17.716	69.091	68	66.926
19.305	19	18.700	70.107	69	67.910
20.321	20	19.684	71.123	70	68.894
21.337	21	20.668	72.139	71	69.879
22.353	22	21.653	73.155	72	70.863
23.369	23	22.637	74.171	73	71.847
24.385	24	23.621	75.187	74	72.831
25.401	25	24.605	76.204	75	73.815
26.417	26	25.589	77.220	76	74.800
27.433	27	26.574	78.236	77	75.784
28.449	28	27.558	79.252	78	76.768
29.465	29	28.542	80.268	79	77.752
30.481	30	29.526	81.284	80	78.736
31.497	31	30.510	82.300	81	79.721
32.514	32	31.495	83.316	82	80.705
33.530	33	32.479	84.332	83	81.689
34.546	34	33.463	85.348	84	82.673
35.562	35	34.447	86.364	85	83.658
36.578	36	35.431	87.380	86	84.642
37.594	37	36.416	88.396	87	85.626
38.610	38	37.400	89.412	88	86.610
39.626	39	38.384	90.428	89	87.594
40.642	40	39.368	91.444	90	88.579
41.658	41	40.352	92.460	91	89.563
42.674	42	41.337	93.476	92	90.547
43.690	43	42.321	94.492	93	91.531
44.706	44	43.305	95.508	94	92.515
45.722	45	44.289	96.524	95	93.500
46.738	46	45.273	97.541	96	94.484
47.754	47	46.258	98.557	97	95.468
48.770	48	47.242	99.573	98	96.452
49.786	49	48.226	100.589	99	97.436
50.802	50	49.210	101.605	100	98.421

METRIC TONNES (1000 kg)—SHORT TONS (2000 lb)

metric tonnes		short tons	metric tonnes		short tons
0.90718	1	1.1023	46.2664	51	56.2178
1.8144	2	2.2046	47.1736	52	57.3201
2.7216	3	3.3069	48.0808	53	58.4224
3.6287	4	4.4092	48.9880	54	59.5247
4.5359	5	5.5115	49.8952	55	60.6271
5.4431	6	6.6139	50.8024	56	61.7294
6.3503	7	7.7162	51.7095	57	62.8317
7.2575	8	8.8185	52.6167	58	63.9340
8.1647	9	9.9208	53.5239	59	65.0363
9.0719	10	11.0231	54.4311	60	66.1386
9.9791	11	12.1254	55.3383	61	67.2409
10.8863	12	13.2277	56.2454	62	68.3432
11.7934	13	14.3300	57.1526	63	69.4455
12.7006	14	15.4323	58.0596	64	70.5478
13.6078	15	16.5347	58.9670	65	71.6502
14.5150	16	17.6370	59.8742	66	72.7525
15.4222	17	18.7393	60.7814	67	73.8548
16.3294	18	19.8416	61.6886	68	74.9571
17.2365	19	20.9439	62.5958	69	76.0594
18.1437	20	22.0462	63.5029	70	77.1612
19.0501	21	23.1485	64.4101	71	78.2640
19.9581	22	24.2508	65.3173	72	79.3663
20.8653	23	25.3531	66.2245	73	80.4686
21.7724	24	26.4554	67.1317	74	81.5709
22.6796	25	27.5578	68.0389	75	82.6733
23.5868	26	28.6601	68.9460	76	83.7756
24.4940	27	29.7624	69.8532	77	84.8779
25.4012	28	30.8647	70.7604	78	85.9802
26.3084	29	31.9669	71.6676	79	87.0825
27.2156	30	33.0693	72.5748	80	88.1848
28.1228	31	34.1716	73.4820	81	89.2871
29.0299	32	35.2739	74.3891	82	90.3894
29.9371	33	36.3762	75.2963	83	91.4917
30.8443	34	37.4785	76.2035	84	92.5940
31.7515	35	38.5809	77.1107	85	93.6964
32.6587	36	39.6832	78.0179	86	94.7987
33.5658	37	40.7855	78.9251	87	95.9010
34.4730	38	41.8878	79.8323	88	97.0033
35.3802	39	42.9901	80.7395	89	98.1056
36.2874	40	44.0924	81.6466	90	99.2079
37.1946	41	45.1947	82.5538	91	100.3102
38.1018	42	46.2970	83.4610	92	101.4125
39.0089	43	47.3993	84.3682	93	102.5148
39.9161	44	48.5016	85.2754	94	103.6171
40.8233	45	49.6040	86.1826	95	104.7195
41.7305	46	50.7063	87.0897	96	105.8218
42.6377	47	51.8086	87.9969	97	106.9241
43.5449	48	52.9109	88.9041	98	108.0264
44.4521	49	54.0132	89.8113	99	109.1287
45.3592	50	55.1155	90.7185	100	110.2310

KILOGRAMS—LONG TONS, HUNDREDWEIGHTS, ETC.

kg	qr	lb	kg	cwt	qr	lb
1		2.2	60	1	0	20.3
2		4.4	70	1	1	14.3
3		6.6	80	1	2	8.4
4		8.8	90	1	3	2.4
5		11.0	100	1	3	24.5
6		13.2	200	3	3	20.9
7		15.4	300	5	3	17.4
8		17.6	400	7	3	13.8
9		19.8	500	9	3	10.3
10		22.0	600	11	3	6.8
20	1	16.1	700	13	3	3.2
30	2	10.4	800	15	2	27.7
40	3	4.2	900	17	2	24.1
50	3	26.2	1000	19	2	20.6

kg	tons	cwt	qr	lb
2000	1	19	1	13.2
3000	2	19	0	5.8
4000	3	18	2	26.4
5000	4	18	1	19.0
6000	5	18	0	11.6
7000	6	17	3	4.2
8000	7	17	1	24.8
9000	8	17	0	17.4
10000	9	16	3	10.0
15000	14	15	1	1.0
20000	19	13	2	20.0
25000	24	12	0	11.0
50000	49	4	0	22.0

KILOGRAMS—SHORT TONS (2000 lb),
HUNDREDWEIGHTS (100 lb) AND POUNDS

kg	cwt	lb	kg	cwt	lb
1	—	2.2	60	1	32.3
2	—	4.4	70	1	54.3
3	—	6.6	80	1	76.4
4	—	8.8	90	1	98.4
5	—	11.0	100	2	20.5
6	—	13.2	200	4	40.9
7	—	15.4	300	6	61.4
8	—	17.6	400	8	81.8
9	—	19.8	500	11	2.3
10	—	22.0	600	13	22.8
20	—	44.1	700	15	43.2
30	—	66.1	800	17	63.7
40	—	88.2	900	19	84.1
50	1	10.2			

kg	short tons	cwt	lb
1000	1	2	4.6
2000	2	4	9.2
3000	3	6	13.9
4000	4	8	18.5
5000	5	10	23.1
6000	6	12	27.7
7000	7	14	32.3
8000	8	16	37.0
9000	9	18	41.6
10000	11	0	46.2
15000	16	10	69.3
20000	22	0	92.4
25000	27	11	15.5
50000	55	2	31.0

LITRES—IMPERIAL GALLONS

litres		Imp gal	litres		Imp gal
4.5459631	1	0.219975	231.844	51	11.2187
9.092	2	0.4400	236.390	52	11.4387
13.638	3	0.6599	240.936	53	11.6587
18.184	4	0.8799	245.482	54	11.8787
22.730	5	1.0999	250.028	55	12.0986
27.276	6	1.3199	254.574	56	12.3186
31.822	7	1.5398	259.120	57	12.5386
36.368	8	1.7598	263.666	58	12.7586
40.914	9	1.9798	268.212	59	12.9785
45.460	10	2.1998	272.758	60	13.1985
50.006	11	2.4197	277.304	61	13.4185
54.552	12	2.6397	281.850	62	13.6385
59.098	13	2.8597	286.396	63	13.8584
63.643	14	3.0797	290.942	64	14.0784
68.189	15	3.2996	295.488	65	14.2984
72.735	16	3.5196	300.034	66	14.5184
77.281	17	3.7396	304.580	67	14.7383
81.827	18	3.9596	309.125	68	14.9583
86.373	19	4.1795	313.671	69	15.1783
90.919	20	4.3995	318.217	70	15.3983
95.465	21	4.6195	322.763	71	15.6182
100.011	22	4.8395	327.309	72	15.8382
104.557	23	5.0594	331.855	73	16.0582
109.103	24	5.2794	336.401	74	16.2782
113.649	25	5.4994	340.947	75	16.4981
118.195	26	5.7194	345.493	76	16.7181
122.741	27	5.9393	350.039	77	16.9381
127.287	28	6.1593	354.585	78	17.1581
131.833	29	6.3793	359.131	79	17.3780
136.379	30	6.5993	363.677	80	17.5980
140.925	31	6.8192	368.223	81	17.8180
145.471	32	7.0392	372.769	82	18.0380
150.017	33	7.2592	377.315	83	18.2579
154.563	34	7.4792	381.861	84	18.4779
159.109	35	7.6991	386.407	85	18.6979
163.655	36	7.9191	390.953	86	18.9179
168.201	37	8.1391	395.499	87	19.1379
172.747	38	8.3591	400.045	88	19.3578
177.293	39	8.5790	404.591	89	19.5778
181.839	40	8.7990	409.137	90	19.7978
186.384	41	9.0190	413.683	91	20.0178
190.930	42	9.2390	418.229	92	20.2377
195.476	43	9.4589	422.775	93	20.4577
200.022	44	9.6789	427.321	94	20.6777
204.568	45	9.8989	431.866	95	20.8977
209.114	46	10.1189	436.412	96	21.1176
213.660	47	10.3388	440.958	97	21.3376
218.206	48	10.5588	445.504	98	21.5576
222.752	49	10.7788	450.050	99	21.7776
227.298	50	10.9988	454.596	100	21.9975

U.S. Gallons—Imperial Gallons

U.S. gal		Imp gal	U.S. gal		Imp gal
1.20095	1	0.83267	61.248	51	42.466
2.402	2	1.665	62.449	52	43.299
3.603	3	2.498	63.650	53	44.132
4.804	4	3.331	64.851	54	44.964
6.005	5	4.163	66.052	55	45.797
7.206	6	4.996	67.253	56	46.630
8.407	7	5.829	68.454	57	47.462
9.608	8	6.661	69.655	58	48.295
10.809	9	7.494	70.856	59	49.128
12.010	10	8.327	72.057	60	49.960
13.210	11	9.159	73.258	61	50.793
14.411	12	9.992	74.459	62	51.626
15.612	13	10.825	75.660	63	52.458
16.813	14	11.657	76.861	64	53.291
18.014	15	12.490	78.062	65	54.124
19.215	16	13.323	79.263	66	54.956
20.416	17	14.155	80.464	67	55.789
21.617	18	14.988	81.665	68	56.622
22.818	19	15.821	82.866	69	57.454
24.019	20	16.653	84.067	70	58.287
25.220	21	17.486	85.267	71	59.120
26.421	22	18.319	86.468	72	59.952
27.622	23	19.151	87.669	73	60.785
28.823	24	19.984	88.870	74	61.618
30.024	25	20.817	90.071	75	62.450
31.225	26	21.649	91.272	76	63.283
32.426	27	22.482	92.473	77	64.116
33.627	28	23.315	93.674	78	64.948
34.828	29	24.147	94.875	79	65.781
36.029	30	24.980	96.076	80	66.614
37.229	31	25.813	97.277	81	67.446
38.430	32	26.645	98.478	82	68.279
39.631	33	27.478	99.679	83	69.112
40.832	34	28.311	100.880	84	69.944
42.033	35	29.143	102.081	85	70.777
43.234	36	29.976	103.282	86	71.610
44.435	37	30.809	104.483	87	72.442
45.636	38	31.641	105.684	88	73.275
46.837	39	32.474	106.885	89	74.108
48.038	40	33.307	108.086	90	74.940
49.239	41	34.139	109.286	91	75.773
50.440	42	34.972	110.487	92	76.606
51.641	43	35.805	111.688	93	77.438
52.842	44	36.637	112.889	94	78.271
54.043	45	37.470	114.090	95	79.104
55.244	46	38.303	115.291	96	79.936
56.445	47	39.135	116.492	97	80.769
57.646	48	39.968	117.693	98	81.602
58.847	49	40.801	118.894	99	82.434
60.047	50	41.634	120.095	100	83.267

Litres—U.S. Gallons

litres		U.S. gal	litres		U.S. gal
3.78530	1	0.26418	193.050	51	13.473
7.571	2	0.5284	196.836	52	13.737
11.356	3	0.7925	200.621	53	14.002
15.141	4	1.0567	204.406	54	14.266
18.926	5	1.3209	208.192	55	14.530
22.712	6	1.5851	211.977	56	14.794
26.497	7	1.8493	215.763	57	15.058
30.282	8	2.1134	219.548	58	15.322
34.068	9	2.3776	223.332	59	15.587
37.853	10	2.6418	227.118	60	15.851
41.638	11	2.9060	230.903	61	16.115
45.424	12	3.1702	234.689	62	16.379
49.209	13	3.4343	238.474	63	16.643
52.994	14	3.6985	242.259	64	16.907
56.780	15	3.9627	246.044	65	17.172
60.565	16	4.2265	249.830	66	17.436
64.350	17	4.4911	253.615	67	17.700
68.135	18	4.7552	257.401	68	17.964
71.921	19	5.0194	261.186	69	18.228
75.706	20	5.2835	264.971	70	18.493
79.491	21	5.5478	268.757	71	18.757
83.277	22	5.8120	272.542	72	19.021
87.062	23	6.0761	276.327	73	19.285
90.847	24	6.3403	280.112	74	19.549
94.633	25	6.6045	283.898	75	19.813
98.418	26	6.869	287.683	76	20.078
102.203	27	7.133	291.468	77	20.342
105.988	28	7.397	295.254	78	20.606
109.774	29	7.661	299.039	79	20.870
113.559	30	7.926	302.824	80	21.134
117.344	31	8.190	306.610	81	21.399
121.129	32	8.454	310.395	82	21.662
124.915	33	8.718	314.181	83	21.927
128.700	34	8.982	317.965	84	22.191
132.485	35	9.246	321.751	85	22.455
136.271	36	9.510	325.536	86	22.719
140.056	37	9.775	329.321	87	22.984
143.842	38	10.039	333.107	88	23.248
147.627	39	10.303	336.892	89	23.512
151.412	40	10.567	340.677	90	23.776
155.197	41	10.831	344.463	91	24.040
158.986	42	11.096	348.248	92	24.305
162.768	43	11.360	352.033	93	24.569
166.553	44	11.624	355.818	94	24.833
170.339	45	11.888	359.604	95	25.097
174.124	46	12.152	363.389	96	25.361
177.909	47	12.416	367.174	97	25.625
181.695	48	12.681	370.960	98	25.980
185.480	49	12.945	374.745	99	26.154
189.265	50	13.209	378.530	100	26.418

KILOGRAMS PER SQUARE MILLIMETRE—LONG TONS PER SQUARE INCH

kg/mm^2		long tons/in^2	kg/mm^2		long tons/in^2
1.57488	1	0.63497	80.32	51	32.38
3.15	2	1.2700	81.89	52	33.02
4.72	3	1.90	83.47	53	33.65
6.30	4	2.54	85.04	54	34.29
7.87	5	3.17	86.62	55	34.92
9.45	6	3.81	88.19	56	35.56
11.02	7	4.44	89.77	57	36.19
12.60	8	5.08	91.34	58	36.83
14.17	9	5.71	92.92	59	37.46
15.75	10	6.35	94.49	60	38.10
17.32	11	6.98	96.07	61	38.73
18.90	12	7.62	97.64	62	39.37
20.47	13	8.25	99.22	63	40.00
22.05	14	8.89	100.79	64	40.64
23.62	15	9.52	102.37	65	41.27
25.20	16	10.16	103.94	66	41.91
26.77	17	10.79	105.52	67	42.54
28.35	18	11.43	107.09	68	43.18
29.92	19	12.06	108.67	69	43.81
31.50	20	12.70	110.24	70	44.45
33.07	21	13.33	111.82	71	45.08
34.65	22	13.97	113.39	72	45.72
36.22	23	14.60	114.97	73	46.35
37.80	24	15.24	116.54	74	46.99
39.37	25	15.87	118.12	75	47.62
40.95	26	16.51	119.69	76	48.26
42.52	27	17.14	121.27	77	48.89
44.10	28	17.78	122.84	78	49.53
45.67	29	18.41	124.42	79	50.16
47.25	30	19.05	125.99	80	50.80
48.82	31	19.68	127.57	81	51.43
50.40	32	20.32	129.14	82	52.07
51.97	33	20.95	130.72	83	52.70
53.55	34	21.59	132.29	84	53.34
55.12	35	22.22	133.86	85	53.97
56.70	36	22.86	135.44	86	54.61
58.27	37	23.49	137.01	87	55.24
59.85	38	24.13	138.59	88	55.88
61.42	39	24.76	140.16	89	56.51
63.00	40	25.40	141.74	90	57.15
64.57	41	26.03	143.31	91	57.78
66.14	42	26.67	144.89	92	58.42
67.72	43	27.30	146.46	93	59.05
69.29	44	27.94	148.04	94	59.69
70.87	45	28.57	149.61	95	60.32
72.44	46	29.21	151.19	96	60.96
74.02	47	29.84	152.76	97	61.59
75.59	48	30.48	154.34	98	62.23
77.17	49	31.11	155.91	99	62.86
78.74	50	31.75	157.49	100	63.50

Kilograms per Square Millimetre—Short Tons per Square Inch

kg/mm^2		short tons/in^2	kg/mm^2		short tons/in^2
1.41	1	0.71	71.71	51	36.19
2.81	2	1.42	73.11	52	36.90
4.22	3	2.13	74.52	53	37.61
5.62	4	2.84	75.93	54	38.32
7.03	5	3.55	77.33	55	39.03
8.44	6	4.26	78.74	56	39.74
9.84	7	4.97	80.14	57	40.45
11.25	8	5.68	81.55	58	41.16
12.65	9	6.39	82.96	59	41.87
14.06	10	7.10	84.36	60	42.58
15.47	11	7.81	85.77	61	43.29
16.87	12	8.52	87.18	62	44.00
18.28	13	9.23	88.58	63	44.71
19.68	14	9.94	89.99	64	45.42
21.09	15	10.65	91.39	65	46.13
22.50	16	11.35	92.80	66	46.84
23.90	17	12.06	94.21	67	47.55
25.31	18	12.77	95.61	68	48.26
26.71	19	13.83	97.02	69	48.97
28.12	20	14.19	98.42	70	49.68
29.53	21	14.90	99.83	71	50.39
30.93	22	15.61	101.24	72	51.10
32.34	23	16.32	102.64	73	51.81
33.75	24	17.03	104.05	74	52.52
35.15	25	17.74	105.45	75	53.23
36.56	26	18.45	106.86	76	53.94
37.96	27	19.16	108.27	77	54.64
39.37	28	19.87	109.67	78	55.35
40.78	29	20.58	111.08	79	56.06
42.18	30	21.29	112.48	80	56.77
43.59	31	22.00	113.89	81	57.48
44.99	32	22.71	115.30	82	58.19
46.40	33	23.42	116.70	83	58.90
47.81	34	24.13	118.11	84	59.61
49.21	35	24.84	119.51	85	60.32
50.62	36	25.55	121.92	86	61.03
52.02	37	26.26	122.33	87	61.74
53.43	38	26.97	123.73	88	62.45
54.84	39	27.68	125.14	89	63.16
56.24	40	28.39	126.54	90	63.87
57.65	41	29.10	127.95	91	64.58
59.05	42	29.81	129.36	92	65.29
60.46	43	30.52	130.76	93	66.00
61.87	44	31.23	132.17	94	66.71
63.27	45	31.94	133.57	95	67.42
64.68	46	32.64	134.98	96	68.13
66.08	47	33.35	136.39	97	68.84
67.49	48	34.06	137.79	98	69.55
68.90	49	34.77	139.20	99	70.26
70.30	50	35.48	140.60	100	70.97

KILOGRAMS PER SQUARE CENTIMETRE—POUNDS FOR SQUARE INCH

kg/cm²		lb/in²	kg/cm²		lb/in²
0.0703	1	14.2233	3.5857	51	725.39
0.1406	2	28.45	3.6560	52	739.61
0.2109	3	42.67	3.7263	53	753.84
0.2812	4	56.89	3.7966	54	768.06
0.3515	5	71.12	3.8669	55	782.28
0.4218	6	85.34	3.9372	56	796.51
0.4921	7	99.56	4.0075	57	810.73
0.5625	8	113.79	4.0778	58	824.95
0.6328	9	128.01	4.1481	59	839.18
0.7031	10	142.23	4.2184	60	853.40
0.7734	11	156.46	4.2887	61	867.62
0.8437	12	170.68	4.3590	62	881.85
0.9140	13	184.90	4.4293	63	896.07
0.9843	14	199.13	4.4996	64	910.29
1.0546	15	213.35	4.5699	65	924.52
1.1249	16	227.57	4.6403	66	938.74
1.1952	17	241.80	4.7106	67	952.96
1.2655	18	256.02	4.7809	68	967.19
1.3358	19	270.24	4.8512	69	981.41
1.4061	20	284.47	4.9215	70	995.63
1.4764	21	298.69	4.9918	71	1009.86
1.5467	22	312.91	5.0621	72	1024.08
1.6171	23	327.14	5.1324	73	1038.30
1.6874	24	341.36	5.2027	74	1052.53
1.7577	25	355.58	5.2330	75	1066.75
1.8280	26	369.81	5.3433	76	1080.97
1.8983	27	384.03	5.4136	77	1095.20
1.9686	28	398.25	5.4839	78	1109.42
2.0389	29	412.48	5.5543	79	1123.64
2.1092	30	426.70	5.6246	80	1137.87
2.1795	31	440.92	5.6949	81	1152.09
2.2498	32	455.15	5.7652	82	1166.31
2.3201	33	469.37	5.8355	83	1180.54
2.3904	34	483.59	5.9058	84	1194.76
2.4607	35	497.82	5.9761	85	1208.98
2.5310	36	512.04	6.0464	86	1223.21
2.6014	37	526.26	6.1167	87	1237.43
2.6717	38	540.49	6.1870	88	1251.65
2.7420	39	554.71	6.2573	89	1265.88
2.8123	40	568.93	6.3276	90	1280.10
2.8826	41	583.16	6.3979	91	1294.32
2.9529	42	597.38	6.4682	92	1308.55
3.0232	43	611.60	6.5385	93	1322.77
3.0935	44	625.83	6.6089	94	1336.99
3.1638	45	640.05	6.6792	95	1351.22
3.2341	46	654.27	6.7495	96	1365.44
3.3044	47	668.50	6.8198	97	1379.66
3.3747	48	682.72	6.8901	98	1393.89
3.4450	49	696.94	6.9604	99	1408.11
3.5153	50	711.17	7.0307	100	1422.33

TABLE OF STRESS VALUES

American	Metric	British
lb/in^2	kg/mm^2	tons/in^2
250	0.175	0.112
500	0.350	0.222
1000	0.700	0.446
2000	1.400	0.893
3000	2.101	1.339
4000	2.801	1.786
5000	3.515	2.232
10000	7.031	4.464
15000	10.546	6.696
20000	14.062	8.929
25000	17.577	11.161
30000	21.092	13.393
35000	24.608	15.652
40000	28.123	17.875
45000	31.639	20.089
50000	35.154	22.321
55000	38.670	24.554
60000	42.185	26.786
65000	45.700	29.018
70000	49.216	31.250
75000	52.731	33.482
80000	56.247	35.594
85000	59.762	37.946
90000	63.277	40.179
95000	66.793	42.411
100000	70.308	44.643

THE FOUNDRYMAN'S HANDBOOK

TEMPERATURE CONVERSIONS

Albert Sauveur type of table. Look up reading in middle column; if in degrees Centigrade, read Fahrenheit equivalent in right-hand column; if in degrees Fahrenheit, read Centrigrade equivalent in left-hand column. Values as printed in "Bethlehem Alloy Steels".

\-459.4 to 0			0 to 100			
C	F		C		F	C		F
−273	−459.4		−17.8	0	32	10.0	50	122.0
−268	−450		−17.2	1	33.8	10.6	51	123.8
−262	−440		−16.7	2	35.6	11.1	52	125.6
−257	−430		−16.1	3	37.4	11.7	53	127.4
−251	−420		−15.6	4	39.2	12.2	54	129.2
−246	−410		−15.0	5	41.0	12.8	55	131.0
−240	−400		−14.4	6	42.8	13.3	56	132.8
−234	−390		−13.9	7	44.6	13.9	57	134.6
−229	−380		−13.3	8	46.4	14.4	58	136.4
−223	−370		−12.8	9	48.2	15.0	59	138.2
−218	−360		−12.2	10	50.0	15.6	60	140.0
−212	−350		−11.7	11	51.8	16.1	61	141.8
−207	−340		−11.1	12	53.6	16.7	62	143.6
−201	−330		−10.6	13	55.4	17.2	63	145.4
−196	−320		−10.0	14	57.2	17.8	64	147.2
−190	−310		− 9.4	15	59.0	18.3	65	149.0
−184	−300		− 8.9	16	60.8	18.9	66	150.8
−179	−290		− 8.3	17	62.6	19.4	67	152.6
−173	−280		− 7.8	18	64.4	20.0	68	154.4
−169	−273	−459	− 7.2	19	66.2	20.6	69	156.2
−168	−270	−454	− 6.7	20	68.0	21.1	70	158.0
−162	−260	−436	− 6.1	21	69.8	21.7	71	159.8
−157	−250	−418	− 5.6	22	71.6	22.2	72	161.6
−151	−240	−400	− 5.0	23	73.4	22.8	73	163.4
−146	−230	−382	− 4.4	24	75.2	23.3	74	165.2
−140	−220	−364	− 3.9	25	77.0	23.9	75	167.0
−134	−210	−346	− 3.3	26	78.8	24.4	76	168.8
−129	−200	−328	− 2.8	27	80.6	25.0	77	170.6
−123	−190	−310	− 2.2	28	82.4	25.6	78	172.4
−118	−180	−292	− 1.7	29	84.2	26.1	79	174.2
−112	−170	−274	− 1.1	30	86.0	26.7	80	176.0
−107	−160	−256	− 0.6	31	87.8	27.2	81	177.8
−101	−150	−238	0.0	32	89.6	27.8	82	179.6
− 96	−140	−220	0.6	33	91.4	28.3	83	181.4
− 90	−130	−202	1.1	34	93.2	28.9	84	183.2
− 84	−120	−184	1.7	35	95.0	29.4	85	185.0
− 79	−110	−166	2.2	36	96.8	30.0	86	186.8
− 73	−100	−148	2.8	37	98.6	30.6	87	188.6
− 68	− 90	−130	3.3	38	100.4	31.1	88	190.4
− 62	− 80	−112	3.9	39	102.2	31.7	89	192.2
− 57	− 70	− 94	4.4	40	104.0	32.2	90	194.0
− 51	− 60	− 76	5.0	41	105.8	32.8	91	195.8
− 46	− 50	− 58	5.6	42	107.6	33.3	92	197.6
− 40	− 40	− 40	6.1	43	109.4	33.9	93	199.4
− 34	− 30	− 22	6.7	44	111.2	34.4	94	201.2
− 29	− 20	− 4	7.2	45	113.0	35.0	95	203.0
− 23	− 10	14	7.8	46	114.8	35.6	96	204.8
− 17	0	32	8.3	47	116.6	36.1	97	206.6
			8.9	48	118.4	36.7	98	208.4
			9.4	49	120.2	37.2	99	210.2
						37.8	100	212.0

TABLES AND GENERAL DATA

100 to 1000					1000 to 2000				
C		F	C		F	C		F	
38	*100*	212	260	*500*	932	538	*1000*	1832	816	*1500*	2732
43	*110*	230	266	*510*	950	543	*1010*	1850	821	*1510*	2750
49	*120*	248	271	*520*	968	549	*1020*	1868	827	*1520*	2768
54	*130*	266	277	*530*	986	554	*1030*	1886	832	*1530*	2786
60	*140*	284	282	*540*	1004	560	*1040*	1904	838	*1540*	2804
66	*150*	302	288	*550*	1022	566	*1050*	1922	843	*1550*	2822
71	*160*	320	293	*560*	1040	571	*1060*	1940	849	*1560*	2840
77	*170*	338	299	*570*	1058	577	*1070*	1958	854	*1570*	2858
82	*180*	356	304	*580*	1076	582	*1080*	1976	860	*1580*	2876
88	*190*	374	310	*590*	1094	588	*1090*	1994	866	*1590*	2894
93	*200*	392	316	*600*	1112	593	*1100*	2012	871	*1600*	2912
99	*210*	410	321	*610*	1130	599	*1110*	2030	877	*1610*	2930
100	*212*	413	327	*620*	1148	604	*1120*	2048	882	*1620*	2948
104	*220*	428	332	*630*	1166	610	*1130*	2066	888	*1630*	2966
110	*230*	446	338	*640*	1184	616	*1140*	2084	893	*1640*	2984
116	*240*	464	343	*650*	1202	621	*1150*	2102	899	*1650*	3002
121	*250*	482	349	*660*	1220	627	*1160*	2120	904	*1660*	3020
127	*260*	500	354	*670*	1238	632	*1170*	2138	910	*1670*	3038
132	*270*	518	360	*680*	1256	638	*1180*	2156	916	*1680*	3056
138	*280*	536	366	*690*	1274	643	*1190*	2174	921	*1690*	3074
143	*290*	554	371	*700*	1292	649	*1200*	2192	927	*1700*	3092
149	*300*	572	377	*710*	1310	654	*1210*	2210	932	*1710*	3110
154	*310*	590	382	*720*	1328	660	*1220*	2228	938	*1720*	3128
160	*320*	608	388	*730*	1346	666	*1230*	2246	943	*1730*	3146
166	*330*	626	393	*740*	1364	671	*1240*	2264	949	*1740*	3164
171	*340*	644	399	*750*	1382	677	*1250*	2282	954	*1750*	3182
177	*350*	662	404	*760*	1400	682	*1260*	2300	960	*1760*	3200
182	*360*	680	410	*770*	1418	688	*1270*	2318	966	*1770*	3218
188	*370*	698	416	*780*	1436	693	*1280*	2336	971	*1780*	3236
193	*380*	716	421	*790*	1454	699	*1290*	2354	977	*1790*	3254
199	*390*	734	427	*800*	1472	704	*1300*	2372	982	*1800*	3272
204	*400*	752	432	*810*	1490	710	*1310*	2390	988	*1810*	3290
210	*410*	770	438	*820*	1508	716	*1320*	2408	993	*1820*	3308
216	*420*	788	443	*830*	1526	721	*1330*	2426	999	*1830*	3326
221	*430*	806	449	*840*	1544	727	*1340*	2444	1004	*1840*	3344
227	*440*	824	454	*850*	1562	732	*1350*	2462	1010	*1850*	3362
232	*450*	842	460	*860*	1580	738	*1360*	2480	1016	*1860*	3380
238	*460*	860	466	*870*	1598	743	*1370*	2498	1021	*1870*	3398
243	*470*	878	471	*880*	1616	749	*1380*	2516	1027	*1880*	3416
249	*480*	896	477	*890*	1634	754	*1390*	2534	1032	*1890*	3434
254	*490*	914	482	*900*	1652	760	*1400*	2552	1038	*1900*	3452
			488	*910*	1670	766	*1410*	2570	1043	*1910*	3470
			493	*920*	1688	771	*1420*	2588	1049	*1920*	3488
			499	*930*	1706	777	*1430*	2606	1054	*1930*	3506
			504	*940*	1724	782	*1440*	2624	1060	*1940*	3524
			510	*950*	1742	788	*1450*	2642	1066	*1950*	3542
			516	*960*	1760	793	*1460*	2660	1071	*1960*	3560
			521	*970*	1778	799	*1470*	2678	1077	*1970*	3578
			527	*980*	1796	804	*1480*	2696	1082	*1980*	3596
			532	*990*	1814	810	*1490*	2714	1088	*1990*	3614
			538	*1000*	1832				1093	*2000*	3632

2000 to 3000

C	F	C	F	C	F	C	F	C	F
1093	2000	3632	1232	2250	4082	1371	2500	4532	1510	2750	4982
1099	2010	3650	1238	2260	4100	1377	2510	4550	1516	2760	5000
1104	2020	3668	1243	2270	4118	1382	2520	4568	1521	2770	5018
1110	2030	3686	1249	2280	4136	1388	2530	4586	1527	2780	5036
1116	2040	3704	1254	2290	4154	1393	2540	4604	1532	2790	5054
1121	2050	3722	1260	2300	4172	1399	2550	4622	1538	2800	5072
1127	2060	3740	1266	2310	4190	1404	2560	4640	1543	2810	5090
1132	2070	3758	1271	2320	4208	1410	2570	4658	1549	2820	5108
1138	2080	3776	1277	2330	4226	1416	2580	4676	1554	2830	5126
1143	2090	3794	1282	2340	4244	1421	2590	4694	1560	2840	5144
1149	2100	3812	1288	2350	4262	1427	2600	4712	1566	2850	5162
1154	2110	3830	1293	2360	4280	1432	2610	4730	1571	2860	5180
1160	2120	3848	1299	2370	4298	1438	2620	4748	1577	2870	5198
1166	2130	3866	1304	2380	4316	1443	2630	4766	1582	2880	5216
1171	2140	3884	1310	2390	4334	1449	2640	4784	1588	2890	5234
1177	2150	3902	1316	2400	4352	1454	2650	4802	1593	2900	5252
1182	2160	3920	1321	2410	4370	1460	2660	4820	1599	2910	5270
1188	2170	3938	1327	2420	4388	1466	2670	4838	1604	2920	5288
1193	2180	3956	1332	2430	4406	1471	2680	4856	1610	2930	5306
1199	2190	3974	1338	2440	4424	1477	2690	4874	1616	2940	5324
1204	2200	3992	1343	2450	4442	1482	2700	4892	1621	2950	5342
1210	2210	4010	1349	2460	4460	1488	2710	4910	1627	2960	5360
1216	2220	4028	1354	2470	4478	1493	2720	4928	1632	2970	5378
1221	2230	4046	1360	2480	4496	1499	2730	4946	1638	2980	5396
1227	2240	4064	1366	2490	4514	1504	2740	4964	1643	2990	5414
									1649	3000	5432

HYDROMETER CONVERSION TABLES
Showing the Relation between Density (CGS) and the Baumé and
Twaddell Scales for Densities above Unity

Density	Degrees Baumé	Degrees Twaddell	Density	Degrees Baumé	Degrees Twaddell
1.00	0.00	0	1.41	42.16	82
1.01	1.44	2	1.42	42.89	84
1.02	2.84	4	1.43	43.60	86
1.03	4.22	6	1.44	44.31	88
1.04	5.58	8	1.45	45.00	90
1.05	6.91	10	1.46	45.68	92
1.06	8.21	12	1.47	46.36	94
1.07	9.49	14	1.48	47.03	96
1.08	10.74	16	1.49	47.68	98
1.09	11.97	18	1.50	48.33	100
1.10	13.18	20	1.51	48.97	102
1.11	14.37	22	1.52	49.60	104
1.12	15.54	24	1.53	50.23	106
1.13	16.68	26	1.54	50.84	108
1.14	17.81	28	1.55	51.45	110
1.15	18.91	30	1.56	52.05	112
1.16	20.00	32	1.57	52.64	114
1.17	21.07	34	1.58	53.23	116
1.18	22.12	36	1.59	53.80	118
1.19	23.15	38	1.60	54.38	120
1.20	24.17	40	1.61	54.94	122
1.21	25.16	42	1.62	55.49	124
1.22	26.15	44	1.63	56.04	126
1.23	27.11	46	1.64	56.58	128
1.24	28.06	48	1.65	57.12	130
1.25	29.00	50	1.66	57.65	132
1.26	29.92	52	1.67	58.17	134
1.27	30.83	54	1.68	58.69	136
1.28	31.72	56	1.69	59.20	138
1.29	32.60	58	1.70	59.71	140
1.30	33.46	60	1.71	60.20	142
1.31	34.31	62	1.72	60.70	144
1.32	35.15	64	1.73	61.18	146
1.33	35.98	66	1.74	61.67	148
1.34	36.79	68	1.75	62.14	150
1.35	37.59	70	1.76	62.61	152
1.36	38.38	72	1.77	63.08	154
1.37	39.16	74	1.78	63.54	156
1.38	39.93	76	1.79	63.99	158
1.39	40.68	78	1.80	64.44	160
1.40	41.43	80	—	...	...

Metric and Decimal Equivalents of Fractions
(Of 1 in)

mm							in
0.3969						$\frac{1}{64}$	0.015625
0.79375					$\frac{1}{32}$		0.03125
1.1906						$\frac{3}{64}$	0.046875
1.5875				$\frac{1}{16}$			0.0625
1.9844						$\frac{5}{64}$	0.078125
2.38125					$\frac{3}{32}$		0.09375
2.7781						$\frac{7}{64}$	0.109375
3.1750			$\frac{1}{8}$				0.125
3.5719						$\frac{9}{64}$	0.140625
3.96875					$\frac{5}{32}$		0.15625
4.3656						$\frac{11}{64}$	0.171875
4.7625				$\frac{3}{16}$			0.1875
5.1594						$\frac{13}{64}$	0.203125
5.55625					$\frac{7}{32}$		0.21875
5.9531						$\frac{15}{64}$	0.234375
6.3500		$\frac{1}{4}$					0.25
6.7469						$\frac{17}{64}$	0.265625
7.14375					$\frac{9}{32}$		0.28125
7.5406						$\frac{19}{64}$	0.296875
7.9375				$\frac{5}{16}$			0.3125
8.3344						$\frac{21}{64}$	0.328125
8.73125					$\frac{11}{32}$		0.34375
9.1281						$\frac{23}{64}$	0.359375
9.5250			$\frac{3}{8}$				0.375
9.9219						$\frac{25}{64}$	0.390625
10.31875					$\frac{13}{32}$		0.40625
10.7156						$\frac{27}{64}$	0.421875
11.1125				$\frac{7}{16}$			0.4375
11.5094						$\frac{29}{64}$	0.453125
11.90625					$\frac{15}{32}$		0.46875
12.3031						$\frac{31}{64}$	0.484375
12.7000	$\frac{1}{2}$						0.5
13.0969						$\frac{33}{64}$	0.515625
13.49375					$\frac{17}{32}$		0.53125
13.8906						$\frac{35}{64}$	0.546875
14.2875				$\frac{9}{16}$			0.5625
14.6844						$\frac{37}{64}$	0.578125
15.08125					$\frac{19}{32}$		0.59375
15.4781						$\frac{39}{64}$	0.609375
15.8750				$\frac{5}{8}$			0.625
16.2719						$\frac{41}{64}$	0.640625
16.66875					$\frac{21}{32}$		0.65625
17.0656						$\frac{43}{64}$	0.671875
17.4625				$\frac{11}{16}$			0.6875
17.8594						$\frac{45}{64}$	0.703125
18.25625					$\frac{23}{32}$		0.71875
18.6531						$\frac{47}{64}$	0.734375

Metric and Decimal Equivalents of Fractions
(Of 1 in)

mm							in
19.0500		$\frac{3}{4}$					0.75
19.4469						$\frac{49}{64}$	0.765625
19.84375					$\frac{25}{32}$		0.78125
20.2406						$\frac{51}{64}$	0.796875
20.6375				$\frac{13}{16}$			0.8125
21.0344						$\frac{53}{64}$	0.828125
21.43125					$\frac{27}{32}$		0.84375
21.8281						$\frac{55}{64}$	0.859375
22.2250			$\frac{7}{8}$				0.875
22.6219						$\frac{57}{64}$	0.890625
23.01875					$\frac{29}{32}$		0.90625
23.4156						$\frac{59}{64}$	0.921875
23.8125				$\frac{15}{16}$			0.9375
24.2094						$\frac{61}{64}$	0.953125
24.60625					$\frac{31}{32}$		0.96875
25.0031						$\frac{63}{64}$	0.984375
25.4000	1						1.0

Areas and Circumferences of Circles
(See note at end of tables)

Diameter	Circumference	Area	Diameter	Circumference	Area
$\frac{1}{16}$	0.1964	0.0031	$\frac{5}{8}$	8.2467	5.4119
0.1	0.3143	0.0078	2.7	8.4823	5.7255
$\frac{1}{8}$	0.3929	0.0123	$\frac{3}{4}$	8.6394	5.9396
$\frac{3}{16}$	0.5893	0.0276	.8	8.7964	6.1575
0.2	0.6286	0.0314	$\frac{7}{8}$	9.0321	6.4918
$\frac{1}{4}$	0.7857	0.0491	.9	9.1106	6.6052
0.3	0.9429	0.0707			
$\frac{5}{16}$	0.9821	0.0767	3.0	9.4248	7.0686
$\frac{3}{8}$	1.1786	0.1105	.1	9.7389	7.5476
0.4	1.2571	0.1257	$\frac{1}{8}$	9.8175	7.6699
$\frac{7}{16}$	1.3750	0.1504	.2	10.1531	8.0424
$\frac{1}{2}$	1.5714	0.1964	$\frac{1}{4}$	10.2102	8.2958
$\frac{9}{16}$	1.7679	0.2486	.3	10.3672	8.5530
0.6	1.8857	0.2829	$\frac{3}{8}$	10.6029	8.9462
$\frac{5}{8}$	1.9643	0.3069	.4	10.6814	9.0792
$\frac{11}{16}$	2.1607	0.3714	$\frac{1}{2}$	10.9956	9.6211
0.7	2.2000	0.3850	.6	11.3197	10.178
$\frac{3}{4}$	2.3571	0.4419	$\frac{5}{8}$	11.3883	10.321
0.8	2.5143	0.5029	.7	11.6239	10.752
$\frac{13}{16}$	2.5536	0.5187	$\frac{3}{4}$	11.7810	11.045
$\frac{7}{8}$	2.7500	0.6016	.8	11.9380	11.341
0.9	2.8286	0.6364	$\frac{7}{8}$	12.1737	11.793
$\frac{15}{16}$	2.9464	0.6906	.9	12.2522	11.945
1.0	3.1416	0.7854	4.0	12.5664	12.566
.1	3.4557	0.9503	.1	12.8805	13.202
$\frac{1}{8}$	3.5343	0.9940	$\frac{1}{8}$	12.9591	13.364
.2	3.7699	1.1309	.2	13.2947	13.854
$\frac{1}{4}$	3.9270	1.2272	$\frac{1}{4}$	13.3518	14.186
.3	4.0840	1.3273	.3	13.5088	14.522
$\frac{3}{8}$	4.3197	1.4849	$\frac{3}{8}$	13.7445	15.033
.4	4.3982	1.5393	.4	13.8230	15.205
$\frac{1}{2}$	4.7124	1.7671	$\frac{1}{2}$	14.1372	15.904
.6	5.0265	2.0106	.6	14.4513	16.619
$\frac{5}{8}$	5.1051	2.0739	$\frac{5}{8}$	14.5299	16.800
.7	5.3407	2.2698	.7	14.7555	17.349
$\frac{3}{4}$	5.4978	2.4053	$\frac{3}{4}$	14.9226	17.720
.8	5.6548	2.5446	.8	15.0796	18.095
$\frac{7}{8}$	5.8905	2.7612	$\frac{7}{8}$	15.3153	18.665
.9	5.9690	2.8352	.9	15.3938	18.857
2.0	6.2832	3.1416	5.0	15.7080	19.635
.1	6.5973	3.4636	.1	16.0221	20.428
$\frac{1}{8}$	6.6759	3.5456	$\frac{1}{8}$	16.1007	20.629
.2	6.9115	3.8013	.2	16.3363	21.237
$\frac{1}{4}$	7.0686	3.9761	$\frac{1}{4}$	16.4934	21.647
.3	7.2256	4.1547	.3	16.6504	22.061
$\frac{3}{8}$	7.4613	4.4301	$\frac{3}{8}$	16.8861	22.690
.4	7.5398	4.5239	.4	16.9646	22.902
$\frac{1}{2}$	7.8540	4.9087	$\frac{1}{2}$	17.2788	23.758
.6	8.1681	5.3093	.6	17.5929	24.630

AREAS AND CIRCUMFERENCES OF CIRCLES
(See note at end of tables)

Diameter	Circumference	Area	Diameter	Circumference	Area
$\frac{5}{8}$	17.6715	24.850	8.7	27.331	59.446
5.7	17.9071	25.517	$\frac{3}{4}$	27.489	60.132
$\frac{3}{4}$	18.0642	25.967	.8	27.646	60.821
.8	18.2212	26.420	$\frac{7}{8}$	27.881	61.862
$\frac{7}{8}$	18.4569	27.108	.9	27.960	62.211
.9	18.5354	27.339			
			9.0	28.274	63.617
6.0	18.8496	28.274	.1	28.588	65.038
.1	19.1637	29.224	$\frac{1}{8}$	28.667	65.396
$\frac{1}{8}$	19.2423	29.464	.2	28.902	66.476
.2	19.4779	30.190	$\frac{1}{4}$	29.059	67.200
$\frac{1}{4}$	19.6350	30.679	.3	29.216	67.929
.3	19.7930	31.272	$\frac{3}{8}$	29.452	69.029
$\frac{3}{8}$	20.0277	31.919	.4	29.531	69.397
.4	20.1062	32.169	$\frac{1}{2}$	29.845	70.882
$\frac{1}{2}$	20.4204	33.183	.6	30.159	72.382
.6	20.7345	34.212	$\frac{5}{8}$	30.237	72.759
$\frac{5}{8}$	20.8131	34.471	.7	30.473	73.398
.7	21.0487	35.256	$\frac{3}{4}$	30.630	74.662
$\frac{3}{4}$	21.2058	35.784	.8	30.787	75.429
.8	21.4628	36.316	$\frac{7}{8}$	31.023	76.588
$\frac{7}{8}$	21.5985	37.122	.9	31.101	76.977
.9	21.6770	37.392			
			10.0	31.416	78.540
7.0	21.991	38.484	.1	31.730	80.118
.1	22.305	39.592	$\frac{1}{8}$	31.808	80.515
$\frac{1}{8}$	22.383	39.871	.2	32.044	81.713
.2	22.619	40.715	$\frac{1}{4}$	32.201	82.516
$\frac{1}{4}$	22.776	41.282	.3	32.358	83.323
.3	22.933	41.853	$\frac{3}{8}$	32.594	84.540
$\frac{3}{8}$	23.169	42.718	.4	32.672	84.948
.4	23.247	43.008	$\frac{1}{2}$	32.986	86.590
$\frac{1}{2}$	23.562	44.178	.6	33.300	88.247
.6	23.876	45.364	$\frac{5}{8}$	33.379	88.664
$\frac{5}{8}$	23.954	45.663	.7	33.615	89.920
.7	24.190	46.566	$\frac{3}{4}$	33.772	90.762
$\frac{3}{4}$	24.347	47.173	.8	33.929	91.609
.8	24.504	47.783	$\frac{7}{8}$	34.164	92.885
$\frac{7}{8}$	24.740	48.707	.9	34.243	93.313
.9	24.818	49.016			
			$11\frac{1}{4}$	36.343	99.40
8.0	25.132	50.265	$\frac{1}{2}$	36.128	103.87
.1	25.446	51.530	$\frac{3}{4}$	36.913	108.44
$\frac{1}{8}$	25.525	51.848			
.2	25.761	52.810	$12\frac{1}{4}$	38.484	117.86
$\frac{1}{4}$	25.918	53.456	$\frac{1}{2}$	39.270	122.72
.3	26.075	54.106	$\frac{3}{4}$	40.055	127.68
$\frac{3}{8}$	26.310	55.088			
.4	26.389	55.417	$13\frac{1}{4}$	41.626	137.89
$\frac{1}{2}$	26.703	56.745	$\frac{1}{2}$	42.411	143.14
.6	27.017	58.088	$\frac{3}{4}$	43.197	148.49
$\frac{5}{8}$	27.096	58.426			

Areas and Circumferences of Circles
(SSee note at end of tables)

Diameter	Circumference	Area	Diameter	Circumference	Area
$14\tfrac{1}{4}$	44.767	159.49	$18\tfrac{1}{4}$	57.334	261.59
$\tfrac{1}{2}$	45.553	165.13	$\tfrac{1}{2}$	58.119	268.80
$\tfrac{3}{4}$	46.338	170.87	$\tfrac{3}{4}$	58.905	276.12
$15\tfrac{1}{4}$	47.909	182.65	$19\tfrac{1}{4}$	60.475	291.04
$\tfrac{1}{2}$	48.694	188.69	$\tfrac{1}{2}$	61.261	298.65
$\tfrac{3}{4}$	49.480	194.83	$\tfrac{3}{4}$	62.046	306.36
$16\tfrac{1}{4}$	51.051	207.39	$20\tfrac{1}{4}$	63.617	322.06
$\tfrac{1}{2}$	51.836	213.82	$\tfrac{1}{2}$	64.402	330.06
$\tfrac{3}{4}$	52.621	220.35	$\tfrac{3}{4}$	65.188	338.16
$17\tfrac{1}{4}$	54.192	233.71	21.0	65.973	346.36
$\tfrac{1}{2}$	54.978	240.52			
$\tfrac{3}{4}$	55.763	247.45	22.0	69.115	380.13

Notes

From these tables circumferences and areas of circles with a diameter of any whole number (in, mm, etc.) up to 110 can be obtained. Thus, the circumference of 74 is read from the value opposite 7.4 and is 232.47, the decimal point being moved one place to the right. The corresponding area is 4 300.8, the point being moved two places to the right.

Diameters are given in 8ths and 4ths, as well as 10ths, so that British units up to 21 can be dealt with. The whole numbers not given, e.g., 11, 12, etc., are read from the first part of the table, since it will be clear that the figures, for instance, opposite the last diameters, 21 and 22, are identical with those opposite 2.1 and 2.2, except for the positions of the decimal points.

Approximate Weights of Materials in Pounds Per Cubic Foot

Material	Weight	Material	Weight
Aluminium Cast	160	Lead	710
Aluminium Wrought	167	Limestone	158–168
Aluminium Bronze	475		
Ashes	37	Mercury	847
		Monel	554
Brass Rolled	524	Nickel Cast	516
Brass Swarf	157	Nickel Silver	516
Babbit Metal	454		
Brick, Common	85–118	Oak Wood	53.5
Bronze	534		
		Phosphor Bronze Cast	536
Cast Iron (mean)	450	Pine Wood	30
Cast Iron Turnings	140	Pig Iron (mean)	300
Cement (loose)	85	Pig Iron and Scrap as	
Chalk	140–145	charged into the cupola	336
Charcoal (lump)	18	Plumbago (graphite)	137–145
Clay	120–135		
Coal	60–80	Sand (moulding), fairly	
Coal Dust	53	dry and unrammed	75–85
Coke	28	Sand (silica)	85–90
Concrete (set cement)	140	Silver Pure Cast	656
Copper Cast	548	Steel (mean)	490
Cupola Slag	150		
		Tin Pure	453
Fire Clay	90	Water—as ice	58.7
Fire brick (in mortar)	112–125	Water at 32°F	62.4
French Chalk	162	Water at 212°F	59.6
		Wrought Iron	480
Gold (cast pure)	1,200	Zinc Cast	428
Gold 22 carat	1,090	Zinc Rolled	448

SOLDERS AND FUSIBLE ALLOYS
Fusible Alloys

Trade name or type of alloy	Composition (%) Sn	Bi	Pb	Cd	Others	Solidus °C	°F	Liquidus °C	°F
Binary Eutectic	99.25	—	—	—	0.75 Cu	227	441	227	441
Binary Eutectic	96.5	—	—	—	3.5 Ag	221	430	221	430
Binary Eutectic	—	—	—	17.0	83.0 Tl	203	397	203	397
Binary Eutectic	92.0	—	—	—	8.0 Zn	199	390	199	390
Binary Eutectic	—	47.5	—	—	52.5 Tl	188	370	188	370
Binary Eutectic	62.0	—	38.0	—	—	183	361	183	361
Binary Eutectic	67.0	—	—	33.0	—	176	349	176	349
Binary Eutectic	56.5	—	—	—	43.5 Tl	170	338	170	338
	40.0	—	42.0	18.0	—	145	293	160	320
Ternary Eutectic	51.2	—	30.6	18.2	—	145	293	145	293
Binary Eutectic	—	60.0	—	40.0	—	144	291	144	291
	48.8	10.2	41.0	—	—	142	288	166	331
Binary Eutectic—Cerrotru	43.0	57.0	—	—	—	138	281	138	281
	41.6	57.4	1.0	—	—	134	273	135	275
Ternary Eutectic	40.0	56.0	—	—	4.0 Zn	130	266	130	266
Ternary Eutectic	46.0	—	—	17.0	37.0 Tl	128	262	128	262
Binary Eutectic—Cerrobase	—	55.5	44.5	—	—	124	255	124	255
Binary Eutectic	48.0	—	—	—	52.0 In	117	243	117	243
Cerroseal-35	50.0	—	—	—	50.0 In	117	243	127	260
	1.0	55.0	44.0	—	—	117	243	120	248
Cerromatrix	14.5	48.0	28.5	—	9.0 Sb	103	217	227	440
	34.5	44.5	—	21.0	—	103	217	120	248
	25.0	50.0	—	25.0	—	103	217	113	235
Ternary Eutectic	25.9	53.9	—	20.2	—	103	217	103	217
	33.0	34.0	33.0	—	—	96	205	143	289
Malotte's	34.2	46.1	19.7	—	—	96	205	123	253
Rose's	22.0	50.0	28.0	—	—	96	205	110	230
D'Arcet's	25.0	50.0	25.0	—	—	96	205	98	208
Newton's	18.8	50.0	31.2	—	—	96	205	97	207
Onion's or Lichtenberg's	20.0	50.0	30.0	—	—	96	205	100	212
Ternary Eutectic	15.5	52.5	32.0	—	—	96	205	96	205
Ternary Eutectic	—	51.7	40.2	8.1	—	92	198	92	198
	15.4	38.4	30.8	15.4	—	70	158	97	207
Cerrosafe	11.3	42.5	37.7	8.5	—	70	158	90	194
	24.5	45.3	17.9	12.3	—	70	158	88	190
	13.0	40.0	37.0	10.0	—	70	158	85	185
	13.0	42.0	35.0	10.0	—	70	158	80	176
Lipowitz's or Cerrobend	13.3	50.0	26.7	10.0	—	70	158	73	163
Wood's	12.5	50.0	25.0	12.5	—	70	158	72	162
Quaternary Eutectic	13.1	49.5	27.3	10.1	—	70	158	70	158
	13.2	49.3	26.3	9.8	1.4 Ga	65	149	66	151
Cerrolow 147	12.77	48.0	25.6	9.6	4.0 In	61	142	65	149
Cerrolow 136B	15.0	49.0	18.0	—	18.0 In	57	136	69	156
Cerrolow 136	12.0	49.0	18.0	—	21.0 In	57	136	57	136
Cerrolow 140	12.6	47.5	25.4	9.5	5.0 In	56	134	65	149
Cerrolow 117	8.3	44.7	22.6	5.3	19.1 In	47	117	47	117
Cerrolow 117B	11.3	44.7	22.6	5.3	16.1 In	47	117	52	126
Cerrolow 105	7.97	42.91	21.7	5.09	18.33 In:4 Hg	38	100	43	110
Binary Eutectic	8.0	—	—	—	92.0 Ga	20	68	20	68

(Courtesy Tin Research Institute)

WEIGHT OF CASTING FROM WEIGHT OF PATTERN

Allowance must be made for the weight of any metal in the pattern. The patterns are without cores.

A pattern weighing 1 lb when made of	Cast iron	Cast steel	Yellow brass	Gunmetal or bronze	Bell bronze	Zinc	Copper	Aluminium
Alder	12.8	14.6	14.3	14.8	15.5	12.2	15.6	4.6
Baywood	8.8	9.8	9.9	10.3	14.0	8.5	10.5	3.2
Beechwood	8.5	9.5	9.5	10.0	12.0	8.2	10.1	3.1
Birch	10.6	11.2	11.9	12.3	12.9	10.2	12.0	3.9
Cedar	12.5	14.5	14.2	14.7	15.3	12.0	15.7	4.5
Cherry	10.7	12.0	12.0	12.6	13.5	10.4	12.8	3.9
Lime	13.4	14.2	15.1	15.6	16.3	12.9	15.3	4.9
Linden	12.0	13.3	13.5	14.1	15.0	11.6	14.3	4.3
Mahogany	8.5	9.5	9.5	10.0	10.5	8.2	10.1	3.1
Maple	9.2	10.3	10.3	10.6	10.9	8.9	11.0	3.2
Oak	9.4	10.4	10.5	10.8	11.0	9.1	11.2	3.4
Pear	10.9	12.1	12.2	12.8	14.0	10.6	13.0	3.9
Pine	14.7	16.3	16.5	16.6	17.3	14.3	17.5	5.3
Whitewood	16.4	18.1	18.4	19.3	20.0	15.9	19.5	5.9
Brass	0.84	0.98	0.95	0.99	1.0	0.81	1.04	0.31
Iron	0.97	1.09	1.09	1.13	1.18	0.93	1.17	0.35
Lead	0.64	0.75	0.72	0.74	0.78	0.61	0.8	0.23
Tin (¼ lead)	0.89	1.1	1.00	1.00	1.12	0.85	1.18	0.32
Zinc	1.00	1.16	1.13	1.17	1.22	0.96	1.18	0.36

SHRINKAGE AND CONTRACTION OF CASTING ALLOYS

Volume Changes on Cooling. Although shrinkage and contraction are more or less synonymous terms they have special application with respect to castings. Three stages of volume changes are recognised.
A. Liquid shrinkage which takes place during the fall in temperature from pouring to freezing.
B. Shrinkage during freezing or solidification which with some alloys may be offset to some extent by an accompanying expansion.
C. Solid shrinkage or contraction which occurs in the solidified casting as it cools to atmospheric temperature.

The following table gives some idea of the reduction in volume which takes place in various casting alloys during cooling from an average pouring temperature to their freezing point.

	Liquid shrinkage vol %
Soft grey iron (3.5% to 4.0% total carbon)	0.6
Aluminium alloy (11% to 13% silicon)	3.5
Copper (deoxidised)	3.8
Aluminium bronze (90%) copper, 10% aluminium)	4.1
Grey iron (3.0% to 3.25% total carbon)	4.2
Manganese bronze (40% zinc, 1.25% iron, 1% aluminium, 0.5% manganese, 0.5% tin)	4.6
Grey iron, high duty (2.5% to 2.75% total carbon)	5.0
Nickel silver (20% nickel, 15% zinc, 65% copper)	5.5
Ni-resist austenitic cast iron	5.6
White iron, chill cast	5.75
Nickel (98% nickel, 1.5% silicon, 0.1% carbon)	6.1
Monel (1.0% silicon)	6.3
Yellow Brass (27% zinc, 2% lead, 1% tin)	6.4
Nickel silver (20% nickel, 15% zinc, 4% tin, 5% lead)	6.5
Aluminium rich alloys not containing silicon	6.5 to 8.0
Steel	7.22
White iron, malleable	7.25
Bearing bronze (10% tin, 10% lead)	7.3

These values are a summation of the total volume change which occurs under headings A and B. Where coring is considerable and with "hard" cores, the expansion of the cores, and to some extent moulds, will neutralise some of the shrinkage tendencies. In some cases liquid metal may exude through the risers or feeders as a result of such expansion.

To reduce shrinkage defects and porosity in castings the liquid and solidification shrinkage (A and B) must be made good by a supply of liquid metal from an adequate system of feeding. Some information on the design of feeding heads is given under a separate heading.

The shrinkage or contraction under heading C is the factor with which the pattern maker is concerned. Pattern dimensions are made to "contraction rule" and not "standard rule". Different casting alloys have different contraction rates. Although average values for unrestricted cooling conditions are available as set out in the following table taken from the *Metal Industry Handbook*, it must be borne in mind that these can vary considerably according to the resistance offered by moulds and cores.

Alloy		Patternmakers' Contraction Allowance
Aluminium alloys		
Commercially pure		
Al–Cu–Si alloys	e.g. LM4	
Al–Si alloys	e.g. LM6, LM28, LM29, LM30	
Al–Cu alloys	e.g. LM12	1.3% or 1/75
Al–Mg 5–10%	e.g. LM5, LM10	
Low expansion Al–Si–Ni	e.g. LM13	
Aluminium bronze		2.0–2.3%
Beryllium-Copper		1.6%
Bismuth		1.3%
Brass, yellow (thick)		1.3%
Brass, yellow (thin)		1.6%
*Cast iron, grey		0.9–1.3%
Cast iron, white		2.0%
Copper		1.6%
Delta Bronze		1.6%
Gunmetal		1.0–1.6%
Lead		2.6%
Magnesium		2.0%
Magnesium alloys (5%, or over of alloys, excluding cadmium, which has no effect on the shrinkage of magnesium)		1.6%
Magnanese bronze		2.0%
Monel		2.0%
Nickel		2.0%
Nickel silver		1.0–1.6%
Phosphor–bronze		1.0–1.6%
Silicon bronze		1.3–1.6%
Steel, carbon		1.6–2.0%
Steel, chromium		2.0%
Steel, manganese		2.6%
Stone's gear wheel bronze		1.0%
Tin		2.0%
White metal		0.6%
Zinc and zinc-base alloys		2.6%

*The amount of contraction in cast iron depends chiefly upon the speed with which the casting cools, greater contraction resulting from more rapid cooling, and vice versa. Large hollow castings contract less in a vertical than in a horizontal direction.

AIR CONSUMPTION OF VARIOUS PNEUMATIC TOOLS

TAKEN FROM *FOUNDRY*

Small spray guns		2–3 ft^3/min						
Jar ram or jolters (platform type)		30–40 ft^3/ton						
Hand grinders—wheel diam., in		$1\frac{1}{2}$	2	4	6	8		
	ft^3/min	12	16	28–35	40–45	50–60		
Hand sanders	pad size, in	7	9					
	ft^3/min	35	45					
Chipping hammers	stroke, in	$1\frac{1}{2}$ to 4 average 20–25 ft^3/min						
Scaling hammers	average	15 ft^3/min						
Air hoists	capacity, lb	500–1000	2000	3000	4000	6000		
	ft/min	0.5	2	3.2	6.3	8.4		
Geared hoists	capacity, tons	1	$1\frac{1}{2}$	2	3	4	5	6	8	10
	ft^3/min per ft lift	3	5	6	8	10	15	20	25	30
Impact wrenches	size, in	$\frac{1}{4}$	$\frac{3}{8}$	$\frac{5}{8}$	$\frac{3}{4}$	$1\frac{1}{4}$	$1\frac{3}{4}$		
	ft^3/min	8–9	12	22	28	38	50		
Jacks	cyl. diam., in	8	10	12	14	16	20	24	
	ft^3/min per lit	1.8	2.8	4.0	5.0	6.9	11.1	16	
Motors	horse power	2	4	5	8	15			
	ft^3/min	40–50	60–70	90–100	140	240			
Rotary steel drills	drill size, in	$\frac{1}{4}$	$\frac{5}{8}$	$\frac{1}{2}$	$\frac{7}{8}$	1	$1\frac{1}{4}$–2		
	ft^3/min	20	28–32	36–42	55	65	85–90		
Wood boring drills	drill size, in	$\frac{1}{2}$	$\frac{7}{8}$	1	$1\frac{1}{4}$				
	ft^3/min	20	32	38	45				

Note

Multiply ft^3/min by 472.0 to obtain cm^3/sec.

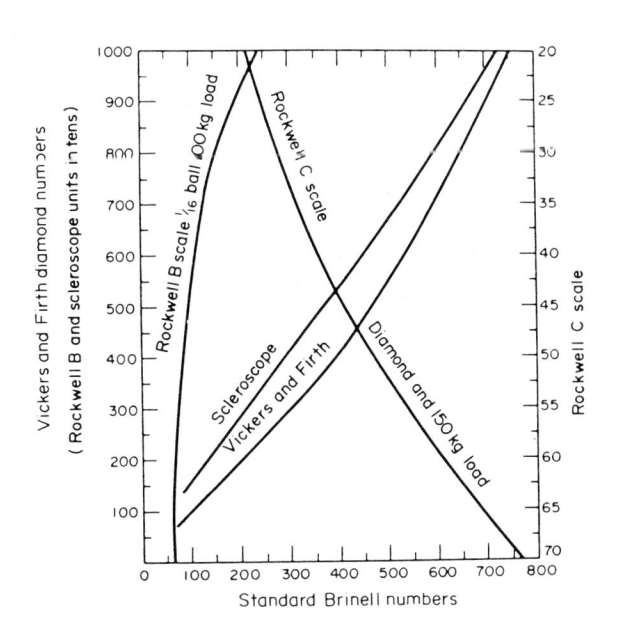

Fig. 1.8. Hardness Conversion Chart. Brinell-Rockwell-Scleroscope-Vickers.

Seger Cones

These are cones made of refractory materials which bend at a definite temperature on heating. The following table gives the bending temperature, in degrees centigrade, for each cone.

Approx. temp. in degrees centrigrade	Cone no.	Approx. temp. in degrees centrigrade	Cone no.	Approx. temp. in degrees centrigrade	Cone no.
600	022	1080	01A	1560	26
650	021	1100	1A	1610	27
670	020	1120	2A	1630	28
690	019	1140	3A	1650	29
710	018	1160	4A	1670	30
730	017	1180	5A	1690	31
750	016	1200	6A	1710	32
790	015A	1230	7	1730	33
815	014A	1250	8	1750	34
835	013A	1280	9	1770	35
855	012A	1300	10	1790	36
880	011A	1320	11	1825	37
900	010A	1350	12	1850	38
920	09A	1380	13	1860	39
940	08A	1410	14	1920	40
960	07A	1435	15	1960	41
980	06A	1460	16	2000	42
1000	05A	1480	17		
1020	04A	1500	18		
1040	03A	1520	19		
1060	02A	1530	20		

Figures supplied by Wenger's Ltd., Etruria.

USUAL THICKNESS OF CHILLS FOR CHILLED ROLLS

Diameter of roll, in	Thickness of chill, in	Diameter of roll, in	Thickness of chill, in	Diameter of roll, in	Thickness of chill, in
3	$2\frac{1}{2}$	13	5	23	$8\frac{1}{2}$
4	$2\frac{1}{2}$	14	$5\frac{1}{4}$	24	9
5	3	15	$5\frac{1}{2}$	25	$9\frac{1}{4}$
6	3	16	6	26	$9\frac{3}{4}$
7	$3\frac{1}{4}$	17	$6\frac{1}{4}$	27	10
8	$3\frac{1}{4}$	18	$6\frac{1}{2}$	28	$10\frac{1}{4}$
9	$3\frac{3}{4}$	19	7	29	$10\frac{1}{2}$
10	$3\frac{3}{4}$	20	$7\frac{1}{4}$	30	11
11	$4\frac{1}{4}$	21	$7\frac{1}{2}$		
12	$4\frac{1}{2}$	22	8		

Multiply figure in inches by 25.4 to obtain metric equivalent in millimetres.

METHOD FOR FINDING THE WEIGHT OF MOLTEN METAL IN POURING LADLES

The diameter of crane ladles is usually greater at the top than the bottom. Thus the sides slope inwards with a uniform taper. As the angle of slope is not very great, the effective diameter may be taken as the average of the top and bottom diameters.

Thus the diagram may be taken to represent the internal dimensions of a ladle 300 mm diameter at the top and 200 mm diameter at the base. Therefore, the average or effective diameter is 250 mm, which is the diameter at half depth.

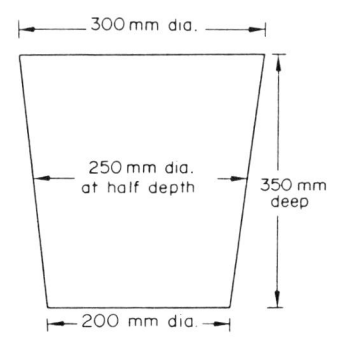

Fig. 1.9.

To obtain the volume of the ladle, the average diameter is squared and multiplied by 0.7854, which gives the average cross-sectional area. This, in turn, is multiplied by the depth so as to get the volume. Keeping to the example illustrated:

$250 \times 250 \times 0.7854 = 49\,087$ mm^2, say 490 cm^2
490×35 (ht in cm) $= 17\,180$ cm^3 approximately

Different metals and alloys have different densities, the more commonly used being as follows:

Weight (g) per cm^3 of some metals and alloys (approximate)

Cast iron	7.0	Aluminium	2.7
Copper	8.9	Steel	7.9
Lead	11.4	Brass and bronze	9.0
Magnesium	1.74		

Then if it is desired to know how much molten cast iron the ladle can hold, multiply its volume in cubic centimetres by the weight of one cubic centimetre of cast iron, thus:

$$17\,180 \times 7.0 = 120\,260 \text{ g}$$

which is 120.26 kilograms.

If we take the average cross-sectional area of the ladle and a layer 1 cm deep, we get 490 cm^3, which is equivalent to a weight of $490 \times 7.0 = 3.43$ kg of cast iron. Therefore, we can say that the ladle holds approximately 3.43 kg of cast iron for each cm of depth.

It is possible, from this, to gauge the weight of any quantity of molten cast iron, in the ladle, simply by measuring its depth with an iron rod bent at right angles. For example, if the depth of liquid iron was found to be 13 cm, the weight of metal in the ladle would be $3.43 \times 13 = 44.59$ kg.

The same procedure is used for any other metal or alloy, but being careful to take its correct weight in grams per cubic centimetre as shown above.

GATING TERMS AND METHODS

Parting Gates

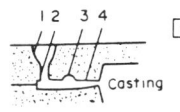

A
1. Sprue cup
2. Sprue
3. Skim bob
4. Gate (in-gate)

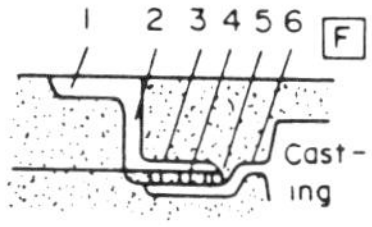

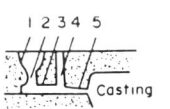

B
1. Sprue cup
2. Sprue
3. Gate
4. Relief sprue
5. Choke (in-gate)

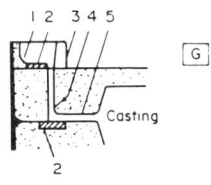

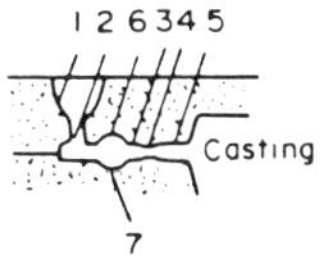

C
1. Sprue cup
2. Sprue
3. Gate
4. Choke
5. Gate
6. Skim bob (cope)
7. Skim bob (drag)

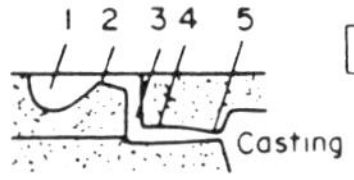

D
1. Pouring basin
2. Dam
3. Sprue
4. Gate
5. Choke

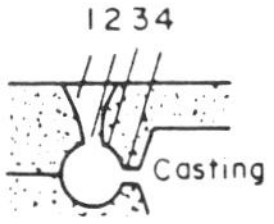

E
1. Sprue cup
2. Sprue
3. Shrink bob
4. Gate

F
1. Pouring basin
2. Sprue
3. Strainer gate (top)
4. Strainer gate core
5. Strainer gate (bottom)
6. Gate

G
1. Pouring basin
2. Splash core
3. Pouring box
4. Sprue
5. Gate

Bottom gate

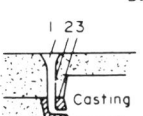

H
1. Sprue cup
2. Sprue
3. Core gate

Branch gate

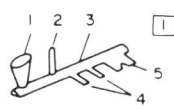

I
1. Sprue
2. Relief sprue
3. Runner
4. Finger gates
5. Flat gate

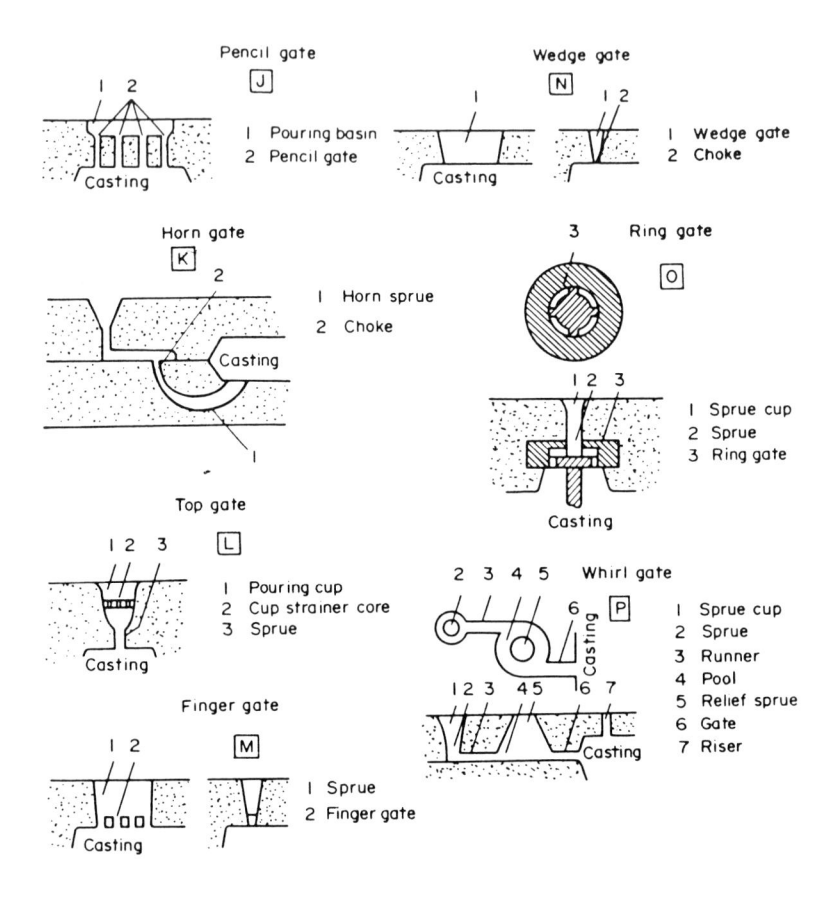

Fig. 1.10. This figure shows the nomenclature recommended by a subcommittee of the American Foundryman's Society (AFS) for the various types of gating arrangements. It is recognised that very many local terms, colloquialisms, and variants are used, but the names given are largely self-explanatory and should be identifiable everywhere.

LINING SYSTEMS FOR POURING LADLES AND HAND SHANKS

Crane Ladle Lining (proportions by weight)

    70% Amorphous crushed silica (ganister rock).
    30% Fireclay.
Add water to give required working consistency when milled. Additional water is added if the mixture is to be used as a brick jointing or facing.

Pouring Ladles and Hand Shanks

Synthetic Mix.
    43% by weight new silica sand.
    43% by weight old silica sand (e.g. old core sand).
    14% by weight bentonite.
Water to give consistency of wet moulding sand after milling.
Natural Mix.
    50% by weight of new red sand (Worksop, Mansfield, etc.).
    50% by weight of ganister mixture as used for cupola patching.
Both the above mixtures should be thoroughly milled in an edge runner type mill.

The steel ladle shells to be lined are first given a coating of wash made by diluting the lining mixture so that it can be painted on. Then straw rope is placed across the base of the ladle shell and up the sides to act as a vent for the escape of steam when drying the lining. For hand shanks, two lengths of straw rope run across the bottom and up the sides at right angles to each other, will be sufficient. Large ladles may need rather more straw rope, but in any case it is usual to drill a number of 5 mm diameter holes in the base of large ladles to act as vents for the escape of steam. After lining, the ladles should be given a thorough drying, finally at a red heat.

The lining mixture, after milling with water to develop the requisite degree of plasticity is placed in position in the bottom of the ladle by hand and then extended up the side walls. It is smoothed off with a water brush prior to drying as described. Thorough drying of the lining is most important. Sometimes the lined ladles are given a coating of plumbago or similar black wash when they are appreciably warm. This helps in keeping them clean and free from adhering dross and "build up".

OTHER LINING METHODS

Following on from the foregoing the refractory manufacturers were not slow to introduce pre-mixes and rammable compositions so that foundries no longer had to obtain and mix their own often less than ideal materials. It was then only necessary for the foundry to obtain an internal former for any given ladle, the lining being rammed up with it in position. Some large ladles could indeed be gunned using the type of refractory gun used for patching cupolas or large furnaces. The larger crane and carrying ladles could also be bricked or sometimes duplexed with a brick backing and a monolithic facing. Whatever the lining method however the ladle always required a lengthy drying period at low heat followed by an extensive pre-heat at high temperature before molten metal could be introduced. Where ladles are used only intermittently—normal for large capacities—the costs of ladle heating could be very high indeed.

Cold Lining Systems

There are important economic advantages in developing some system that avoids extended drying and pre-heating of all types of metal transporting methods crane ladles, pouring ladles and hand shanks, launders, etc. Obviously this involves the development and availability of pre-formed shapes in refractory materials that will withstand the temperature of the molten metal and at the same time ideally have a sufficiently high insulation value as to render pre-heating, or at least a lengthy pre-heat, unnecessary. Such materials are now available and are being continuously improved.

INSURAL

In the aluminium field where metal temperatures usually lie between about 600° and 900°C, pre-cast and pre-dried shapes are available as board, pre-formed launders, hot tops, downspouts, floats, ladle linings, pouring bushes, one-piece furnace and filter box bodies, and a host of other similar applications. Because of the specialised natures of the mixing and drying cycles INSURAL can only be supplied as a shape but INSURAL adhesive and mastic are available to provide bond and sealant respectively for construction work. INSURAL is white in colour, easy to cut and shape, is not wetted by molten aluminium and although not a structural material is reasonably strong.

KALTEK

For cast irons and steel a different approach has been necessary using different materials. To cope with a temperature range of 1200°–1700°C a new non-wettable, non-ceramic and therefore slightly flexible and permeable composition had to be developed. This is known as KALTEK and is available as boards which fit together to form a disposable ladle lining or, for the small hand ladles, as a pre-formed insert lining.

The ladle, shank, tundish or whatever, is furnished with a bricked or rammed permanent lining inside which the Kaltek lining is built up. The space between the two—and there must be some space for venting—is filled with dry sand or crushed refractory. Small gaps between boards or between side and bottom can be filled with KALSEAL, a special air-drying refractory.

For hand shanks a pre-formed insert can be made to sit inside the normal steel shell. Plain inserts are available as are alternatives with dam boards or "teapot" spouts, internally or externally formed.

In the case of INSURAL all shapes are permanent or at least semi-permanent as they are not wetted by molten aluminium, damage being limited to that from erosion or mechanical bruising, etc. KALTEK board ladle linings are however essentially a one-trip product while the one-piece hand shank linings are intermediate lasting from one trip to a full shift depending on conditions. None of these highly insulating lining materials require the fierce pre-heating of their precursors but a gentle drying out with a lazy gas flame is usually a sensible precaution. With steel alloys the great advantages of a one-trip KALTEK disposable lining are, apart from the reduction of temperature loss during holding, the very much cleaner metal obtained when using virgin linings. After casting the ladle is simply inverted, the disposable lining drops out to be immediately replaced for the next heat. Thus each ladle is clean and new and slag-free and the improvement in casting cleanliness is clearly evident, particularly during subsequent machining.

BOTTING PRACTICE FOR CUPOLA OPERATORS

The British Cast Iron Research Association, in their publication, Broadsheet No. 3, state that a refractory mixture, based on fireclay is the most satisfactory botting material.

A botting mixture should combine the qualities of high fusion temperature, medium dry strength and low drying contraction. The BCIRA report that the following mixtures give satisfaction in practice.

Type A

Suitable for small botts and frequent tappings.
Fireclay (30–50% on clay grade)—70–88%.
Coat Dust (11% ash max., superfine grade)—10–20%.
Note: 2–10% medium fine sawdust can be used instead of coal dust if required.

Type B

Suitable where long gathering periods and large tap hole is employed.
Fireclay (as above)—50%.
Coal dust or sawdust (as above)—10–20%.
Black sand—20–38%.

Design of Tap Hole

Of the various forms of tap hole used experience shows that the design shown below gives the most consistently good results. With this design the dried plug can often be teased from the tap hole so that it is flushed away in one piece by the molten metal.

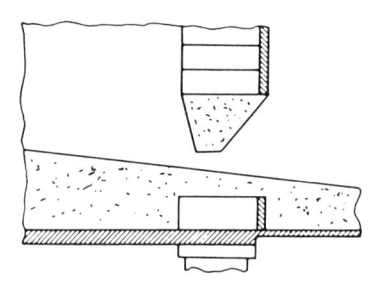

Fig. 1.11.

BOTTING PRACTICE

The mixtures are best mixed dry and water should be added gradually with thorough kneading. It is best to prepare the mixture the day before use, keeping it covered with a damp sack.

After making a conical shaped bott by hand it should be dipped in coal dust. This prevents "splutter" when the bott comes into contact with the metal stream.

The sketch below illustrates some useful features associated with good botting practice. Note the short bott stick which is easier to handle than a long one. The "splutter" shield as shown affixed to the bott stick is an added safeguard. Note also the bracket which holds spare bott sticks and the container of coal dust which is readily available. The bott stick should have a convex surface as shown. This is found to provide a more secure adhesion of the clay bott than a flat metal face.

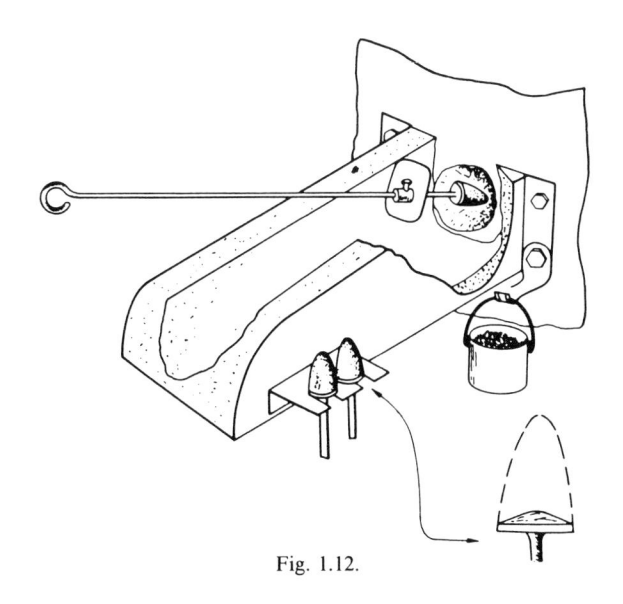

Fig. 1.12.

RAMMING MIX OR PATCHING FOR CUPOLA LINING

Ganister rock—$\frac{5}{16}$ + 20 mesh	40%
Fine silica sand	40%
Fireclay of maximum refractoriness and plasticity	20%
Moisture	7% approx.

When used as a patching or daubing material for cupolas, all the slag from the previous heat must be removed from the brickwork or lining, otherwise when the furnace is in blast the old slag will melt and allow the new patching material to come away.

METAL FILTRATION

The attractive idea of filtering out non-metallic inclusions, slags, etc. from molten metal has been attempted for many years with little real progress being made, mainly due to non-availability of suitable filter materials. Aluminium gravity diecasters in particular have used materials such as glass and steel wool, perforated steel sheet, asbestos string, woven fibre-glass and stainless steel wire, etc., with varying degrees of success and difficulty in application. Strainer cores have been common-place in the cast iron industry but the degree of actual filtering associated with them must be questionable.

Foreign inclusions in castings are a widespread problem. Much is known about their origin and behaviour; the inclusions can arise from several sources, including general melting debris, pieces of refractory, sand grains from moulds and, most commonly, slags from furnaces, oxidation, fluxing or other treatment agents.

These inclusions have numerous undesirable effects, including poor surface appearance, reduction of mechanical properties and machining difficulties. They can in many cases result in the production of scrap castings.

Methods for controlling inclusions by ladle design and preparation, by gating-system design and by use of spinners and strainer cores, are well known and widely applied. Nevertheless, despite great care and the best practice, inclusions still lead to reject castings. In addition, the use of complex gating systems designed to trap slag can be a heavy cost burden in loss of yield and the associated handling and remelting costs. If the gating system could be used only for its primary purpose of delivering metal into the mould cavity, and the task of controlling inclusions carried out by an efficient filter, then there would be the joint benefits of inclusion-free castings and improved yield.

The advent of cellular ceramic foams and the ability to weave or bond cloths made from silica and other temperature resistant filaments has recently permitted a much wider investigation of the principles and confirmed the very considerable benefits obtainable.

Initially, the wrought aluminium industry was the first to institute metal filtration on a commercial scale, using ceramic foam filters based on zirconia 50 mm thick and from 150–450 mm square and handling up to 30 tonnes of metal. Pore sizes ranged from 2.5 mm diameter for the coarser filter, 1.3 mm diameter for the intermediate to 1 mm diameter for the fine. Composite or duplex layer filters are also in use. All such large scale systems and filters must be located in a proper filter box as shown in on p. 66 and sufficient fall in metal levels must be availabled to allow priming and maintain adequate flow even allowing for some slowing down as the filter performs its function.

The logical next step was to manufacture smaller filters for use in sand and gravity die foundries and located in the running systems of individual castings. Very many such applications are now in regular use employing filters of about 15–25 mm thick and in a range of shapes from 50–150 mm square and up to 75 mm diameter. Alumina is also available as the refractory material and application has been extended to most of the copper base alloys.

Similar small individual filters can also be formed with silicon carbide as the principal refractory and these are ideally suited to the filtration of all the cast irons.

Advantages of In-mould Metal Filtration

1. Cleanliness. Of particular application to the ready oxide forming alloys such as all the cast aluminium series, but particularly the magnesium containing members, high tensile brass (Manganese Bronze), aluminium bronze and ductile (S.G.) iron. Slags, drosses, exogenous and endogenous non-metallics are kept out of the mould cavity and there are enormous benefits in casting appearance and machining characteristics.
2. Simplicity. One of the major effects of in-mould filters is the controlling and smoothing effect on metal flow. Downstream metal is clean and non-turbulent with obvious improvements in casting finish. Because of their effect as a super-choke mechanism only the simplest running systems are necessary. Large savings are therefore possible with such metals as aluminium bronze and spheroidal graphite irons where complicated and difficult to mould in-gating systems have hitherto been the order of the day.
3. Economy. The advantages described in 1 and 2 have immediate and direct effects on casting costs and visual appeal to the customer. The savings are both real and aesthetic and together can have a significant effect on customer acceptance value.
4. Improved mechanical properties. Removal of entrained discontinuous inclusions must obviously have a considerable effect on physical properties and such proves to be the case: the effect is particularly marked in flake and spheroidal graphite cast irons. See Fig. 1.13 for the effect on an aluminium alloy.

Methods of Using Ceramic Foam Filters

Use of large rigid ceramic filters for cleaning tonnages of molten aluminium for the wrought industry has previously been referred to (see p. 66) and is now standard procedure. Application to castings has

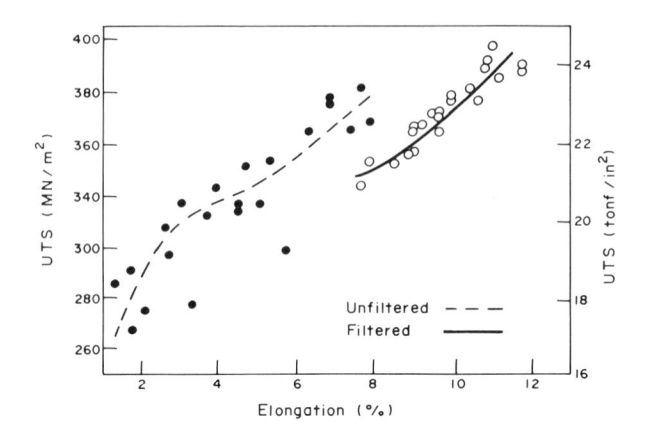

Fig. 1.13. The effects on mechanical properties of solution heat treated aluminium alloy (Cu 4.5% Mg 1.5%) castings after filtration.

necessarily to be more versatile and adaptable and some methods of including ceramic foam filters into running systems are shown in Fig. 1.14. Note that the dimensions refer to a $50 \times 50 \times 22$ mm filter: note also the increase in cross-sectional areas to compensate for any possible restrictive effect on metal flow, especially if the filter begins to clog.

For the average aluminium alloy that is reasonably clean and for most of the copper base alloys a $50 \times 50 \times 22$ mm SIVEX F filter will pass around 50 kg of metal before the flow rate becomes impractically slow. A similar weight of spheroidal graphite iron can usefully be put through a similar sized SEDEX filter but the cleaner grey and malleable irons will usually extend the capacity of a similar filter to around 150 kg. These indications are based on filters of 10 pores/in for SEDEX and 20 p.p.i. for the SIVEX F.

Comments

Positive and effective filtration of all cast alloys, at the moment with the sole exception of steel, is now a feasible and practical proposition. The benefits in terms of improved casting appearance and mechanical/machining properties will more than justify the small relative cost of the filter and there are significant ancillary savings by way of simplified ingating systems and moulding costs.

TABLES AND GENERAL DATA

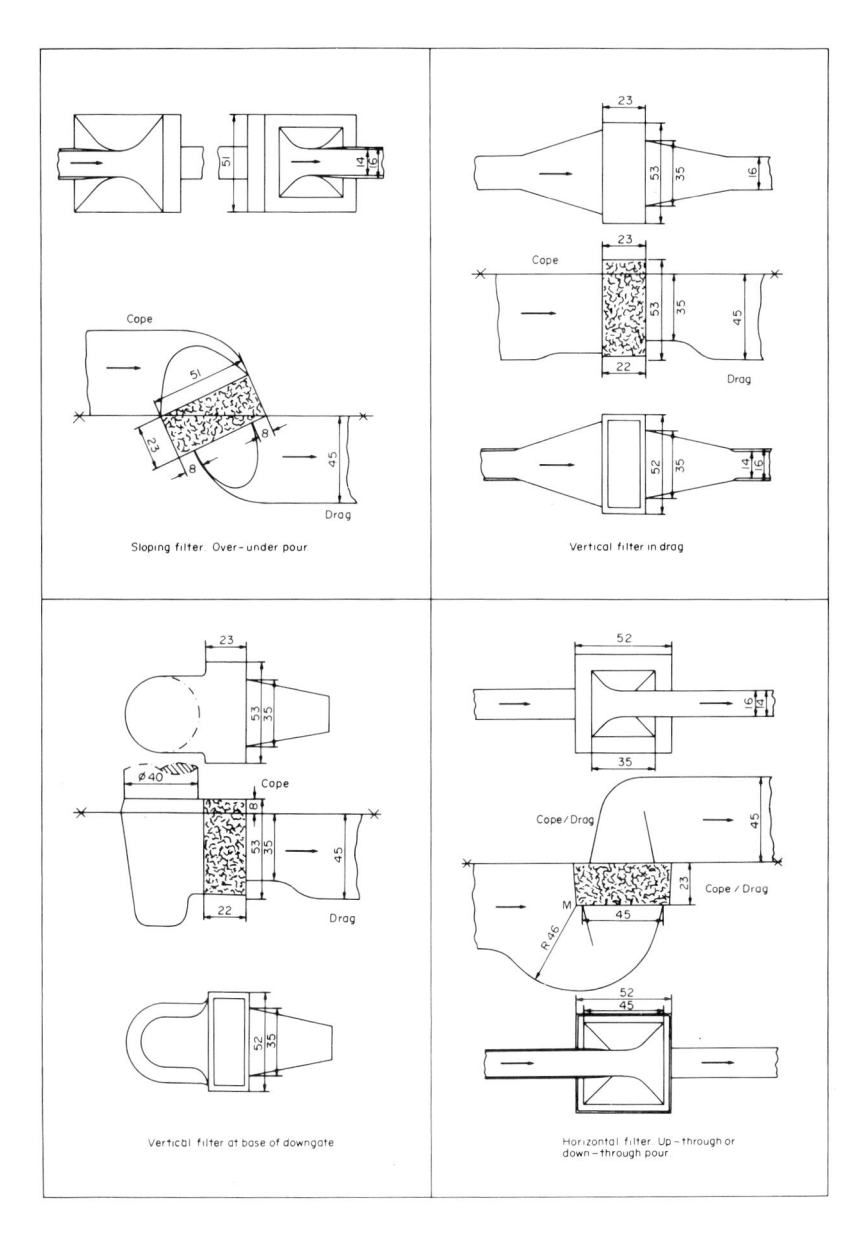

Fig. 1.14. Methods of locating ceramic foam filters in running/ingating systems

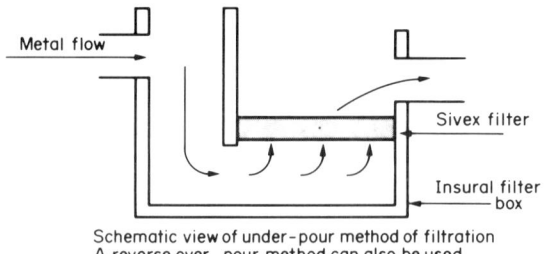

66　　　　　　　　　THE FOUNDRYMAN'S HANDBOOK

Schematic view of under-pour method of filtration. A reverse over-pour method can also be used.

Methods of Using Gauze or Woven Fibre Filters

Provided the gauze or mesh is reasonably rigid this type of filter material can be located at the box joint or even at the runner bush/cope joint if such is available. These situations may not, however, be as close to the casting as is theoretically desirable and there could be opportunities for turbulence effects to build up between the filter and the casting proper. Also the single layer mesh or gauze will only act as a primary mechanical filter and the

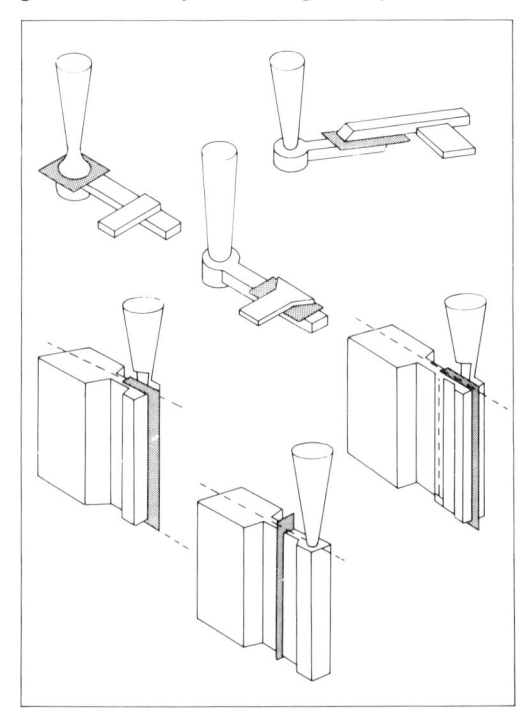

Fig. 1.15. Various applications of filter screens in moulds.

cleaning effect will almost certainly not be as great as a ceramic foam where the impingement effects due to the many internal changes of direction in the metal stream are very important.

Even so, they can have a useful and practical function, usually at appreciably less cost than a ceramic foam alternative. Figure 1.15 illustrates various methods and locations for positioning this type of filter in sand moulds. The concept is also adaptable to shell moulding and gravity diecasting.

Conclusions

The concept of filtering molten metal to improve mechanical properties, visual appearance, machining characteristic and general casting performance is becoming increasingly accepted and the use of filtered metal is beginning to appear in specifications. The major immediate advantages are apparent, as might be expected, in the "dirty" metals such as the ready oxide formers—aluminium bronze, high tensile brass, S.G. (ductile) iron and most, if not all, of the aluminium alloys. Advantages are also beginning to be seen in some specialised applications (sacrificial electrodes for example) in some of the lower melting point metals such as zinc, tin and lead and gradual extension of use into these areas is inevitable. Eventually of course the one outstanding large usage area—that of continuously cast and billet steel and steel sand castings—will also be covered as new materials are developed.

SECTION II

SANDS AND SAND BONDING SYSTEMS, ALTERNATIVE MOULDING SYSTEMS

Sand Test Data which have Proved Satisfactory for Various Castings in Different Casting Alloys

The test results recorded were obtained using the American Foundrymen's Association apparatus and are after H. Dietert.

Casting alloy	Moisture %	Permeability	Green compression lbs/in^2	Clay substance %	Fineness no. (see page 86)
Aluminium castings	6.5 to 8.5	7 to 13	6.5 to 7.5	12 to 18	225 to 160
Brass and Bronze castings	6.0 to 8.0	13 to 20	7.0 to 8.0	12 to 14	150 to 140
Copper-Nickel castings	6.0 to 7.5	37 to 50	6.5 to 8.0	12 to 14	130 to 120
Grey Iron, light stove plate castings	6.5 to 8.5	10 to 15	6.0 to 7.5	10 to 12	200 to 180
Grey Iron, squeeze moulds	6.0 to 7.5	18 to 25	6.2 to 7.5	12 to 14	120 to 87
Grey Iron, medium floor sand	5.5 to 7.0	40 to 60	7.5 to 8.0	11 to 14	86 to 70
Grey Iron, medium, synthetic sand	4.0 to 6.0	50 to 80	7.5 to 8.5	4 to 10	75 to 55
Grey Iron, heavy, green or dry sand	4.0 to 6.5	80 to 120	5.0 to 7.5	8 to 13	61 to 50
Malleable Iron castings, light	6.0 to 8.0	20 to 30	6.5 to 7.5	8 to 13	120 to 92
Malleable Iron castings, heavy	5.5 to 7.5	40 to 60	6.5 to 7.5	8 to 13	85 to 70
Steel, light green sand castings	2.0 to 4.0	125 to 200	6.5 to 7.5	4 to 10	56 to 45
Steel, heavy green sand	2.0 to 4.0	130 to 300	6.5 to 7.5	4 to 10	62 to 38
Steel castings, dry sand	4.0 to 6.0	100 to 200	6.5 to 7.5	6 to 12	60 to 45

GREEN SAND FACING MIXTURES FOR IRON CASTINGS USING NATURALLY BONDED SANDS

Casting Section	New Sand	Floor Sand	Coal Dust
up to 6 mm	6 parts	20 parts	1½ parts
6–13 mm	8 ,,	20 ,,	1¾ ,,
13–18 mm	10 ,,	20 ,,	2½ ,,
18–40 mm	15 ,,	20 ,,	3 ,,
above 40 mm	20 ,,	20 ,,	4 ,,

The above mixtures need to be well mixed or lightly ground in an edge runner type sand mill and used to cover the pattern. In each case the mould can be finished off by dusting with the TERRAPOWDER dressing which can be sleeked if required.

"New Sand" in the mixtures refers to any of the well known naturally bonded sands. In each case the mixture can be applied equally well to non-ferrous casting by omitting the coal dust and replacing it by TERRADUST dressing.

In the case of moulds for non-ferrous work the correct type of dusting dressing is TERRACOTE or TERRADUST.

All facing sand should be used as dry as possible consistent with good working. An average moisture content is about 6%.

SYNTHETIC SAND MIXTURES FOR IRON CASTINGS

Green Sand Facing (proportions by weight).
60% Old silica sand.
34% to 37% New silica sand.
1% to 4% Bentonite (dependent on green strength required).
2% Coal dust.

Core Sand for Average Castings (proportions by weight).
93% Silica sand.
5% Core oil.
2% Bentonite.

Core Sand for Heavy Section Castings (proportions by weight).
83% Silica sand.
9.5% Silica flour.
5.0% Core oil.
2.5% Bentonite.

SAND ADDITIVES FOR IRON CASTINGS

Traditionally most iron castings have been made in natural, semi-synthetic or synthetic green sand to which a very variable addition of coal dust (2–10%) has been made. The inclusion of coal dust in the mix ensures good strip of the sand from the casting and a better surface finish. Opinions differ as to how this is accomplished, but some features are probably important. Combustion of some of the coal dust as the metal enters may ensure oxygen-free reducing conditions in the mould cavity, thus reducing oxide formation and aiding apparent fluidity. Evolution of gas from the heated coal dust can also be said to produce a gas cushioning effect, thus giving finish and polish to the casting. Strong heating of coal dust also ensures the formation and deposition of a type of carbon layer, known as lustrous carbon, which is not wetted by metal and so produces improved surface finish in this way. Certainly the inclusion of certain types of volatile material in the sand can ensure castings with better strip and surface finish and can also materially help to eliminate sand expansion defects such as rat-tails, buckles, scabs, etc. Unfortunately, coal dust is extremely dirty to work with, and mill and shake-out areas can be very dusty indeed; the mixed sand is also dirty to handle.

Many if not all of these objections can be removed by the alternative use of specially blended additives such as BENTOKOL. Where the clay/volatile ratio in the sand is suitable, one or another of the BENTOKOL versions can be substituted for the coal dust. A complete substitution can often be made, but balancing additions of supplementary clay or volatile can be introduced if necessary. The use of BENTOKOL often means that a "one-shot" sand addition is sufficient and always improves working conditions. Fume after casting is reduced, and the whole foundry is cleaner and pleasanter while casting finish is maintained or often enhanced.

BENTOKOL is basically a blend of natural clays to which have been added balanced essential volatiles to produce a light coloured, virtually dust-free product. In addition, a number of other, powerful, ingredients are present to provide special properties having a far-reaching effect on the chemistry of the whole clay/volatile system. The result is a composite green sand conditioner which provides both bond and volatile content for iron foundry moulding sands, often eliminating the need for other additives.

EXAMPLE OF ROUTINE SAND CONTROL TESTS

Sample of Typical Mansfield Moulding Sand (unmilled)

Moisture	5.0 to 7.0%
Permeability (AFS)	25.0 to 30.0
Green Compression Strength (AFS)	4.5 to 6.0 lb/in^2
Clay substance	10.0 to 12.0%

Sieve Grading—British Standard Sieves (BSS 410–31).

Sieve no.	Per cent retained
8	0.04
10	0.03
16	0.10
22	0.22
30	0.55
44	2.23
60	2.22
100	10.02
150	34.70
200	28.35
300	7.31
Pan	4.05
Clay Substance	10.18

Interpretation of Sand Test Results

Although no hard and fast rules can be laid down with respect to sand properties and grain scatter, there are general guiding principles.

Excessive moisture should be avoided, as this will tend to lower the permeability and increase the possibility of blown castings. Plasticity and deformation of a sand also increase with the moisture content. Low permeability and compression strength encourage the entrapment of gas and the washing away of sand by the flow of the molten metal.

The free venting properties or the permeability of a sand will depend largely on the distribution of the sand grains on different sieves. The less scattered the grains, the higher the permeability. Therefore, maximum permeability would be obtained in sand having all the grains of the same size, that is of a single sieve grading.

In practice, it is possible to maintain a distribution where 60% to 80% of the sand grains are retained on three adjacent sieves. If the sand were composed entirely of rounded grains of uniform grading it would most likely give trouble due to scabbing and if the grains were fairly coarse, metal penetration could be expected.

Fines or silt in a moulding or core sand may be regarded as all material passing the 200 mesh sieve. Burnt out clay, coal dust and ruptured quartz grains help to form silt. Too much silt seriously lowers the permeability. On the other hand, some silt is of advantage in strengthening the bond and preventing metal penetration and scab formation. Apart from the clay substance, which has a grading equal to silt, about 5% of minus 200 mesh grade is about right.

The test values given for the specimen sand, although they show some range of properties, should be taken as a general example only. Each type of casting requires a sand with properties which vary according to the metal or alloy concerned and to the points most desired in the cast components. For example, decorative castings need fine detail and good as-cast appearance. Fine grained sands must therefore be employed, and green compression strength is obtained at the expense of permeability.

Alternatively, heavy sectioned castings call for high permeability sands. Values between 40 and 60 or even higher may be specified. Such "open" sands tend to possess a low green compression strength which may have to be made up by using additions such as bentonite.

Sands with a high silt fraction are often not workable with moisture contents of less than 7%. Coarse, open sands with grains mainly contained on the larger mesh sieves will bond with less moisture. Incidentally, they also tend to dry out more rapidly in use, and are not so easily patched, in the case of hand moulding.

ART CASTINGS

For art castings produced by sand moulding methods, a very fine grade of sand is desirable, especially in the case of small work and where detail is very intricate, a natural clay bonded sand such as Erith is suitable, but a synthetic mixture based on Ryarsh silica sand is appended:

>150 kg of Ryarsh silica sand.
>2.75 kg of bentonite.
>2.75 kg of Terradust No. 2.
>0.9 kg of coregum powder.
>4 litres of water.

First, mix all the ingredients together dry, and then add the water gradually and finish milling. The milled sand is put through an aerator.

Alternatively, use can be made of water-less, high definition, sand mixtures such as LUTRON sand that is available in the ready for use form. These are based on a very fine silica sand and are organically/oil bonded so no tempering with water is required. They are ideal for fine detail art castings, plaques, name plates, etc, in aluminium and thin section copper base alloys. Small thin section iron castings are also possible provided the mould is well vented.

The pattern is rammed up in the usual way using LUTRON sand to form the complete mould or it may be used as a facing backed up with green sand, carbon dioxide process sand or any self setting medium. After casting a thin adhering layer of burnt sand is lifted with the casting and discarded leaving the remainder to be used again. Facing sand relics are usually absorbed into the return sand (where such a system is used) without harm. It is also possible to obtain LUTRON binder on its own for incorporation into an individual dry sand mix if required. These new sand premixes are excellent for high definition work and ideal for the amateur as no mixing or tempering is required.

MOULDING SAND MIXTURES FOR COPPER-BASE ALLOYS

A variety of sand mixtures are used, both naturally and synthetically bonded with mineral clay such as bentonite. In Great Britain the naturally occurring clay bonded red sands are common and some facing sand mixtures recommended by R. F. Hudson are given below.

Green Sand Mixture

10% New Red Sand.
90% Old Floor Sand (comprising old red sand).
Add approximately 6% of water and mill for 5 minutes. If desired, 0.5% of pelleted pitch can be added to provide an improved "as-cast" surface to the casting.

Dry Sand Mixture

12% New Red Sand.
88% Old Red Sand.
Add 8% of water, mill and finally aerate.

Properties of a Typical Natural Moulding Sand

Moisture content	5.0–7.0%
Permeability No. (AFS)	25–30
Compression strength (AFS)	4.5–6.0 lb/in^2
Clay content (below 20 microns)	10.0–12.0%
Ignition loss on dried sample (volatile matter content)	Normally less than 0.5%
Flowability (AFS)	65%

Sieve Analysis: British Standard Sieve Nos.

Retained on BSS 8-mesh sieve	Nil
Retained on BSS 10-mesh sieve	0.1%
Retained on BSS 16-mesh sieve	0.3%
Retained on BSS 22-mesh sieve	0.6%
Retained on BSS 30-mesh sieve	0.8%
Retained on BSS 44-mesh sieve	1.2%
Retained on BSS 60-mesh sieve	4.3%
Retained on BSS 100-mesh sieve	18.6%
Retained on BSS 150-mesh sieve	32.9%
Retained on BSS 200-mesh sieve	15.6%
Through 200-mesh sieve	14.4%
Clay grade	11.2%

Loam Mixture

44% Old Red Sand.
44% New Red Sand.

7.5% Horse Manure.
3.5% Sawdust.
1.0% Cowhair.
Add 20% water and mix thoroughly.

Core Sand Mixtures

A suitable mixture based on natural sand and bentonite mineral clay is as follows:

80% Old Red Sand.
18% New Red Sand.
2% Bentonite.
Add $7\frac{1}{2}$ to 8% of water and mill thoroughly.

Oil Sand Cores

Core sand mixtures bonded by linseed oil and cereals, or blended semi-solid oil type binders are in widespread use and a number of suitable mixtures can be found on page 77. See also the section (page 82) concerned exclusively with core binders.

MOULDING AND CORE SAND MIXTURES FOR STEEL CASTINGS

Mixtures developed by the U.S.A. Navy Yard Foundry. The basis is silica sand of 50 to 60 mesh grading.

Green Sand Facing for Steel Castings (proportions by weight)

95% Silica sand.
4% Western bentonite.
1% Cereal type binder.
3% to 4% Water.

The sand and binders are milled dry for five minutes. Then the water is added and milling continued for an additional ten minutes. Old silica sand may be riddled and used to replace new silica sand in proportion up to 50%.

This facing has a green compression strength of about 5 lb/in^2 and is suitable for the majority of small and medium weight castings.

Backing Sand for Steel Castings (proportions by weight).

A suitable backing sand contains:

>95% Floor sand.
>5% Bentonite.
>2.5% to 4% Water.

Dry Sand Facing for Heavy Section Steel Castings (proportions by weight).

>70% Silica sand.
>23% Silica flour.
>5.5% Bentonite.
>1.5% Molasses.
>plus 6% to 7% Water.

Dry Sand Facing for Light Section Steel Castings (proportions by weight).

>75% Silica sand.
>20% Silica flour.
>4% Bentonite.
>1% Molasses.
>plus 6% to 7% Water.

Core Sand Mixture for Light Section Steel Castings (proportions by weight).

>90% Silica sand.
>6% Silica flour.
>2% Core oil.
>1% Bentonite.
>1% Cereal binder.

Core Sand for Small Cores in Average Metal Section (proportions by weight).

>80% Silica sand.
>16% Silica flour.
>2.5% Core oil.
>1.5% Bentonite.

Core Sand for Heavy Section Steel Castings
(proportions by weight).

> 43.5% Silica sand.
> 48.5% Silica flour.
> 4.0% Bentonite.
> 4.0% Core oil.

NOTE.—All sand mixtures after milling should be put through some form of aerator or disintegrator such as a Royer, before passing to the moulder. This not only improves the permeability but also increases its homogeneity, especially with respect to uniform distribution of moisture. It is also advisable that sand mixtures be prepared some hours before using, as this encourages full development of the bonding properties.

RAMMED REFRACTORY LININGS

ACID STEEL PROCESS

Rammed Lining for Side Blown Convertors

Silica ganister graded minus 15 mm mesh	50%
Silica ganister graded minus 5 mm mesh	30%
Silica flour graded 140 mesh or finer	14%
Western bentonite	6%
Water—by weight	7%

The sieve grading of the dry mixture should be:
Retained on 28 mesh sieve	53–55%
Through 28 and on 65 mesh sieve	12–14½%
Through 65 mesh sieve	33–36½%

Mill the ingredients with the water for eight minutes.

After placing in position in the converter, dry slowly at 315° C. up to 36 hours. Heat slowly from 315° C. to 675° C. in from 3 to 4 hours. Pre-heat from 675° C. to 1,095° C. before charging in the molten iron.

Rammed Lining for Bottom Pour Steel Ladle

> 30% Rock ganister.
> 30% Fireclay.
> 30% Coarse silica sand.

SAND BONDING, ALTERNATIVE MOULDING SYSTEMS 79

This mixture is rammed with pneumatic rammers around a metal covered wooden former, which is afterwards withdrawn. It is important that the ganister mixture is added round the former in small quantities and rammed down hard before adding more. This is to ensure the maximum density of the rammed up material.

The bottom is rammed in next, tying in carefully with the side wall material. A clear spot is left in the nozzle area, the nozzle inserted, and the ganister rammed hard around it. Then the bottom is shaped and smoothed over into the nozzle well. Brush over the completed ladle lining with fireclay slurry, set aside, and allow to air dry for at least 48 hours. After this the ladle is dried on low fire at about 540° C. for 8 hours. It is subsequently heated to a cherry red 875° C. for at least 15 hours before use.

A high quality ganister rock has the following analysis:

Silica	98.82%
Ferrous Oxide	0.09%
Alumina	0.73%
Magnesia	0.01%
Titania	0.02%
Total alkalies	0.03%
Ignition loss	0.30%

COMPARISON OF STANDARD SIEVES

British Standard Sieve No.	American Society for Testing Materials Sieve No.	Tyler Sieve No.	German Din 1171 Sieve No.	BSS Aperture Diameter	ASTM Aperture Diameter	Tyler Aperture Diameter	Din 1171 Aperture Diameter
8	—	9	—	2.06 mm	—	2.00 mm	—
10	12	10	4	1.68 mm	1.68 mm	1.68 mm	1.5 mm
16	18	16	6	1.00 mm	1.00 mm	1.00 mm	1.0 mm
22	25	24	8	0.70 mm	0.71 mm	0.71 mm	0.75 mm
30	35	32	12	0.50 mm	0.50 mm	0.50 mm	0.5 mm
44	45	42	16	0.35 mm	0.35 mm	0.35 mm	0.4 mm
60	60	60	24	0.25 mm	0.25 mm	0.25 mm	0.25 mm
100	100	100	40	0.15 mm	0.15 mm	0.15 mm	0.15 mm
150	140	170	60	0.10 mm	0.10 mm	0.09 mm	0.10 mm
200	170	200	80	0.08 mm	0.09 mm	0.08 mm	0.08 mm
300	270	270	E 110	0.05 mm	0.05 mm	0.05 mm	0.05 mm
350	400	—	E 185	0.03 mm	0.04 mm	—	0.03 mm

COMPARISONS BETWEEN SIEVE ANALYSIS AND INDEX NUMBERS
BATHGATE 60

Old British Standard			New ISO² Series						ISO² Series Converted to Mesh Nos.				
Mesh No.	% Ret'd	Mult	Prod	Apper size	% Ret'd	Mult	Prod	Mesh No.	% Ret'd	Mult	App Product
30	2.87	22	63	500	2.93	600	1758	30	2.93	22	64
44	10.22	30	307	355	10.70	425	4548	44	10.70	30	321
60	34.90	44	1536	250	35.00	300	10500	60	35.00	44	1540
72	15.30	60	918	180	27.04	212	5732	85	27.04	60	1622
100	26.46	72	1905	125	17.64	150	2646	120	17.64	85	1499
150	8.90	100	890	90	6.29	106	667	170	6.29	120	755
200	1.21	150	182	63	0.30	75	23	240	0.30	170	51
PAN	0.02	200	4	PAN	TR	38	0	PAN	TR	240	0
TOTAL	99.88	—	5805	TOTAL	99.90	—	25874				5852

$\dfrac{5805}{99.88} = 58.2$ AFS

$\dfrac{25874}{99.90} = 259$ AGS

$\dfrac{5852}{99.90} = 58.6$ AFS

SOUTHPORT 60

Old British Standard				New ISO² Series					ISO² Series Converted to Mesh No.				
Mesh No.	% Ret'd	Mult	Prod	Apper Size	% Ret'd	Mult	Prod	Mesh No.	% Ret'd	Mult	Product
30	TR	22	0	500	TR	600	0	30	TR	22	0
44	1.38	30	41	355	1.42	425	604	44	1.42	30	43
60	36.44	44	1603	250	36.24	300	10872	60	36.24	44	1595
72	26.56	60	1594	180	40.70	212	8628	85	40.70	60	2442
100	33.58	72	2418	125	20.00	150	3000	120	20.00	85	1700
150	2.00	100	200	90	1.36	106	144	170	1.36	120	163
200	TR	150	0	63	TR	75	0	240	TR	170	0
PAN	TR	200	0	PAN	TR	38	0	PAN	TR	240	0
TOTAL	99.96	—	5856	TOTAL	99.72	—	23248	TOTAL	99.72	—	5943

$\dfrac{5856}{99.96} = 58.6$ AFS

$\dfrac{23248}{99.72} = 233$ AGS

$\dfrac{5943}{99.72} = 59.6$ AFS

RESIN-BONDED SAND

Resins are liquids or gums, natural or synthetic, where the individual molecules have the capacity to polymerise or fuse together to form very long chains. When this happens the resins harden to form a strong solid block of material that is extremely inert and forms a powerful bond with other materials. This polymerising reaction can be triggered by certain chemical reagents, usually a strong acid, and by heat.

The possibility that such a material could be used as a sand bonding medium for foundry work was recognised at an early stage. As a result there are now very many resins and resinous mixtures available to foundries so that a particular choice becomes quite complicated. The following section does not set out to be a complete survey of all the different materials available but hopefully it will provide some introduction to the different types of resin and resin derivatives in addition to suggesting the routes by which development may well proceed. One of the big disadvantages of resins in a modern society is that because they are essentially organic and based on formaldehyde there are considerable environmental difficulties in disposing of waste and used material. The acid catalysts that are used as setting agents also add to these difficulties particularly in atmospheric pollution within the foundry itself when the moulds are cast. It is clear therefore that development was bound to take note of these characteristics and endeavour to alleviate them. Currently resin percentages are as low as around the 1% level by weight of sand and the newer bonds, some coupled with chemically set silicate, mean that reclamation is simpler and environmental impact much reduced. There is no doubt that continuing efforts will be made to further increase the acceptability of sand bonding materials at the same time improving the possibilities and economics of reclaim and re-use of the sand itself.

FURAN Resin Binders

There are basically four types of cold set furan resin binders available, each producing a range of physical properties. They may be classified as follows:

1. Urea Formaldehyde/Furfuryl Alcohol (UF/FA).
2. Phenol Formaldehyde/Furfuryl Alcohol (PF/FA).
3. Urea Formaldehyde/Phenol Formaldehyde/Furfuryl Alcohol (co-polymers).
4. Phenol Formaldehyde (PF, Furfuryl Alcohol-free).

UF/FA TYPES. These vary in FA content between 30% and 90% with varying nitrogen and water contents. These types of binder promote high strengths and good "throughcure" characteristics. However, in certain cases the nitrogen content is fairly high which can create problems with sand reclamation. Suitable for use with aluminium and low grade irons. Some ironfoundries are particularly sensitive to nitrogen contents of the sand in relation to the occurrence of "nitrogen fissuring"—a well defined type of evolved gas defect—in the castings. However, the problem is not universal.

PF/FA TYPES. Have FA contents in the region of 30% and 70%. These types of resin exhibit a lower performance in terms of "throughcure" and cold strength development in comparison with the UF/FA types. Due to their reduced nitrogen contents they are more acceptable for reclamation purposes. Suitable for Heavy Duty Iron, S.G. Iron and Steel.

CO-POLYMER TYPES. Vary in FA contents between 40% and 85%. These resins exhibit low nitrogen levels yet maintain reasonable strength development. Suitable for Heavy Duty Irons, S.G. Iron and Steel.

PF Types

These resins are furfuryl alcohol- and nitrogen-free. They are generally more cost effective over other resin types although higher addition rates are usually required, i.e. 1.5%–2.0%. These types also exhibit a plastic stage during the curing process which facilitates easier stripping from poor quality patterns. The hot strength however is high and can lead to hot tearing and finning defects. Suitable for use with most ferrous metals.

Acid Catalysts for Use with Cold Set FURAN Resins

The most common types of catalysts used with furan resins are:
1. Phosphoric acid (65–85%).
2. Paratoluene Sulphonic acid (PTSA).
3. Xylene Sulphonic acid (XSA).
4. Blended catalysts with additions of sulphuric acid.

Phosphoric acid and phosphoric/sulphuric acid blends are only recommended for use with UF/FA types. Reclamation is not usually possible with these types of catalyst due to a build up of phosphates in the sand leading to inferior cured strength and phosphorous pick up in the metal. Paratoluene sulphonic and xylene sulphonic acids can be used with all types of resin binder. These acids are highly recommended for sand reclamation due to the ease of burn-out with the resin during casting.

In certain instances when more rapid strip times are required these faster, blended, catalysts may be employed.

The terms hot and cold box (see below) are used to describe systems that require either heat as a setting agent for the resin ("hot" box) or an injected gas ("cold" box) to initiate the setting reaction.

Both types of sand are precoated either by the individual foundry or from a commercial supplier. There is a wide range of sand gradings and resin percentages available to accommodate practically any requirement. In the case of hot box and shell moulding sand the coating consists of a resin/catalyst mixture that is inactive at ambient temperature but is triggered by exposure to temperatures of the order of 2–300°C. Dumping the sand on to a heated pattern plate or blowing it into a hot core box will start an exothermic setting reaction that in just a few seconds will give the core, etc. enough strength to be handled and removed from the pattern. The reaction will then proceed to completion independently and develop full mechanical properties in the sand compact.

Cold box sands on the other hand are precoated with a resin/activator layer and are tucked, blown or rammed into the box. The hardener gas is then blown through ("pulsed" gassing is the more modern method), reacts with the resinous layer and induces a set by one chemical means or another.

The heat setting systems are expensive in energy terms and a fairly pungent fume is produced on casting. The cold box systems often utilise a hazardous gas that requires total enclosure at the gassing station although processes based on methyl formate are less dangerous in this respect.

COLD BOX SYSTEMS

SO_2 Process

The SO_2 Process can employ either furan or phenolic resins as the basis of the bond. The sand and resin mixture being mixed with a small proportion of peroxide which supplies oxygen to the subsequent reaction. SO_2 is passed through the sand mixture, the gas reacts with the peroxide to produce SO_3 which combines with water in the binder to produce Sulphuric Acid, promoting rapid curing of the sand bond.

Advantages of this process being rapid cure times (5–15 seconds), high flowability of the sand and good core storage.

Amine/Phenolic Urethane Process

This process has found widespread application in both long series automatic production of cores and in the production of smaller jobbing type cores of modest size.

The bond is based on the use of a 2 part isocyanate, part one being a phenolic resin, and the second part an isocyanate contained in a solvent. The resultant reaction promoted by a reaction with Triethylamine (TEA) or dimethylethylamine (DMEA) vapour is the formation of a solid polyurethane resin.

Phenolic Ester Process

This is a recent cold box technique which is based on an alkali-catalysed phenolic resin which is subsequently hardened by the vapour of a volatile ester, namely Methyl Formate. Advantages of this system are low toxicity, the chemicals used being phenol formaldehyde which has been in use in foundries for some 30 years and Methyl Formate which has a TLV of 100 ppm. Low odour at the coremaking station, excellent surface finish in the final casting, good core storage are some of the advantages that this system offers.

Phenolic Ester Cold Setting System

The process comprises an alkaline phenolic liquid resole type resin which is cured by the addition of specially formulated organic esters.

Environmentally the process with its low fume emission on mixing and casting is most acceptable. In addition the products used are nitrogen and sulphur free.

The system is finding widespread use especially in steel foundries where excellent casting surface finishes are produced. Excellent breakdown after casting is one of the major benefits of this system, overcoming hot tearing and finning problems.

HOT BOX SYSTEM

UF/FA and UF Resins suitable for most metals except steel. FA contents control the ease of breakdown, this being faster as FA percentage decreases. Straight UF Resins have very good breakdown and are usually used with light alloys.

PF Resins—are nitrogen free or very low in nitrogen and are usually used on steels.

UF/PF Resins are suitable for most types of grey iron castings, being low in nitrogen without the problems of brittle bond normally associated with nitrogen free resins.

The Hot Box System is suitable for a wide range of core making applications in ferrous and non-ferrous foundries. Mixed sand has good flowability allowing the blowing of complex shapes with pressures of 80–100 psi. Cure rates of 5 seconds are possible although 10–20 seconds is more typical. After ejection from the box, cure continues until complete through cure is attained.

HOW TO CALCULATE SAND FINENESS NUMBER

The equipment required is a nest of standard laboratory sieves, a sieve vibrating mechanism and a sensitive balance accurate to plus or minus 0.5 g.

Test Procedure

1. Weigh a 100 g sample of the sand when it is perfectly dry.
2. Place the entire sample into the top (coarsest) sieve, place nest of sieves on vibrator, vibrating for 15 minutes.
3. Remove sieves, and, beginning with the top sieve, weigh the quantity of the sample remaining on each sieve.
4. Arrange the respective weights in a column as shown below, adjacent to their sieve mesh number (either the British standard sieve numbers or the AFS American standard sieves may be used).
5. Multiply each separate sieve weight by the **preceding** sieve mesh number.
6. Divide the total product by the total sample weight and this produces the Fineness Number, which is the AFS Fineness Number when calculating by this method.

Sieve No. British Standard	Sand Retained on Sieve (g)	Multiplied by Previous Sieve No.	Product
+10	Nil.	—	—
+16	Nil.	—	—
+22	0.2	0.2 × 16	3.2
+30	0.8	0.8 × 22	17.6
+44	6.7	6.7 × 30	201.0
+60	22.6	22.6 × 44	1104.4
+100	48.3	48.3 × 60	2898.0
+150	15.6	15.6 × 100	1560.0
+200	1.8	1.8 × 150	270.0
Through 200	4.0	4.0 × 200	800.0
		Product =	6854.2

$$\text{AFS Fineness No.} = \frac{\text{Product}}{\text{Weight of Sample,}} \quad \text{or} \quad \frac{6854.2}{100}$$

Which is 68 AFS

SILICATE-BONDED SAND

INTRODUCTION

Silica sand bonded with sodium silicate exhibits some clear advantages over other bonding materials and methods. The original Carbon Dioxide Process utilised the gas carbon dioxide as the hardening agent. Other methods of inducing hardening have been suggested, usually by incorporating a powdered reagent into the mix. These have included such materials as anhydrite, di-calcium silicate, cement, ground blast furnace slag, powdered ferro-silicon (the Nishyama Process), etc. but all have suffered from some disadvantage. Recently a versatile range of organic ester hardeners has superseded almost all other forms of catalyst hardening.

In addition to the true silicate bonded processes, i.e. those where the silicate forms the major or essential part of the bond, there is a number of other proprietary systems of intermediate characteristics. Usually they consist of silicate/phenolic/resole resin mixtures, ester set, exemplified by the Fenotec binder system with its appropriate hardeners. These are described elsewhere in this Handbook.

THE CARBON DIOXIDE PROCESS

Using Carbon Dioxide Gas

The process consists essentially of mixing a clean, dry silica sand with a silicate-based binder, compacting the sand into shape and then hardening it by passing carbon dioxide gas through the sand.

The reaction of carbon dioxide with sodium silicate, which is the main ingredient of binders employed for the process, has frequently been represented by the equation:

$$Na_2O\ SiO_2 aq. + CO_2 \longrightarrow Na_2CO_3 + SiO_2 aq.$$

This equation is an over-simplification of the true reaction. The structure of most soluble silicates consists of ionic groupings or micelles of silicate ions interspersed with the equivalent number of sodium ions. On the introduction of carbon dioxide into the silicate some of the silicate ions are discharged leading to the formation of a silica gel and of carbonate ions. In addition, this results in a growth of the ionic micelles so that they become intermingled, leading to a very marked increase in viscosity of the binder. This increase in viscosity inhibits any further reaction between carbon dioxide and silicate ions, further treatment with carbon dioxide resulting in a secondary reaction, the formation of bicarbonate ions. In practice, this secondary reaction is revealed by the formation of white crystals of sodium bicarbonate on the surface of the sand; there is considerable evidence to suggest that this formation of bicarbonate is responsible for the marked friability of moulds and cores that occurs after prolonged treatment or overgassing with carbon dioxide, and the appearance of a white deposit is a useful indicator of the onset of "overgassing".

Self-setting Silicate Bonds

In this process the liquid setting agent is pre-dispersed into the sand before adding the silicate. For strength reasons the silicates used in the self-setting process are usually of a somewhat higher ratio than those used in the gas hardened process. Once mixed, the hardening reaction will proceed at a rate independent of access to atmosphere but dependent on ambient temperature. It is therefore necessary to have available a range of products to meet seasonal temperature variations and different bench life requirements. The process and the materials are described in greater detail later.

Heat Setting

As described above, the setting process involving sodium silicate can also be regarded as a form of dehydration. It is therefore possible to induce a set by oven baking. Ordinary core stoves and normal temperatures (around 200°C) can be used and fully dried, the method will give the strongest possible mechanical strengths. It is not often used however as there is no economic advantage and maximum strength is not always desirable.

BINDERS

For the Carbon Dioxide Process

A very wide range of both straight silicates and additioned or modified materials is available for gas hardening. Most silicates with ratios ($SiO_2:Na_2O$) between about 2.0:1 and 3.0:1 can be utilised giving varying degrees of bond strength and sensitivity to carbon dioxide gas. One of the main objections to the use of straight silicates has always been poor breakdown after casting, the heat from the metal often only serving to develop maximum bond strength. To counter this difficulty breakdown agents (see section on collapsibility later) may be used or similar, mainly organic, materials can be pre-mixed with the silicate solution. Binders containing varying amounts of agents such as starches, sugars, molasses, etc. are available as CARSIL, GASBINDA or SOLOSIL—see below for details. This type of addition will also function as a surface finish improver and some can in fact act as supplementary bonds or binder extenders thus allowing economy in percentage additions which in turn will have a beneficial effect on post-casting breakdown. The latest and more sophisticated binders such as SOLOSIL will also allow almost indefinite shelf life of stored cores especially in damp conditions, an area where the sugar containing silicates have well-known shortcomings. They are also of low viscosity, incorporating readily with the sand and avoiding the usual cold weather mixing problems. The choice is therefore very wide indeed.

THE CARSIL RANGE

Product	Ratio	Type of Addition	Comments
CARSIL 1	2.2:1	Sugar	The quality grade for core and mould making. Excellent for cast iron, further additions of breakdown agent can be made for aluminium etc.
CARSIL 2	2.2:1	Sugar/Graphite	Additions specifically designed to improve surface finish on light section NF and cast iron castings.
CARSIL 7	2.2:1	Sugar	Similar to CARSIL 1: an economy grade.
CARSIL 9	2.33:1	—	A "straight" economical silicate that may require additions of breakdown agent.
CARSIL M	2:1	Clay	Exhibits a green strength of 0.2–0.4 kg/cm^2 allowing pattern withdrawal before gassing. May require breakdown additions.

THE GASBINDA RANGE

Product	Ratio	Type of Addition	Comments
GASBINDA 500	2.0:1	Molasses	General purpose high viscosity binder for moulds and cores. Some green strength.
GASBINDA 502	2.0:1	Molasses	Low viscosity. Principally for core production.
GASBINDA 507	2.3:1	High Molasses	Mainly for cores. Very high breakdown potential.
GASBINDA 508	2.0:1	Molasses	High viscosity and high breakdown. Cores mainly.
GASBINDA 510	2.0:1	Molasses	Medium viscosity. Suitable for moulds and cores.
GASBINDA 513	2.4:1	Molasses	Low viscosity binder for easy mixing. Moulds and cores.
GASBINDA 515	2.0:1	High sugar	Good breakdown/surface finish. Cores mainly.
GASBINDA 520	2.0:1	High sugar	As GASBINDA 515. Lower viscosity.
GASBINDA 540	2.2:1	Low molasses	Suitable for either cores or moulds.
GASBINDA 542	2.6:1	Low molasses	Cores or moulds where high ratio silicate required.
GASBINDA 560	2.5:1	Sugar	Low viscosity high ratio silicate for cores.
GASBINDA 567	2.2:1	High sugar	Medium viscosity high ratio silicate for cores.

THE SOLOSIL RANGE

Product	Type of Addition	Comments
SOLOSIL 123	Starch/Sugar intermediates etc.	Low viscosity, exceptional breakdown.
SOLOSIL 433	As SOLOSIL 123 with bond supplement and waterproofing.	As SOLOSIL 123. Long shelf life for moulds and cores. Minimum gas requirement. Not temperature sensitive.

The SOLOSIL range represents the latest developments in one-shot silicate binders. Exceptionally fluid and easy to use they are ideal for continuous mixing/core shooting/blowing, are extremely economical in gas usage, cores sit well on the shelf and post casting breakdown is excellent even with thin section aluminium castings.

For the Self-setting Processes

These processes work best with a slightly higher ratio silicate than the gas hardened process, of the order of 2.5:1. Lower ratio silicates can be used but hardening times and strengths are not so reliable and subsequent breakdown may also suffer. Proprietary breakdown additives should not be used with any self-setting process as they may interface with planned setting times, bench life, etc. Where additional breakdown potential beyond that inherent in the binder is required up to 1% of fine coal dust or pelleted pitch can be dispersed in the sand before or alongside the setting agent.

From the list of GASBINDA silicates given above those containing the higher silica:soda ratios can also be used for the chemical setting process. They are:

GASBINDA 507, 513, 540, 542, 560 and 567.

If, for any special reason, a lower ratio is required GASBINDA 502, 508 or 510 may be substituted. From the CARSIL range CARSIL 100 (2.5:1 ratio) has been specially developed for the process and should provide consistent, reliable results in almost any application. The complex SOLOSIL type of silicate does not work well with the self-setting process and the others mentioned will be preferred.

Sand

Silica sands free from clay must be used. It is desirable that the sand used should be dry (0.5% maximum moisture), since excessive moisture may slow down the rate of hardening, harm the bond and also lead to blowing troubles in the foundry.

The sand should also be cool but not cold (between 10 and 40°C) in order to react reasonably quickly to avoid evaporation of the combined water present in the binder and retain adequate bench and shelf lives. Sand temperature is extremely important in the self-setting process (q.v.)

If a large quantity of fines is present it will cause friability and although this does not mean that fine sand cannot be used with the process (provided the percentage of binder is increased) it is better to use only moderately fine grades (AFA fineness No. 52–70).

The relation between optimum binder content and AFA fineness number is shown in Fig. 2.1.

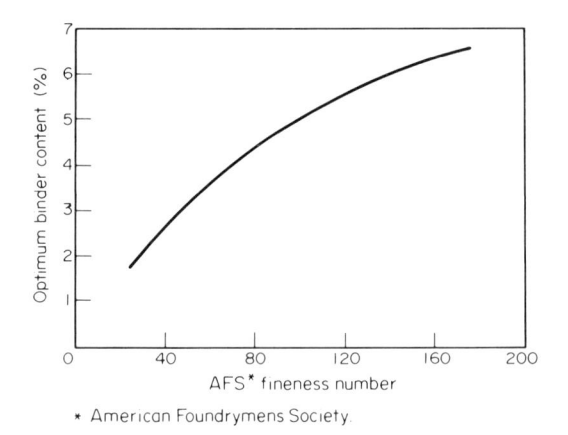

Fig. 2.1.

* American Foundrymens Society.

MIXING PROCEDURE—THE CARBON DIOXIDE PROCESS

Batch Mixing

It is quite satisfactory to mix a CO_2 process mixture in a small batch mill—a "mixing" rather than a "milling" action gives the best results—provided that it is not overmixed and the sand is transferred without delay to a plastic bag, dustbin, covered bin or other reasonably airtight container.

Measure the sand (see under **Sand** above) into the mixer and add the DEXIL or other breakdown agent if it is to be used. Premix these additions for about 60 seconds, minimum, before adding the *correct* (weigh or measure, do not guess) amount of binder, spreading the addition around the mixer as much as possible. Continue mixing for a further 1–3 minutes (depending on mixer efficiency) and then discharge. Overmixing may lead to heating up and will shorten bench life resulting in weak friable cores, etc.

Refer to Fig. 2.1 for amount of binder required for the grading of the sand being used.

Continuous or On-demand Screw-type Mixers

If an external powder type breakdown agent is to be used separate hoppering and measured feed arrangements must be available. The breakdown agent should be metered into the trough as early as possible in the cycle and some degree of mixing should take place before the silicate is introduced. Where extra pumping is available the liquid type DEXIL 60 will be ideal, but again as much pre-mixing as possible should be arranged. Ensure that the trough is cleared at each switch-off to avoid delivery of partially air hardened sand at the next start-up.

Troughs and screw ploughs of this type of machine can be kept clean by an initial and then a regular coating of STRIPCOTE AL. Mixed sand will not stick to it and the trough becomes almost self cleaning.

Again refer to Fig. 2.1 for the amount of binder required and to the appropriate leaflet for quantities if a breakdown agent is used.

PHYSICAL PROPERTIES OF CARBON DIOXIDE PROCESS SANDS

Green Strength

After mixing, the green strength of the sand is comparatively low, i.e. around 35–105 g/cm^2 (0.5–1.5 lb/in^2), but this is of little consequence as moulds and cores are normally hardened while still on pattern plates or in core boxes. In the case of CARSIL M, a higher green compression strength is achieved which will make it possible to draw the pattern before hardening the mould.

Compression Strength and Hardness

Compression strengths of 14–21 kg/cm^2 (200–300 lb/in^2) can be obtained after gassing. In addition, it has been shown that the hardened compression strength increases on standing. Scratch hardness tests have shown that cores having a 50/55 hardness immediately after the passage of CO_2, increased in hardness to 75/80 after standing for half an hour. SOLOSIL exhibits a more rapid response to gassing than the other binders and the set strengths are also higher.

Effect of Clay Content

Sands should be specified as clay free. Even very small quantities of clay will affect the strength of cores after standing for a short while. The presence of small quantities of bentonite has a very detrimental effect and similar results, although much less severe, are obtained with clays of the ball and fireclay type. Consequently, the incorporation of very small quantities of ballclay in the sand (up to about 1%) is tolerable when it is desired to develop green strength; bentonite should never be used for this purpose.

Collapsibility

The extent to which a core will break down after casting can vary from acceptable to almost impossible depending on a number of factors. The lower temperature metals such as aluminium often only inject sufficient heat into a core to develop optimum bond strength which can be very high indeed—of the order of 800 p.s.i. in some cases. As a matter of principle, the minimum amount of binder that can be tolerated should be used and advantage should be taken of additional breakdown agents such as DEXIL if necessary. Organic additions such as coal dust, pelleted pitch, plumbago, starches, sugars, etc, have all been used but may have distinct disadvantages from the point of view of gas evolution or metal/mould reaction.

A change to a different GASBINDA or CARSIL having more (or less) breakdown potential may also be indicated. In general terms a high breakdown potential will mean a reduced shelf life except when using SOLOSIL binders.

The varieties of DEXIL breakdown agent available are as follows:

DEXIL 11   A balanced organic/inorganic additive to improve cold core strength and surface finish. Used at around 1% addition.

DEXIL 34BNF   Made to a formula developed by the British Non-Ferrous Metals Research Association. Also acts as a binder extender so reducing silicate requirement. Application rate 0.5–1.5%.

DEXIL 35	As for DEXIL 34BNF but with additions to improve shelf life of cores in adverse, damp conditions.
DEXIL 60	A pumpable liquid organic intermediate particularly suitable for continuous or on-demand mixers. Use at 0.5–1.0% of sand weight.

Note that all external breakdown agents are added to the sand and pre-dispersed *before* adding the silicate. While addition rates up to about 2% are occasionally used, it should not be necessary to exceed the amounts suggested for fear of inducing friability and undue gas evolution. Rather, evaluate the possibility of a reduced binder addition or an alternative silicate of different ratio.

Surface Finish

Provided the core boxes/patterns are in good condition and the grading of the sand is well chosen an extremely good, almost glass-like, finish can be obtained. Casting surface finish will similarly be of a very high standard. Wherever possible avoid "rubbing" the original finish as this opens the way for penetration and casting finish will deteriorate.

Where metal sections are heavier or the sand more open, flammable, air drying or electrostatically applied dressings may be used to ensure a good finish to the casting. It is not usual to apply water-based dressings on a silicate bond since water penetrating the sand structure may drastically weaken the bond. It may occasionally be permissible to apply a very thick water based coating such as TERRAPAINT 55 but this should not be diluted below 80° Baume (3 vol paste to 1 vol water). Allow at least 12 hours air drying, or torch or stove adequately before casting.

Conventional based dressings that can be used are as follows:

Alloy	Air Dry	Flash Dry
Steel and Ni base	ZEROTHERM 4	ISOMOL 100 and 185 MOLDCOTE 33
Cast Iron	ZEROTHERM 1, 2	MOLDCOTE 6, MOLCO 216, ISOMOL 100, 185, 290
Cu. Base	ZEROTHERM 1, 2, 4	MOLDCOTE 6 or 11, ISOMOL 100, 185, 290
Aluminium	ZEROTHERM 2	MOLDCOTE 6 or 9

The electrostatically applied coating, suitable for all metals, is TRIBONOL. It must be used via a specialised "spray" unit that induces an electrical charge on each refractory particle as it is discharged from the gun. This charge attracts the particle to the oppositely charged mould or core surface where it is held. Light torching or infrared heating then develops a permanent bond. This method of coating application involves no solvents, lends itself to automatic operation and gives full coverage with no "shaded" areas.

GASSING WITH CARBON DIOXIDE

Having produced a sand mix, and after tucking or lightly ramming it into the core or moulding box, it remains to harden it by passing carbon dioxide gas. Incidentally, some silicate mixes are very sticky and it will probably be necessary to use a more sophisticated parting agent than the usual foundry powder or liquid agents. Silicones or PTFE are sometimes useful but the best results will probably be obtained with STRIPCOTE AL, an aluminised self-drying liquid that is very effective with these difficult sand mixtures.

The early gassing techniques were based on a simple tube probe, probably protruding through a backing plate to give depth control and ensure some degree of dispersion, or a plain rubber edged hood placed over the whole of the top (open) side of the box. (See Fig. 2.2.) Gas was passed (at about 10–20 p.s.i.) until the sand could be felt to have hardened or some white crystals appeared (indicating overgassing and the production of sodium bicarbonate). Strategic venting of boxes, patterns, etc, to prevent dead spots and equalise gas flow was, and always will be, extremely important. Because there is no dimensional change in the sand during gas hardening it is usual to rap the box before gassing as slight burrs, rough spots or sand flash can prevent easy stripping afterwards.

More modern methods of gassing will certainly incorporate additional equipment, probably heaters to ensure gas at a constant temperature, flow meters and timers, etc, most likely located at a specialised gassing station so that the whole operation becomes virtually automatic. Further refinements are available as pulsed gassing or vacuum assistance—all indicative of the benefits and economics that close control can give. Carbon dioxide can be expensive if it is not used efficiently. Most of the equipment mentioned is available from or via the gas suppliers who should be consulted.

Very accurate control over time, temperature and pressure of gassing assumes greater importance with the more advanced silicates and core manufacturing methods. CO_2 process cores are now commonly blown and gassed in a single operation, shooting is no problem provided the silicate is

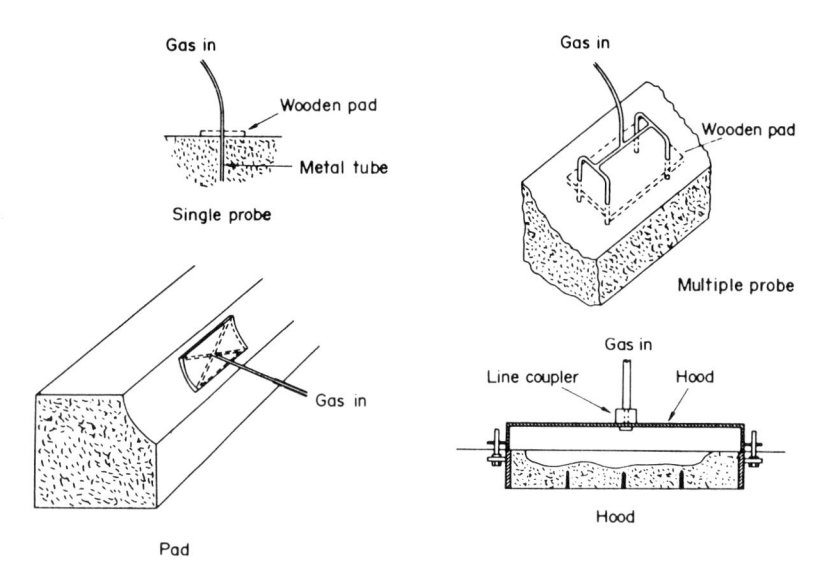

Fig. 2.2. Methods of Gas Application

carefully chosen and high speed production from, for example, a Corebelter (using VELOSET, a high speed self-setting agent—see later) is a recognised process. The complex silicates such as SOLOSIL are so sensitive to carbon dioxide that an exposure of a few seconds only is all that is usually required to develop adequate handling properties.

With each of the various types of silicate it is important that overgassing is avoided, any error being always towards undergassing.

SUPPLY OF CARBON DIOXIDE GAS

For foundry use carbon dioxide is usually supplied conforming to BS.4105:1967 (gas for industrial use) and contains a maximum of 0.2% residual other gas, usually air, which of course does not interfere with its use in the CO_2 process in any way.

It is available in three forms:

(a) in normal gas cylinders. The standard cylinder holds 13 kg (28 lb) of carbon dioxide as a liquid under high pressure—usually around 750–800 p.s.i. (51–55 bar). At room temperature and pressure 28 lb of liquid carbon dioxide will be equivalent to about 224 ft^3 (6.34 m^3) of gas. From the high pressure cylinder gas must obviously be put through a reducing valve to the line pressure of 15–20 p.s.i. (1–1.4 kg/cm^2).

Pressure reductions of this magnitude also mean that temperature is considerably reduced and it is quite common for reducing valves, etc. to frost or freeze if any large volume of gas is passed. Since cold carbon dioxide will obviously adversely affect the speed of the hardening reaction, it is usual to incorporate some form of heater to meet high production demands.

Individual gas cylinders are transportable and may be used singly in that manner: the alternative is to connect up a central battery and pipe the carbon dioxide to the gassing station.

(b) as a frozen solid known as "dry ice". Solid CO_2 is supplied in blocks of 11.5 kg (25 lb) nominal which are placed in a refrigerated converter, the evolved gas being led off, again through suitable pressure reducing valve-work and heaters, to the gassing station. One block will produce about 870 ft^3 (25 m^3) of gas at room temperature and pressure.

(c) as a liquid cooled to $-20°C$ and pressurised to 270 p.s.i. (19 bar) delivered by road tanker. It can be pumped direct into refrigerated storage containers, in this case without any interruption of supply to the foundry, as is the case with the other methods.

The supply, storage and distribution of carbon dioxide gas at the correct temperature and pressure is well developed and a wide range of equipment is available.

All the suppliers offer a consumer technical service and it is strongly recommended that reference is made to them, in the first instance, so that the various safety precautions, etc, are complied with. Carbon dioxide is not toxic and is colourless and odourless. It is, however, slightly denser than air and may accumulate in enclosed areas or deep moulds leading to oxygen exclusion or displacement and hence suffocation. Gassing stations should therefore be provided with adequate ventilation.

The main U.K. suppliers of carbon dioxide gas are:

The Distillers Company Limited.
British Oxygen Company Limited.
Air Products Limited.
The Carbon Dioxide Company Limited.

GENERAL COMMENTS ON THE CO_2 PROCESS

It is true that silicate bonding has certain limitations, but it possesses many advantages over conventional methods when correctly used. It is not more economical than green sand moulding unless castings are required to much closer tolerances, thereby allowing a reduction in machining allowances, but it does have clear cost and environmental advantages over most other methods of bonding sand.

Any comparison with other bonding methods should take into account the following proved advantages of the process:

1. Reduction in fuel costs as a result of the elimination of stoves and driers.
2. Increased output per unit area of working space, achieved as a result of installing production units in areas formerly occupied by stoves, etc.
3. Reduction in the number of moulding boxes required to achieve a given output.
4. Reduction in handling and transport costs as a result of hardening moulds and cores *in situ*.
5. The greater accuracy obtained by hardening cores in core boxes and moulds with patterns in position leading to a reduction in machining allowances. With suitable equipment, certain castings can be produced to tolerances of the same order as those obtained with shell moulding and at lower cost.
6. Reduction in the number of core irons, lifters, and sprigs required and the elimination of core carriers.
7. Reduction in the quantity of sand required in coremaking by wider application of hollow cores.
8. The low cost of the equipment required.
9. In many cases, existing pattern equipment can be used.
10. No unpleasant gas or fume given off when casting.
11. Lower sand disposal costs due to non-hazardous nature.

TWELVE SIMPLE STEPS TO THE CO_2 PROCESS

1. Select dry, clay free sand (see page 92).
2. Select correct quantity of binder (see page 92).
3. Select breakdown agent (see page 94).*
4. Prepare sand mixture in mill (see page 93).
5. Keep mixed sand away from contact with air.*
6. Use an efficient parting agent such as STRIPCOTE AL.
7. Ram sand into core or moulding box with flat end rammer.
8. Insert gassing vents.*
9. Rap core box or pattern sharply to facilitate stripping.
10. Apply carbon dioxide gas (see page 96).*
11. Remove core from core box or pattern from mould.
12. Sand is now rigid and ready for final assembly of mould. For improved surface finish apply air or flash drying dressing (see page 95).

*Steps marked thus are eliminated when using the self-setting process.

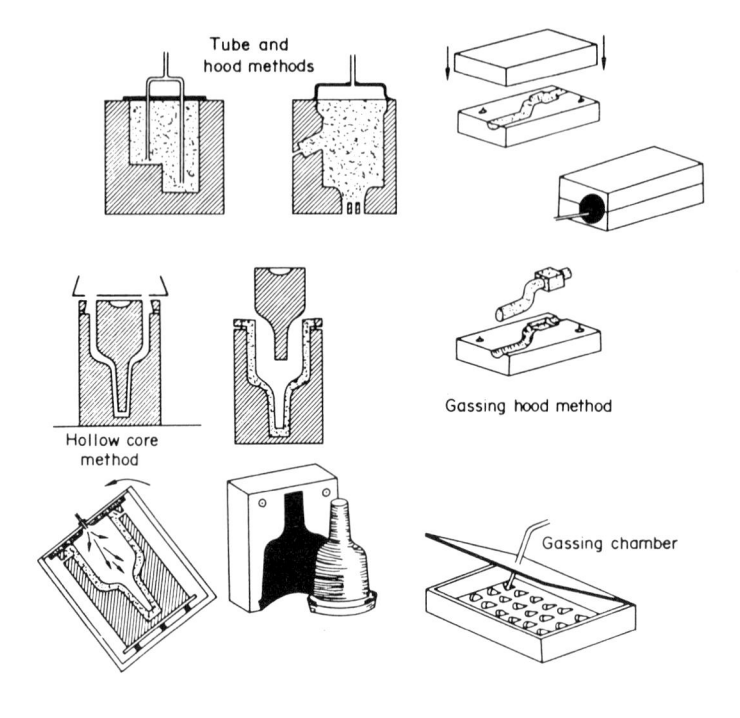

Fig. 2.3. Typical Methods of Core Production by CO_2 Process.

THE SELF-SETTING PROCESSES

The self or chemical hardening processes may be classified into two main groups according to the speed of setting. The fastest process based on the VELOSET group of hardeners will give a sand working life of between 1 and 11 minutes suggesting stripping times of roughly twice that interval. This group overlaps with the main ester based group of products and these SILISET and CARSET ranges can, depending on which product is selected, carry useful sand life well beyond an hour. A critical factor in this process (the selection of a suitable grade of silicate is also important—see later) is the sand temperature as discharged from the mixer. To accommodate normal seasonal fluctuations, etc, a number of grades within each product group must be available to ensure constant setting times winter and summer.

All the VELOSET, SILISET and CARSET materials are clear, mobile liquids of similar consistency to water. They are non-corrosive and non-hazardous and all mix readily with sand especially in the context of continuous or on-demand mixing where the main application clearly lies.

THE PRODUCTS

VELOSET

Bench Life (minutes at 20°C)

Silicate Binder Ratio	VELOSET 1 (Clear)	VELOSET 2 (Red)	VELOSET 3 (Blue)
2.2:1	11	6	4
2.4:1	7	5	3
2.6:1	4	2	1

For most practical purposes the stripping time for a VELOSET system can be assessed as twice the bench life. The amount of VELOSET used will usually lie between 10 and 15% of the silicate addition.

SILISET and CARSET

The SILISET S range are colour identified clear liquids and represent a fully economic general purpose approach to the self-setting silicate process. An accompanying SILISET T range comprises essentially high performance agents where maximum and consistent through-curing, minimum sag, high mechanical strength and minimal pattern or varnish attack are required. This T group are all clear, colourless liquids.

Gel Time—Minutes

Grade	Colour	GASBINDA 240 5°C	GASBINDA 240 20°C	GASBINDA 513 5°C	GASBINDA 513 20°C	GASBINDA 542 5°C	GASBINDA 542 20°C
			SILISET S				
S1100	Clear		95		75		38
S1104	Violet		29		30	53	15
S1109	Blue	110	24	70	21	31	9
S1113	Yellow	76	15	49	13	20	7
S1119	Green	43	10	35	9	12	5
S1121	Orange	39	8	28	8	8	3
S1125	Brown	35	7	27	6	7	2
			SILSET T				
T1150							123
T1154			117		92		41
T1158	All		105		53		21
T1162	Clear	96	56	120	32		14
T1166	Liquids	56	19	60	15		9
T1170		50	12	49	13		8
T1174		26	8	30	9		6
T1178		18	6	27	7		4

NOTE: The gel time is the time taken for gelling to occur when a quantity of silicate binder is mixed with an appropriate amount of setting agent. The results may not be repeated exactly when sand is present, due to the possibility of impurities, etc., but it does provide a good independent guide to probable bench life.

The CARSET range of setting agents is essentially similar to the SILISET T group and for all practical purposes the following equivalents are valid:

CARSET 511	is equivalent to SILISET	T1174
CARSET 522	is equivalent to SILISET	T1170
CARSET 533	is equivalent to SILISET	T1166
CARSET 544	is equivalent to SILISET	T1158
CARSET 555	is equivalent to SILISET	T1154

TYPES OF SILICATE

While almost all foundry grade silicates, including the complex or additioned varieties, can be set to varying degrees by chemical hardeners the more practical foundry applications are in general met by those of a higher ratio. The usual Carbon Dioxide Process (gas hardening) binders of around 2.0:1 ratio will set but the setting process may be unreliable, through-cure can be slow and core sag becomes a problem. Subsequent breakdown after casting may also be adversely affected.

Silicates for use with any of the self-setting processes should be selected from the higher ratios—usually from about 2.2:1 and upwards. Where good post casting breakdown is required, the presence of a breakdown agent is acceptable but further additions of external agents should not be made. Selection of the correct grade of silicate becomes more important as the required setting time is shortened so that with VELOSET a special high ratio silicate blend—CARSIL 100—should be used.

For the CARSET and SILISET processes silicate selection can be from:

GASBINDA 507	2.3:1 ratio
GASBINDA 508	2.0:1 ratio
GASBINDA 513	2.4:1 ratio
GASBINDA 540	2.2:1 ratio
GASBINDA 542	2.6:1 ratio
GASBINDA 560	2.5:1 ratio
GASBINDA 567	2.2:1 ratio

The use of SOLOSIL with any self-setting process is not recommended.

MIXING PROCEDURES

The self-setting silicate process is ideally suited to continuous screw-type mixer-slinger and other mix on demand machines in view of the ease with which the hardener can be incorporated. Special pumps and piping are not necessary because of the non-corrosive nature of the liquid catalyst. Calibration should be carried out as indicated below.

The process can also be used with the conventional pan or blade mixers (a tumbling, mixing action is preferred to milling) but here close control over additions and mixing time is very necessary because of the direct effect on setting times, etc, of even small variations in mix temperature and composition.

A recommended general purpose binder is CARSIL 100 which is a 2.5:1 ratio silicate with additions and which will provide optimum results in terms of correct hardening and good subsequent breakdown. Other silicates can also be used ranging from the normal foundry grades of 2.0:1 ratio to special high ratios of 2.65:1 or even higher. Higher ratio silicates are slightly cheaper and give improved through-curing with marginally better breakdown: on the other hand set strength is slightly lower and mould and core surfaces are not quite so hard. The effect of substitution of different silicate ratios and compositions is indicated in some of the tables below.

INSTRUCTIONS FOR USE

One decision and one measurement must be taken first of all.

(a) Decide bench life required in the mixed sand.
(b) Measure temperature of sand at discharge from mill or mixer after a normal mixing time (approximately 3–4 minutes). Any clean dry silica sand that can be used successfully with the Carbon Dioxide Process will be suitable for use with a chemically set process.

Refer to graph (Fig. 2.4) and read off bench life required horizontally against sand temperature vertically. Where these lines intersect indicates the grade of setting agent to use together with the amount required. Note that the percentage of setting agent indicated refers to a percentage of the weight of silicate.

The amount of silicate—CARSIL or GASBINDA—required is again dependent on the grain size of the sand and will normally fall between $2\frac{1}{2}$–$3\frac{1}{2}\%$ by weight of sand for AFA sand fineness Nos. 50–70.

PROCEDURE

Batch Mixers

(1) Add setting agent to sand in the mill and mix until uniformly dispersed, normally $\frac{1}{2}$–$1\frac{1}{2}$ minutes. Both sand and agent MUST be measured accurately.

(2) Add silicate binder and continue mixing until homogeneous—1-2 minutes. DO NOT OVERMILL. Measure silicate accurately.
(3) Discharge mill and ensure that sand is used within the bench life period used for calculation.

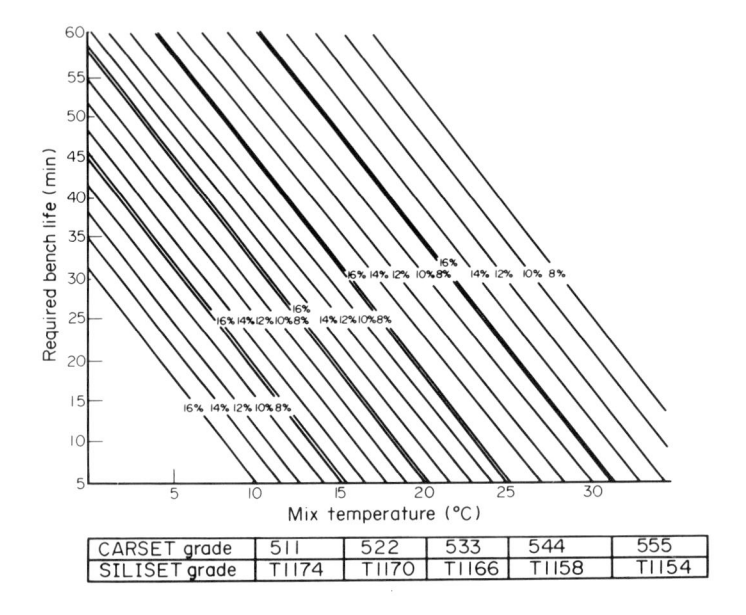

Fig. 2.4. SILISET T and CARSET 500 SERIES. A guide to selection of hardener addition level using CARSIL 100 or an equivalent GASBINDA. Percentages relate to weight of binder.

A graph similar to the above can be derived for the SILISET S range with any suitable silicate and any particular mixer under constant conditions. It is recommended that such a graph is constructed so that seasonal adjustments, etc, can be made.

Continuous or Screw-type Mixers

(1) Clean and calibrate mixer and pumps to deliver quantities of sand, silicate and setting agent as calculated above.
(2) Position of entry of silicate binder and agent will be largely dictated by machine design, but wherever possible introduce the ester prior to the binder.
(3) Note that if the machine is used intermittently and is not self-cleaning the sand remaining in the trough may be wasted if its bench life is exceeded before starting up again.

NOTES

1. REMEMBER that reproducibility of results depends on accuracy of measurement of sand, binder, ester and sand temperature.
2. If coal dust is used as a breakdown agent in conjunction with special silicate it should be added after the setting agent, but before the binder. A breakdown agent is not usually required with CARSIL 100, or GASBINDA silicates.
3. Normal spirit-based dressings can be used to improve surface finish if necessary.
4. Stripping time is generally 2–3 times bench life for simple cores. Up to 8 times may be necessary for large, complex cores with lifting irons, i.e. with a bench life of 10 minutes most patterns can be stripped 20–30 minutes after ramming.
5. Excessive use of liquid parting agents should be avoided, it is better to use STRIPCOTE AL as previously described. Boxes and patterns should be in good condition and rapped before expiry of bench life period.
6. Choose the bench life that suits the work in hand best and then vary either the type of setting agent or the amount used according to how sand temperature at discharge varies with the season.
7. Moulds, etc, may be cast as soon as satisfactorily set.

ANCILLARY NOTES

Bench Life. Can be varied by selection of a suitable grade of hardener. Will be affected by variations in mix temperature—higher sand temperature will reduce bench life—and by alteration of silicate ratio. The use of higher ratios will decrease bench life of the sand. All mixed sand should be used within its predicted bench life.

Ramming. Freshly mixed sand is free flowing and easily rammed by any conventional method. Even, consistent ramming is conducive to good casting surface finish. Interfaces between separate mixes should be trowelled smooth to avoid areas of soft compaction.

Mould and Core Dressings. Those proposed for the Carbon Dioxide Process are quite suitable.

Patterns, etc. Coreboxes, etc, should be in good condition and coated with a paint or varnish resistant to the hardeners, e.g. polyurethane. All esters will attack and soften ordinary pattern paints very severely. Use STRIPCOTE AL as a parting agent but excessive application should be avoided.

Hazard and Metal/Mould Reaction. No toxic or unpleasant fumes are evolved on casting and the relatively inert binder system ensures that any tendency towards metal/mould reaction, pinholing, etc. is minimal.

ADVANTAGES ASSOCIATED WITH THE SELF-SETTING PROCESS

1. Total binder and catalyst costs are significantly lower than those of conventional air and cold setting resins and oils.
2. Mixing may be performed satisfactorily in either batch or continuous mixers.
3. Mixer and equipment clean-down is facilitated by the water solubility of the uncured system.
4. The mixed sand is free from objectionable odour, dermatitic and toxic hazards to operators are negligible. No hydrogen evolved during setting.
5. The desired bench life and stripping time can be accurately achieved and controlled.
6. Complete through-cure is obtained.
7. No heat is generated during curing and thus there is no expansion to cause problems of dimensional accuracy after pattern removal.
8. Gas evolution during casting is small and no unpleasant fumes are evolved.
9. A nitrogen-free system eliminates trouble from subsurface pinholes with cast irons.
10. The incidence of finning and metal penetration defects is minimised.
11. The storage life of moulds and cores is excellent.
12. Moulds and cores are dimensionally stable during casting; therefore, minimal dilation and more accurate reproduction of pattern dimensions in the castings is possible.
13. Casting strip is excellent, easy shakeout after casting is obtainable and the used sand may be easily reclaimed.
14. Since the system is carbonate-free contamination of green sand is not detrimental.
15. It is possible to incorporate low cost, shell containing sea sand with this process if other considerations permit.

SAND RECLAMATION

The escalating cost of new, clean, dry, clay-free sand in conjunction with enhanced environmental requirements, transport and tipping charges has meant that greater urgency has been given to possible recovery and re-use of sands used in the silicate bonded processes. The original wet scrubber systems, although relatively effective, were awkward to operate, were expensive in energy terms and suffered from effluent problems. More recent wet systems are better in all respects so that currently four sand reclamation systems can be employed for recovery of silicate (or resin, or clay) bonded sands:

1. Wet reclamation, using some form of hydraulic scrubbing.
2. Dry attrition.
3. Thermal reclamation.
4. Pneumatic scrubbing.

Wet Reclamation

The basis of most wet systems is the scrubbing of a sand/water mixture in attrition cells by means of abrasion resistant high speed paddles. The treated sand is de-watered, dried and re-graded before re-use. Even in its simplest form the difficulties of effluent disposal and drying of large quantities of sand are major problems.

More recently an enclosed single chamber treatment system has been developed by S.C.R.A.T.A.* that is claimed to be able to handle CO_2 process and ester hardened sands. The novel engineering concept is said to be unique in that after crushing to grain size the sand is processed completely in the single vessel. Each stage of washing, de-watering, drying, cooling, fines removal and sand discharge is easily controlled. Sand is delivered to a storage hopper, fines to a collector and effluent is said to be acceptable for sewer discharge.

Dry Attrition

A number of suppliers now offer attrition units which by some mechanical means break down the sand lumps to grain size and then hurl them on a target or otherwise abrade them thus shattering any adhering film bond. Subsequent dust/fines removal and grading will then leave a re-usable sand of low loss on ignition value. This type of recovery unit is probably more suited to the more brittle resin bonds than to silicate but it is said that up to 50% reclaim sand can be used under some conditions.

Thermal Reclamation

Again more suited to the combustible organic bonds than to inorganic silicates these systems rely on exposing the sand to relatively high temperatures of the order of 3–500°C to burn or crack off any residual bond. The rate of heating must be controlled in order to avoid shattering individual grains unduly and the treatment chamber is often a fluidised bed where a degree of grain to grain abrasion is also helpful. This type of

---

* Steel Castings Research and Trade Association, Sheffield.

reclaimer can produce very little fume, even when used to remove resinous films, but its use with silicate bonds which are largely non-combustible is probably less effective.

It has been suggested that pre-coating the sand with a carbon layer (by pre-mixing just enough—about 0.25%—of a readily available carbonaceous liquid) before adding the ester and/or silicate provides a combustible layer rendering the sand more amenable to these thermal reclamation processes.

Pneumatic Scrubbing

Most of the systems in this section use the principle, after the return sand has been crushed to approximately individual grain, of entraining the sand in a high volume low pressure air stream and passing it through a series of (usually vertical) scrubbing cells so that impingement of the grains on each other and on various targets in the resultant "sand-storm" has the combined effect of removing the tenacious silicate film and also applying a little beneficial rounding to the sand grain without undue shattering. The latest equipment in this field often utilises separate treatment stages some with rapidly rotating grinding or abrasive discs against which the sand is allowed to fall, with fines removal and sand classification at strategic locations. They can be very effective and an expected design efficiency of 75% reclaim sand has frequently been exceeded often by a considerable margin.

Multi-stage machines capable of handling from 2–20 tonnes per hour throughput are on offer and are capable of treating all furane, phenolic, urethane and phenolic ester systems in addition to Carbon Dioxide Process and all the ester silicate systems. These dry attrition machines avoid the wet effluent problems associated with the water-based methods and are capable of providing a reclaim material of residual soda content (in the case of silicate bonds) that allows very considerable proportions of re-use.

ALTERNATIVE MOULDING SYSTEMS—THE REPLICAST PROCESSES

Despite all the recent advances that have taken place in the resin and silicate fields, the most popular method of forming a mould remains with "green" sand—a natural or synthetic clay bonded, water tempered, silica sand with or without certain additions, such as starch, coal dust, pelleted pitch, wetting agents, etc. The resin and silicate bonded systems are described in other chapters (q.v.) but there is a number of other processes that have been proposed and in fact used at one time or another with varying degrees of success. They may be briefly outlined as follows, all are subject to licensing arrangements.

The "V" or Vacuum Process

A pattern is covered with a polythene film and the box filled with unbonded dry silica sand. The top is struck off level and covered with another layer of special polythene. A powerful vacuum is then drawn in the sand and maintained during pattern withdrawal, core placement where necessary and subsequent casting.

The "F" Process

Wet, clay-free silica sand is tucked into a moulding box and the whole is then "gassed" with air at the temperature of liquid nitrogen so that the water film round the sand grain is frozen. Casting must obviously take place before melting of the ice bond.

The EPS (Expanded Polystyrene) or Full Mould Process

Patterns are formed from a special foundry grade of expanded polystyrene and are assembled complete with heading and running systems, etc. They are then "rammed" up in loose sand (cores are integral) or a self-setting silicate or resin bonded material and the metal cast. The EPS will vaporise on contact and the metal will take the shape of the pattern.

The Fluid Sand Process

Really a variant of a silicate bonded process, the base sand is mixed with a silicate with a hardening agent and a surface active or foam producing ingredient. Vigorous mixing produces a lightweight foamed slurry that can be poured as a liquid into the moulding box and then left to set before removing the pattern, coating and casting.

The Randupson or Cement Bonding Process

Although not new, this process is still used for some very large castings, for example for large propellers for ships, etc. A clean silica sand is mixed with a refractory cement and water, placed and vibrated into position and allowed to set. Very strong moulds can be produced.

Recent developments of the EPS (Expanded Polystyrene) or Full Mould Process, have led to two new subsidiary Moulding Systems of some promise. They are proprietary and known as the Replicast FM and Replicast CS Processes.

REPLICAST FM PROCESS

The original EPS or Full Mould Process, described briefly above, suffered from some drawbacks. The use of unbonded sand sometimes gave rise to "lift" during casting, penetration was occasionally a difficulty and there was always a danger of mould collapse if the pouring rate was not exactly right.

The reliability of the EPS process has recently been transformed by development work carried out by Steel Castings Research and Trade Association culminating in the REPLICAST FM process. Success lay in a combination of:

(1) a refractory coating (developed by Foseco) applied to the EPS pattern rather like a core coating;
(2) controlled compaction of the loose sand by vibration;
(3) application of vacuum to the open topped mould during casting.

Appreciation and application of these factors together with improvement of the quality of the EPS pre-form opened up the possible applications to the process enormously to include quite small castings. The EPS pattern must be very well made—it is blown on a machine similar to a coreshooter—with a very good, dense surface finish and of high dimensional accuracy. Quality of the EPS is usually defined by density and bead size and the slight shrinkage of the mouldings over the first few days of storage can be compensated for when the die is made. Dies are relatively cheap (aluminium is quite suitable) and there is practically no process wear. Complicated patterns may be built up from a number of dowelled and glued EPS multi-piece parts. Running and risering systems may be incorporated in the original pattern or they may be glued on as additions as required.

After application of the coating—usually by a pour-over technique since the buoyancy of EPS makes dipping difficult—it is dried by warm air not above about 40°C. A number of coatings based on different refractories is available. The completed and dried assembly is located in a steel moulding box which is then filled with sand (usually around 50–55 AFS grading) and the vibrating table switched on for about 30 seconds. The correct amount of vibration is critical in order to fill cavities, undercuts and hollows and reach the correct degree of overall compaction. Before casting the vacuum is applied via a plenum chamber in the base of the moulding box so that a pressure differential exists between open top and bottom. The consequent air flow removes any fume on casting and also renders it unnecessary to weight the box. It also appears to lock the sand grains in position preventing movement or collapse during the casting process.

A sand cooler-classifier will remove shattered grain fines from the largely re-usable sand and will also cool it to below 40°C for re-use. Sand can be re-cycled many times with only small losses from drag-out needing to be made good. Castings require only a light shot blast to remove any adherent coating.

REPLICAST CS PROCESS

This is a development from the Full Mould Process described above. It utilises similar equipment with the principal difference being the quality of the coating applied to the EPS pattern. In this case a ceramic shell (hence the CS) is built up of four or five coats of slurry/stucco applied by dipping (slurry) and fluidised bed or rain booth (stucco) and hardened as in the true investment process. This "shell" is then fired at about 1000°C for a few minutes so vaporising the EPS and leaving a thin, lightweight rigid shell. Up to this point the EPS pattern has mirrored the function of the wax pattern in the true investment process but the subsequent difficulties of de-waxing and firing are dispensed with. The resulting thin, carbon free shell requires a little care in handling but is suitable for casting any metal including the very low carbon high alloy steels. Any openings in the shell can be sealed with thin plastic film (cling film) and the completed unit is then located in a moulding box, surrounded with loose sand, carefully vibrated and a vacuum applied before casting exactly as described for the REPLICAST FM Process.

REPLICAST CS Process characteristics may be summarised as follows:

STEP 1.	**Pattern Production**
	Blow partially expanded polystyrene beads into aluminium tool at low pressure and complete expansion by injecting steam.
	● High density patterns, dimensionally accurate and with good surface finish.
	● Lightweight and readily handled.

STEP 2.	**Ceramic Shell Production**
	(a) Apply successive coats of refractory slurry and stucco.
	● Only 3-4 coats required.
	● Relative light and manual handling acceptable even on large components.
	● Possible to automate.
	(b) Heat in furnace to 1000°C for 5 minutes to vaporise EPS pattern and to fire binder.
	● Lightweight shells readily handled when cold.

STEP 3.	**Mould Assembly and Casting**
	Bed the ceramic shell in loose sand and vibrate to maximum bulk density: vacuum optionally applied during pouring.
	● Adequate support ensures good dimensional accuracy in the casting and prevents metal breakout from the shell.

Benefits of the Process have been proposed as:

Improvement in as-cast quality. Finish and dimensional accuracy are significantly improved by the ceramic shell.

Elimination of core making. The EPS route permits one-piece patterns.

No sand binders required. Using vibration and vacuum the shells are adequately supported.

Reduced fettling costs. No joint line flash and runners and risers can be placed for easy access.

Better yield. Components can be clustered round common runners/feeders.

Reduced machining. No draft angles and high dimensional accuracy.

Improved foundry environment. Low fume, dust and noise levels. All processes are clean and lightweight.

Low capital outlay. Maintenance costs, energy requirements and manning levels are low as is the original equipment cost.

The REPLICAST CS Process is recommended for casting up to about 100 kg and gives a quality approaching that of lost-wax investment in terms of surface finish. Costs are said to be comparable with those of conventional moulding process.

NOTE: Licensing arrangements for the REPLICAST FM and CS Processes are vested in Foseco to whom enquiries should be directed in the first instance.

SECTION III

MOULD AND CORE COATING

INTRODUCTION

There are many occasions when the application of a refractory or other type of dressing to a mould or core surface is very necessary as a means of achieving a required standard of finish to the casting. This despite the considerable advances that have taken place over recent years in binder and sand technology giving the user enormous choice and control of these basic foundry raw materials. It might be supposed that since casting finish is largely a function of sand particle grading a suitable selection of a particular grade of sand would be all that is required to obtain a desired casting appearance. Frequently, however, there are other considerations, such as ability to vent away gases produced during casting, economic use of a binder, non-availability of sand of a certain grading, etc., that means the use of a suitable surface dressing is the best practicable approach.

Types of Dressing Available

Every conceivable method of presenting mould and core dressings to the user has at some time been utilised. These range from dry powders for soft camel hair brush application, through powders and pastes for dispersion in both water and various flammable solvents, ready-to-use suspensions in the same solvents, air-drying versions and electrostatically deposited powders, to inclusion of the powder dressing in the sand mix itself.

Similarly, all possible methods of placing the dressing on the mould or core surface have been explored—swabbing, brushing, spraying, dipping and flow-coating of liquids with electrostatic spray, shaker-bag and soft brush for powders.

Dressing Development

For many years foundries made up their own dressings as occasion required. Mostly these were based on talc, mica, graphite, whiting or some similar readily available refractory material and dispersed in water using simple bucket and stick methods, the more sophisticated establishments adding a little clay, lye, bentonite, etc. to improve suspension and adhesion. Gradually the benefits of a more scientific approach began to be appreciated and specialist coating suppliers emerged bringing an expertise and consistency the average foundry could not achieve.

Carrying the development programme to its logical conclusion saw the production of ready-mixed or virtually ready-to-use dispersions, often tailor-made to meet the requirements of a particular foundry. Many of these highly specialised coatings have survived and continue to provide excellent results despite the pre-requirement of exceptionally good suspension characteristics and the counter argument of difficult handling and cost of transporting large volumes of water or solvent.

Following the lead of the decorative paint producers some mould and core dressings are now available in thixotropic suspensions even accommodating such dense and difficult to suspend refractories as zircon (zirconium silicate). At the same time more sophisticated bonding agents, organic and inorganic, have been incorporated largely eliminating the difficulties associated with the clay bonds and suspensions previously used. Natural clays have absorption and swelling (gelling) characteristics that were originally very useful but as quality requirements advanced it became increasingly difficult to ensure the requisite levels of consistency and reliability from a naturally occurring and therefore variable material. Even from the same source, different batches of clay can have slightly different properties that will affect dressings made from them. Improvement in fundamental properties can usually be obtained by adding a source of sodium ions (as soda ash or sodium carbonate) so that some base exchange, as it is called takes place. Whilst such treatments can help they are not always certain and the clay may still be affected by other incidentals such as, for example, the hardness or softness of the water used to mix the dressing.

Turning to the spirit-based dressings, suspension agents have largely centred around the bentones, a group of organically dispersed clays. These have suspension and dry bond characteristics similar to the water tempered types but are generally rather less effective and give off fairly large volumes of gas when heated. In turn the bentones have now been displaced by more recent developments that in addition to a more certain performance do not evolve so much gas when the metal is poured.

Present Position

As always in a progressing situation as outlined above some users have stuck with a particular type of dressing that suits their particular purpose or meets individual cost/efficiency criteria. All types of dressing will therefore be found in use somewhere and suppliers have, within certain economic constraints, been under some pressure to keep supplying despite the later developments.

These notes therefore constitute an endeavour to suggest how this wide variety and complexity of dressing types should be stored, mixed, used and dried to give the best possible results. For the sake of convenience dressings will be divided into three principal groups, as follows:

1. Water-based dressings.
2. Spirit-based dressings (including all solvent bases except water), and
3. Others.

Distinctions are not drawn between dressings for use on moulds and on cores as with very few exceptions they are largely non-specific in this context and may be used for both applications.

Because of the continually changing structure of the foundry dressings field, the arrival of new paints and the demise of out-dated or uneconomic formulations, only a general indication of the numerous types available, can be suggested. A summary is given at the end of this section.

WATER-BASED DRESSINGS

Within this general heading there are three sub-groups, viz:

(a) Powders for mixing with water.
(b) Pastes or slurries for dilution with water.
(c) Ready-to-use (or virtually so) suspensions.

Powders for Mixing with Water

This was the original and simplest way of supplying dressings. Almost exclusively they were clay suspended with, as required, supplementary bonding additions such as dextrine, starches, sulphite lye, etc. The range of refractory fillers used was equally as varied with graphite/plumbago or crushed coke (the so-called "blackings") forming the major proportions but whiting, mica, talc, etc. also being used alone or in conjunction with the carbonaceous materials. The relative inefficiency of the available clay suspension systems virtually precluded the use of more dense fillers such as zircon, chronite, alumina, etc. although some dressings based on olivine were quite successful.

As might be expected, the carbonaceous dressings are intended for use primarily on cast irons although there is an additional application to some copper base alloys where the "polish" resulting from the graphite and its associated gas cushioning effect can be beneficial. The other fillers, of generally lower refractoriness, are usually recommended for use with either thin sectioned iron castings, the copper based alloys or the lower melting point aluminium alloy group. Mixed refractories can also have some attractive properties, particularly in the field of partially fusible dressings and their use to prevent or overcome the problem of "finning" often encountered in automobile and similar castings where the core is almost totally enclosed or thermal gradients in the sand are very steep during casting. Because of the sensitivity of steel castings to carbon pick-up (from the dressing) and the requirement for high dressing refractoriness—zircon is most commonly used—there is normally little call for this type of coating.

Special types of powder dressings: Powders for mixing with water are also supplied for two other rather specialised areas of use, i.e. refractory coatings for tools, etc. and the manufacture of wet applied dressings for centrifugal or spun castings and chilled rolls.

The former will usually require a specially strong bond and as a consequence powdered sodium silicates are often used. Clay as such is not often present as its hot strength is not great and dressings tend to crack or spall on sudden heating. Organic materials burn out too quickly. The centrifugal and chilled roll dressings are essentially similar to the normal dressings for use on a sand substrate but here, because of the impermeable nature of the mould material, there is often a requirement for very low or closely controlled gas contents together with some lateral dry coating permeability.

Pastes, Slurries and Ready-for-use Paints

These will mostly utilise similar refractories, etc. to those available in the dry powder varieties but in these forms the supplier has greater freedom to incorporate more specialised suspension systems, bonding agents, preservatives, etc. than would be possible in a dry powder presentation. Practically the preparation of a dressing from a paste or slurry is very much easier than from a dry, dusty powder where the waterproofing effect of fine particle size can be a problem despite the use of surfactants, etc. This is the group of greatest commercial appeal.

STORAGE

Dry Powder Dressings

These present no problems, the only requirement being that they are kept dry to prevent deterioration of the clays, starches, etc. present. Under good conditions, shelf life can be virtually indefinite but for practical purposes stock turn round should be arranged on a twelve month maximum interval.

Pastes, slurries and ready-for-use paints: All these variants will normally be subjected to a recommended storage period and the manufacturer should be consulted. Pastes and slurries will usually exhibit a longer shelf life than ready-for-use dressings and three months and one month respectively seems reasonable. Where possible there may well be some merit in inverting the containers at weekly intervals to prevent hard packing of any segregates.

Obviously, all water-based coatings must be protected from extremes of temperature and should be stored between 10 and 20°C. Freezing will destroy any gel structure and the refractory particles can settle and pack hard. Overheating if stored near furnaces or in summer sun can promote fermentation of any organics present and the dressing will thicken, bubble and stink. Preservatives are usually added by the manufacturer but most can only be effective up to a certain point and there may be health and safety or other limits to the amount that can be tolerated.

MIXING

Whatever the type or composition of the coating there will be a preferred way of preparing it for use and the manufacturer's recommendations should be obtained and followed. Always measure all ingredients including water.

Dry Powder Dressings

Except where these are sleeked, as a powder, on to a mould or core surface by means of a soft brush, they will require mixing with water. In general terms the requisite quantity of powder should be added to half the amount of water required and stirred briskly to form a cream. This should then be allowed to stand for an hour or two to "age" before adding the remainder of the water, again stirring thoroughly.

Pastes, Slurries and Ready-for-use Paints

Paints supplied as "ready-for-use" may not always be exactly so and could require slight further dilution: pastes and slurries always do. The correct proportions of paint and water should be mixed together until homogeneous.

DO
1. Measure or weigh all ingredients including water.
2. Premix pastes and slurries before removing required amounts.
3. Always use *clean* utensils, tools, etc.
4. Thoroughly wash out used containers.
5. Remix regularly during use. Use mechanical agitation or "bubblers" if available.
6. Add powder to water—never vice versa.

DON'T 1. Mix large amounts of coating so that the excess stands about for more than 24 hours, especially in warm weather.
2. Add new paste, etc. to previously mixed paint.

SPECIAL NOTE FOR MIXING THIXOTROPIC SUSPENSIONS

After storage or transport even a thixotropic suspension may be affected to the extent that some refractory separates but such segregates are usually soft and relatively easily re-incorporated. If a dressing is being made to a given consistency—most conveniently controlled by means of a Baume reading—a constant time interval must be allowed between mixing and measuring. This is to allow the gel structure (which is broken down when "work" is put into the mixture by shear stress) to recover and as much as 10–15 minutes may be required in some instances. The same recovery time must be used before every measurement.

USING

Apart from the dry sleeking with a soft brush already referred to the main application methods for prepared water-based paints are swabbing or brushing, spraying and dipping or flow-coating.

Swabbing and Brushing are self explanatory and are usually reserved for the very thick paints applied to large moulds and heavy section castings. Viscosity of the paint is not critical but is usually kept quite high, say above 80° Baume, so that an adequate layer of paint is laid down. These methods are inclined to leave application marking on the casting unless carefully carried out.

Spraying is widely practised and there are many types of applicators available from simple hand spray equipment to specialised pressure pot dispensers. Each method can have a preferred viscosity for optimum effect and careful control of paint quality should be ensured at all times.

Dipping and Flow-Coating. High production foundries may well install a core coating station where cores can be prepared by hand or in bulk on frames or trays for a simple dipping or a flowover enrobing operation, both being followed by a drain and return section. Coating consistency is obviously critically dependent on control of dressing quality. Regular checks on viscosity and/or density, etc. are necessary as is some form of continuous agitation to maintain homogeneity. A simple bubbler or propellor type stirrer is often sufficient. Once the desired dressing consistency has been found it must be held within very close limits.

Importance of Cleanliness. Particularly with the continuous methods of application great care should be taken especially in hot weather, to clean down very thoroughly at regular intervals—at least once a week. Steam clean if this facility is available or use a fairly strong disinfectant. A coating that is fermenting will usually thicken up considerably or froth, both of which will affect dried finish and casting surface. Most FOSECO dressings employ a rota of preservatives so that bacterial colonies of immunity are not built up with time.

DRYING

Only occasionally can residual heat from a previously dried core or mould be relied upon to dry out a water-based coating completely. Nor should a coating be dried by fierce or localised torching—very few coatings will stand up to this sort of treatment. They may craze or crack off the substrate due to differential expansion characteristics. Some modern or clay-free compositions (TERRAPAINT 55, HOLCOTE 110) will, however, withstand a very great deal of this sort of maltreatment.

A gentle, well-ventilated, form of drying should be used such as a return to a normal "up and over" core store or ventilated batch oven operating at between 180°–220°C. Time at temperature should be long enough to ensure complete drying. Some deep-seated penetration may take longer than expected to dry out especially if the core is more or less completely covered by a relatively impermeable coating.

SPIRIT-BASED DRESSINGS

Technical considerations usually mean that there are really only two viable sub-groups in this sector—(a) Pastes and slurries for dilution with solvent and (b) ready-to-use suspensions. Powders for mixing with spirit may still be available in some areas but the technical constraints on this method of presentation are such that successful formulations with commercially satisfactory properties etc, are unlikely. They will not be considered further.

PASTES AND SLURRIES FOR FURTHER DILUTION

This has always been a popular method of presenting dressings, as it gives the supplier the opportunity to include suspension and surface active agents, binders, preservatives, etc. of his choice and the user the freedom to dilute to his own liking. By far the largest proportion of dressings in this group are flammable and are based on iso-propyl alcohol as the solvent, perhaps with an addition of methanol or other additive to increase burning characteristics if necessary.

All the refractory fillers and mixtures mentioned in the "Water-Based Dressings" section are again utilised but zircon containing dressings, either as the sole filler or as a mixture, are much more common. Indeed the most recent advances have taken place in the field of thixotropic and virtually ready for use suspensions of zircon and other dense and difficult refractory materials. Gas contents are much lower, solids are higher and coating performance is much improved.

Storage

Most flammable dressings, because they are required to burn out, will fall within the Petroleum Regulations in terms of flash point and both the dressing paste/slurry and the solvent used for dilution must be stored and handled accordingly. Most will also settle out a little after transport vibration or lengthy storage and any clear liquid or thinners appearing on the top should be thoroughly remixed with the bulk before taking any part for use or for further dilution. Similar comments apply to any settled sediment.

Use

The importance of determining a correct consistency for the application has already been stressed for water-based dressings and can be repeated here. A further complication is that dressing consistency could also affect the burning or drying properties.

All the usual methods of dressing application previously discussed can be used. When brushing or spraying, however, because of the volatile nature of the solvent, large mould or core areas should be dressed and flashed off in smaller sections. It is necessary that a coating should burn correctly in order to cure the resin bond.

READY-TO-USE SUSPENSIONS

This section will also include a group of air-drying coatings based on non-flammable solvents. Some alcohol-based dressings may also be allowed to air dry where sufficient time is available but there may be problems with explosive vapour/air mixtures if suitable precautions are not taken.

The major problem with all coatings in this section has always been sedimentation, vigorous remixing before and during use being necessary. However, some of the more recent heavy duty flammable slurry presentations based on zircon are exceptionally good showing only minor settlement after one month. Even then there is no hard packing of solids and redispersion is relatively easy. There is no doubt that improvement in this area will be continuous.

Storage

Flammable coatings will fall within the Petroleum Acts and must be stored and used accordingly. The air-drying dressings will probably utilise 1.1.1 trichlorethane (Trade names Genklene or Chlorothene) as being perhaps the most acceptable solvent from a Health and Safety point of view. Although technically non-flammable it should not be stored or used in an area of high intensity electrical sparking or other high value energy or temperature locations.

Use

Essentially as for water based dressings above. Care must be exercised, particularly when dipping cores, that all alcohol vapour has been removed before completing a mould and casting metal. It is sometimes possible that unsatisfactory application and combustion can drive vapour deep into the sand, this vapour subsequently creeping back to condense on the sand side of the coating layer. As with water vapour, discussed previously, the way is then open to blown castings and metal/mould reaction.

With air dried coatings arrangements must be made to disperse the vapour. Alcohol-based coatings may build up a critical explosive concentration. The non-flammable coatings produce a very heavy vapour that will collect in deep moulds, pockets, subways, cellars, etc. and can easily provide a hazard through oxygen displacement. Exhaust ventilation should be low level.

PROPRIETARY COATING TYPES

Powders for mixing with water:
TERRACOTE
FIRIT

Powders for electrostatic deposition:
TRIBONOL

Water-based pastes, etc:
TERRAPAINT
HOLCOTE
DYCOTE
TELLURIT
FRACTON

Spirit-based (flammable) dressings:
MOLDCOTE
ISOMOL
MOLCO
FLAMCO
CERAMOL
TELLURIT
HARDCOTE
INGOTOL

Air-drying dressings:
CHILCOTE
ZEROTHERM
HARDCOTE

The generic name as given above is nearly always followed by an identifying number, e.g. MOLDCOTE 9, specifying the particular type of dressing. For further details of dressings for almost any application please contact your local FOSECO organisation.

SECTION IV
LIGHT CASTING ALLOYS

ALUMINIUM CASTING ALLOYS—LM SERIES

USES AND GENERAL REMARKS

LM0   Mainly used for sand castings for electrical, chemical, and food applications.

LM2   One of the two most widely used alloys for all types of die castings.

LM4   The most versatile of the alloys; has very good casting characteristics and is used for a very wide range of applications. Strength and hardness can be greatly increased by heat treatment.

LM5   Suitable for sand and chill castings requiring maximum resistance to corrosion, e.g. marine applications.

LM6   Suitable for large, intricate, and thin-walled castings in all types of moulds; also used where corrosion resistance or ductility is required.

LM9   Used for applications especially low-pressure die castings requiring the characteristics of LM6 but higher tensile properties following heat-treatment.

LM10   Mainly used for sand and chill castings requiring high strength and shock resistance. Requires special foundry technique; heat-treated.

LM12   Mainly used where a very good machined surface finish and hardness is required.

LM13   Mainly used for pistons.

LM16   Suitable where high mechanical properties are desired in fairly intricate sand or chill castings. Requires heat-treatment.

LM18   Combines good foundry characteristics with high resistance to corrosion.

LM20   Mainly used for die castings. Similar to LM6 but a little better machinability and hardness.

LM21   Generally similar to LM4-M in characteristics and applications but better machinability and higher proof strength.

LM22   Used for chill castings requiring good foundry characteristics with good ductility. Requires heat treatment.

LM24   One of the two most widely used alloys for all types of die castings.

LM25   Suitable where good resistance to corrosion combined with high strength is required.

LM26   Mainly used for pistons as alternative to LM13.

LM27   A versatile sand and chill casting alloy introduced as an alternative to LM4 and LM21.

LM28   Piston alloy with lower coefficient of expansion than LM13. Requires special foundry technique.

LM29   As LM28 but lower coefficient of expansion.

LM30 For unlined die-cast cylinder blocks with low expansion and excellent wear resistance.

The Tables appearing on pages 127–138 inclusive are based on data given in The Properties and Characteristics of Aluminium Casting Alloys, published by the Association of Light Alloy Refiners and Smelters (Alar), London.

CASTING CHARACTERISTICS

BS 1490	Sand casting	Chill casting	Die casting	Fluidity	Resistance to hot-tearing	Pressure tightness
LM0	F	F	F	F	P	F
LM2	G*	G*	E	G	E	G
LM4	G	G	G	G	G	G
LM5	F	F	F*	F	F	P
LM6	E	E	G	E	E	E
LM9	G	E	G*	G	E	G
LM10	F	F	F*	F	G	P
LM12	F	G	U*	F	G	G
LM13	G	G	F*	G	E	F
LM16	G	G	G*	G	G	G
LM18	G	G	G*	G	E	E
LM20	E*	E	G	E	E	E
LM21	G	G	G*	G	G	G
LM22	G*	G	G*	G	G	G
LM24	F*	F*	E	G	G	G
LM25	G	E	G*	G	G	G
LM26	G	G	F*	G	G	F
LM27	G	E	G*	G	G	G
LM28	P	F	—	F	G	F
LM29	P	F	—	F	G	F
LM30	*	F	G	G	G	F

*Not normally used in this form.
E—Excellent G—Good F—Fair P—Poor U—Unsuitable

Chemical Composition (%)*

BS	Cu	Mg	Si	Fe	Mn	Ni	Zn	Pb	Sn	Ti	Others
1490	0.03	0.03	0.30	0.40	0.03	0.03	0.07	0.03	0.03	—	Al 99.50 min
LM0	0.7–2.5	0.30	9.0–11.5	1.0	0.5	0.5	2.0	0.3	0.2	0.2	—
LM2	2.0–4.0	0.15	4.0–6.0	0.8	0.2–0.6	0.3	0.5	0.1	0.1	0.2	—
LM4	0.1	3.0–6.0	0.3	0.6	0.3–0.7	0.1	0.1	0.05	0.05	0.2	—
LM5	0.1	0.10	10.0–13.0	0.6	0.5	0.1	0.1	0.1	0.05	0.2	—
LM6	0.1	0.2–0.6	10.0–13.0	0.6	0.3–0.7	0.1	0.1	0.05	0.05	0.2	—
LM9	0.1	9.5–11.0	0.25	0.35	0.10	0.10	0.10	0.05	0.05	0.2[b]	—
LM10	9.0–11.0	0.2–0.4	2.5	1.0	0.6	0.5	0.8	0.1	0.1	0.2	—
LM12	0.7–1.5	0.8–1.5	10.0–12.0	1.0	0.5	1.5	0.5	0.05	0.1	0.2	—
LM13	1.0–1.5	0.4–0.6	4.5–5.5	0.6	0.5	0.25	0.1	0.1	0.05	0.2[b]	—
LM16	0.1	0.10	4.5–6.0	0.6	0.5	0.1	0.1	0.1	0.05	0.2	—
LM18	0.4	0.2	10.0–13.0	1.0	0.5	0.1	0.2	0.1	0.1	0.2	—
LM20	3.0–5.0	0.1–0.3	5.0–7.0	1.0	0.2–0.6	0.3	2.0	0.2	0.1	0.2	—
LM21	2.8–3.8	0.05	4.0–6.0	0.6	0.2–0.6	0.15	0.15	0.1	0.05	0.2	—
LM22	3.0–4.0	0.1	7.5–9.5	1.3	0.5	0.5	3.0	0.3	0.2	0.2	—
LM24	0.1	0.20–0.45	6.5–7.5	0.5	0.3	0.1	0.1	0.1	0.05	0.2[b]	—
LM25	2.0–4.0	0.5–1.5	8.5–10.5	1.2	0.5	1.0	1.0	0.2	0.1	0.2	—
LM26	1.5–2.5	0.3	6.0–8.0	0.8	0.2–0.6	0.3	1.0	0.2	0.1	0.2	—
LM27	1.3–1.8	0.8–1.5	17–20	0.7	0.6	0.8–1.5	0.2	0.1	0.1	0.2	Cr 0.6; Co 0.5
LM28	0.8–1.3	0.8–1.3	22–25	0.7	0.6	0.8–1.3	0.2	0.1	0.1	0.2	Cr 0.6; Co 0.5
LM29	4.0–5.0	0.4–0.7	16–18	1.1	0.3	0.1	0.2	0.1	0.1	0.2	—

*—single figures in this table are maxima. b—0.2% in castings

LIGHT CASTING ALLOYS

AND OTHER PROPERTIES

Machin-ability	Resistance to corrosion	Specific gravity	Coefficient of linear expansion* ($C \times 10^6$)	Thermal conductivity[2] at 25°C (c.g.s. units)	Electrical conductivity** at 20°C (% IACS)	BS 1490
F	E	2.70	24	0.50	57	LM0
F	G	2.74	20	0.24	26	LM2
G	G	2.75	21	0.29	32	LM4
G	E	2.65	23	0.33	31	LM5
F	E	2.65	20	0.34	37	LM6
F	E	2.68	22	0.35	38	LM9
G	E	2.57	25	0.21	20	LM10
E	P	2.94	22	0.31	33	LM12
F	G	2.70	19	0.28	29	LM13
G	G	2.70	23	0.34	36	LM16
F	E	2.69	22	0.34	37	LM18
F	G	2.68	20	0.37	37	LM20
G	G	2.81	21	0.29	32	LM21
G	G	2.77	21	0.29	32	LM22
F	G	2.79	21	0.23	24	LM24
F	E	2.68	22	0.36	39	LM25
F	G	2.76	21	0.25	26	LM26
G	G	2.75	21	0.37	27	LM27
P	G	2.68	18	0.32	—	LM28
P	G	2.65	16	0.30	—	LM29
P	G	2.73	18	0.32	20	LM30

* Relates to temperature range 20–100°C.
** These values are approximate and will vary with the condition of the casting.

TYPICAL MECHANICAL PROPERTIES

Alloy BS 1490* 1970	0.2% proof stress (N/mm^2)	Sand cast Tensile stress (N/mm^2)	Elongation (%)	Brinell hardness number
LM0-M	30	80	30	25
LM2-M	—	—	—	—
LM4-M	100	150	2	70
LM4-TF	250	280	1	105
LM5-M	90	170	5	60
LM6-M	70	170	8	55
LM9-M	—	—	—	—
LM9-TE	120	180	2	70
LM9-TF	220	250	—	100
LM10-TB	180	310	15	85
LM12-M	—	—	—	—
LM13-TE	—	—	—	—
LM13-TF	200	200	—	115
LM13-TF7	140	150	1	75
LM16-TB	130	210	3	80
LM16-TF	240	280	1	100
LM18-M	70	120	5	40
LM20-M	—	—	—	—
LM21-M	130	180	1	85
LM22-TB	—	—	—	—
LM24-M	—	—	—	—
LM25-M	90	140	2.5	60
LM25-TE	130	170	1.5	70
LM25-TB7	100	170	3	65
LM25-TF	220	250	1	105
LM26-TE	—	—	—	—
LM27-M	90	150	2	75
LM28-TE	—	—	—	—
LM28-TF	120	130	0.5	120
LM29-TE	120	130	0.3	120
LM29-TF	120	130	0.3	120
LM30-M	—	—	—	—
LM30-TS	—	—	—	—

* Suffix letters indicate the condition of castings: M—as cast; TE—precipitation treated; TB—solution treated; TB7—solution treated and stabilised; TF—solution and precipitation treated; TF7—fully heat-treated and stabilised; TS—stress relieved.

LIGHT CASTING ALLOYS

TYPICAL MECHANICAL PROPERTIES—(continued)

0.2% proof stress (N/mm²)	Tensile strength (N/mm²)	Chill cast Elongation (%)	Brinell hardness number	Alloy BS 1490[1] 1970
30	80	40	25	LM0–M
90	180	2	80	LM2–M
100	200	3	80	LM4–M
250	310	3	110	LM–4TF
90	230	10	65	LM5–M
80	200	13	60	LM6–M
—	200	3	—	LM9–M
150	250	2.5	80	LM9–TE
280	310	1	110	LM9–TF
180	360	20	95	LM10–TB
150	180	—	95	LM12–M
—	220	1	105	LM13–TE
280	290	1	125	LM13–TF
200	210	1	75	LM13–TF7
140	250	6	85	LM16–TB
270	310	2	110	LM16–TF
80	150	6	50	LM18–M
80	220	7	60	LM20–M
130	200	2	90	LM21–M
120	260	9	75	LM22–TB
110	200	2	85	LM24–M
90	180	5	60	LM25–M
150	220	2	80	LM25–TE
100	230	8	65	LM25–TB7
240	310	3	105	LM25–TF
180	230	1	105	LM26–TE
100	180	3	80	LM27–M
170	190	0.5	120	LM28–TE
170	200	0.5	120	LM28–TF
170	210	0.3	120	LM29–TE
170	210	0.3	120	LM29–TF
150	150	—	110	LM30–M
160	160	—	110	LM30–TS

Note. The typical properties shown in this table are those of separately cast test bars and may not be obtained in all parts of a casting.

			Related or Approximately	
	BS	U.S.A.	U.S.A.	U.S.A.
BS 1490	aerospace	AA	commercial	SAE
LM0–M	—	—	—	—
LM2–M	—	—	—	—
LM4–M	—	319.0	319	326
LM4–TF	—	319.0	319	326
LM5–M	—	514.0	213	320
LM6–M	L.33	—	—	—
LM9–M	—	A360.0	A360	309
LM9–TE	—	—	—	—
LM9–TF	—	—	—	—
LM10–TB	L.53	520.0	220	324
LM12–M	—	222.0	122	34
LM13–TE	—	A332.0	A132	321
LM13–TF	—	A332.0	A132	321
LM13–TF7	—	A332.0	A132	321
LM16–TB	—	355.0	355	322
LM16–TF	L.78	355.0	355	322
LM18–M	—	443.0	43	35
LM20–M	—	413.0	13	305
LM21–M	—	319.0	319	329
LM22–TB	—	—	—	—
LM24–M	—	A380.0	A380	306
LM25–M	—	356.0	356	323
LM25–TE	—	356.0	356	323
LM25–TB7	—	356.0	356	323
LM25–TF	—	356.0	356	323
LM26–TE	—	F332.0	F132	332
LM27–M	—	—	—	—
LM28–TE	—	—	—	—
LM28–TF	—	—	—	—
LM29–TE	—	—	—	—
LM29–TF	—	—	—	—
LM30–M	—	390.0	390	—
LM30–TS	—	390.0	390	—

Suffix letters indicate the condition of castings: M—as cast; TE—precipitation treated; TB—solution treated; TB7—solution treated and stabilised; TF—solution and precipitation treated; TF7—fully heat-treated and stabilised; TS—stress relieved.

LIGHT CASTING ALLOYS

SIMILAR SPECIFICATIONS

U.S.A. ASTM	Canada commercial	France A57–702/3	Germany DIN 1725/2	Germany Alloy No.	BS 1490
995A	100	A5	—	—	LM0–M
—	—	—	—	—	LM2–M
SC 64D	117	—	—	—	LM4–M
SC 64D	117	—	—	—	LM4–TF
G 4A	B320	A–G3T	G–AlMg	245	LM5–M
S 12C	6290	A–S13	G–AlSi12	230	LM6–M
SG 100A	161	A–S10G	G–AlSi10Mg	230/10	LM9–M
—	161	A–S10G	G–AlSi10Mg	230/10	LM9–TE
—	161	A–S10G	G–AlSi10Mg	230/10	LM9–TF
G 10A	350	—	G–AlMg10	—	LM10–TB
CG 100A	—	A–U10G	—	—	LM12–M
SN 122A	162	A–12UN	—	230/Ni	LM13–TE
SN 122A	162	A–S12UN	—	230/Ni	LM13–TF
SN 122A	162	A–S12UN	—	230/Ni	LM13–TF7
SC 51A	125	—	—	234	LM16–TB
SC 51A	125	—	—	234	LM16–TF
S 5A	123	—	—	—	LM18–M
S 12A/B	—	A–S13Y4	G–AlSi12(Cu)	331	LM20–M
SC 64C	—	—	G–AlSi6Cu4	225	LM21–M
—	C117	—	—	—	LM22–TB
SC 84A	C143	—	G–AlSi8Cu3	333	LM24–M
SG 70A	135	—	G–AlSi7Mg	—	LM25–M
SG 70A	135	—	G–AlSi7Mg	—	LM25–TE
SG 70A	135	—	G–AlSi7Mg	—	LM25–TB7
SG 70A	135	—	G–AlSi7Mg	—	LM25–TF
SC 103A	—	—	—	—	LM26–TE
—	—	—	—	—	LM27–M
—	—	—	—	—	LM28–TE
—	—	—	—	—	LM28–TF
—	—	—	—	—	LM29–TE
—	—	—	—	—	LM29–TF
—	—	—	—	—	LM30–M
—	—	—	—	—	LM30–TS

ALUMINIUM CASTING ALLOYS
Chemical

BS aerospace	Cu	Mg	Si	Fe	Mn
L33	0.1	0.10	10.0–13.0	0.6	0.5
L35	3.5–4.5	1.2–1.7	0.6	0.6	0.6
L51	0.8–2.0	0.05–0.20	1.5–2.8	0.8–1.4	0.1
L52	1.3–3.0	0.5–1.7	0.6–2.0	0.8–1.4	0.1
L53	0.1	9.5–11.0	0.25	0.35	0.10
L78	1.0–1.5	0.4–0.6	4.5–5.5	0.6	0.5
L91	4.0–5.0	0.10	0.25	0.25	0.10
L92	4.0–5.0	0.10	0.25	0.25	0.10
L99	0.10	0.20–0.45	6.5–7.5	0.15**	0.10
DTD361B	4.0–5.0	0.10	0.25	0.25	0.10
DTD716A	0.1	0.3–0.8	3.5–6.0	0.6	0.5
DTD722A	0.1	0.3–0.8	3.5–6.0	0.6	0.5
DTD727A	0.1	0.3–0.8	3.5–6.0	0.6	0.5
DTD735A	0.1	0.3–0.8	3.5–6.0	0.6	0.5
DTD741A	3.5–4.5	1.2–2.5	0.5	0.5	0.1
DTD5008B	0.1	0.5–0.75	0.25	0.5	0.1
DTD5018	0.2	7.4–7.9	0.25	0.35	0.1–0.3
DTD5028	0.10	0.20–0.45	6.5–7.5	0.15**	0.10

* Single figures in this table are maxima.   ** 0.20 in castings

—AEROSPACE SERIES
Composition (per cent)*

Ni	Zn	Pb	Sn	Ti	Others
0.1	0.1	0.1	0.05	—	
1.8–2.3	0.1	0.05	0.05	0.25	Si+Fe 1.0
0.8–1.7	0.1	0.05	0.05	0.25	—
0.5–2.0	0.1	0.05	0.05	0.25	—
0.10	0.10	0.05	0.05	0.25	—
0.25	0.10	0.05	0.05	0.25	—
0.10	0.10	0.05	0.05	0.25	—
0.10	0.10	0.05	0.05	0.25	—
0.10	0.10	0.05	0.05	0.20	—
0.10	0.10	0.05	0.05	—	Ti+Nb 0.05–0.30
0.1	0.1	0.1	0.05	—	Ti+Nb 0.2
0.1	0.1	0.1	0.05	—	Ti+Nb 0.2
0.1	0.1	0.1	0.05	—	Ti+Nb 0.2
0.1	0.1	0.1	0.05	—	Ti+Nb 0.2
0.1	0.1	0.1	0.05	—	Co 0.5–1.0; Nb 0.05–0.3
0.1	4.8–5.7	0.1	0.05	0.25	Cr 0.4–0.6
0.10	0.9–1.4	0.05	0.05	0.25	—
0.10	0.10	0.05	0.05	0.20	—

Aluminium Casting Alloys—Aerospace Series
Related or approximately similar specifications

Aerospace	BS 1490	U.S.A. AA	U.S.A. commercial	U.S.A. SAE	U.S.A. ASTM	Canada commercial	France A57-702/3	Germany DIN 1725/2	Germany alloy no.
L33	LM6-M	—	—	—	S 12C	6290	A-S13	G-AlSi12	230
L35	LM14-WP*	242.0	142	39	CN 42A	218	A-U4NT	—	—
L51	LM23-P*	—	—	—	—	—	A-S2U	—	—
L52	LM15-WP*	—	—	—	—	—	—	—	—
L53	LM10-TB	520.0	220	324	G 10A	350	—	G-AlMg10	234
L73	LM16-TF	355.0	355	322	SC51A	125	—	—	—
L91	LM11-W*	—	—	—	—	226	A-U5GT	G-AlCu4Ti	220
L92	LM11-WP*	—	—	—	—	226	A-U5GT	G-AlCu4Ti	220
L99	—	A356.0	A356	336	SG 70B	B135	—	G-AlSi7Mg	—
DTD361B	LM11-WP*	—	—	—	—	226	A-U5GT	G-AlCu4Ti	220
DTD716A	LM8-M*	—	—	—	—	B116	A-S4G	G-AlSi5Mg	235
DTD722A	LM8-P*	—	—	—	—	B116	A-S4G	G-AlSi5Mg	235
DTD727A	LM8-W*	—	—	—	—	B116	A-S4G	G-AlSi5Mg	235
DTD735A	LM8-WP*	—	—	—	—	B116	A-S4G	G-AlSi5Mg	235
DTD741A	—	—	—	—	—	—	—	—	—
DTD5008B	—	D712.0	40E	310	ZG 61A	432	A-Z5G	—	—
DTD5018	—	—	—	—	—	—	—	—	—
DTD5028	—	A356.0	A356	336	SG 70B	B135	—	G-AlSi7Mg	—
DTD5028	—	A356.0	A356	336	SG 70B	B135	—	G-AlSi7Mg	—

ALUMINIUM CASTING ALLOYS—AEROSPACE SERIES
Minimum mechanical properties

Aerospace	BS 1490	Sand cast 0.2% proof stress[c] (N/mm²)	Sand cast Tensile stress (N/mm²)	Elongation (%)	0.2% proof stress[c] (N/mm²)	Chill cast Tensile stress (N/mm²)	Elongation %
L33	LM6–M	60[c]	162[d]	5	80[c]	185[d]	7
L35	LM14–WP*	210	220	—	230	280	—
L51	LM23–P*	125	160	2	140	200	3
L52	LM15–WP*	245	280	—	295	325	—
L53	LM10–TB	170	280	8	170	310	12
L78	LM16–TF	220	250	—	250	300	—
L91	LM11–W*	165	220	7	165	265	13
L92	LM11–WP*	200	280	4	200	310	9
L99	—	185	230	2	200	280	5
DTD361B	LM11–WP*	310[c]	324[d]	—	360[c]	402[d]	4
DTD716A	LM8–M*	80[c]	124[d]	2	90	162[d]	3
DTD722A	LM8–P*	130[c]	147[d]	1	130[c]	185[d]	2
DTD727A	LM8–W*	100[c]	162[d]	2.5	100[c]	232[d]	5
DTD735A	LM8–WP*	220[c]	232[d]	—	220[c]	278[d]	2
DTD741A	—	250[c]	263[d]	—	260[c]	340[d]	—
DTD5008B	—	150[c]	216[d]	4	180[c]	232[d]	5
DTD5018	—	170[d]	278[d]	5	170[d]	309[d]	10
DTD5028	—	193[ad]	263[ad]	5[a]	201[d]	278[d]	5
DTD5028	—	178[bd]	224[bd]	3[b]	—	—	—

*Obsolete

One Newton per square millimetre (N/mm²) = 0.0645 tons-force per square inch (tonf/in²)
One Ton-force per square inch (ton/in²) = 15.4443 newtons per square millimetre (N/mm²)
([a]) test pieces cut from designated locations of casting; ([b]) test pieces cut from undesignated locations. ([c]) for information only for LM alloys and also for others where indicated; ([d]) conversion from tons per sq. inch.

ALUMINIUM CASTING ALLOYS—AEROSPACE SERIES

BS Aerospace	Specific gravity	Casting characteristics			General properties	
		Sand casting	Chill casting	Die casting	Corrosion resistance	Machinability
L33	2.65	E	E	G	E	F
L35	2.82	F	G	U	F	G
L51	2.77	G	G	G*	G	G
L52	2.75	F	G	U	G	G
L53	2.57	F	F	F*	E	G
L78	2.70	G	G	F*	G	F
L91	2.80	F	P	U*	F	G
L92	2.80	F	P	U*	F	G
L99	2.68	G	E	F*	E	F
DTD361B	2.80	F	P	U*	F	G
DTD716A	2.68	G	G	G*	E	F
DTD722A	2.68	G	G	G*	E	F
DTD727A	2.68	G	G	G*	E	F
DTD735A	2.68	G	G	G*	E	F
DTD741A	2.80	F	G	—	F	G
DTD5008B	2.81	F	P	U*	E	G
DTD5018	2.64	F	F	—	E	G
DTD5028	2.68	G	E	—	E	F

*Not normally used in this form
E—Excellent; G—Good; F—Fair; P—Poor; U—Unsuitable

Aluminium casting alloys

					0	White
				12	11*	Green
			5	30	10	Black
		23*	14*	29	28	Brown
	1*	26	3*	27	—	Blue
6	25	9	13	20	18	Yellow
4	8	24	22	16	21	2	Red
Red	Yellow	Blue	Brown	Black	Green	White

* Obsolete.

STANDARD MELTING AND FLUXING PROCEDURES FOR ALUMINIUM ALLOYS

ALUMINIUM—SILICON—COPPER ALLOYS
e.g. BS 1490, LM4, LM13, LM16, LM21, LM22, LM24 and LM27

Metal Treatment for Sand Castings

Melt initial charge under a cover of 250 g of COVERAL 11 per 50 kg melt. Continue charging until melt is complete and has acquired requisite degree of superheat. When metal temperature begins to fall slowly plunge 0.25% DEGASER 190 or 0.35% DEGASER 450 by weight per melt. Hold until bubbling ceases. These amounts may be increased if the charges are corroded or dirty giving heavy gas contamination in the metal. Increased amounts of DEGASER are best plunged in two stages consecutively.

For certain applications the above combined degassing/grain refining treatment may be superseded by separate treatments with DEGASER 200 or DEGASER 400 in conjunction with NUCLEANT 2 (0.25 by weight), a powerful grain refiner. Whichever method is used, finish off by adding a further 250 g per 50 kg of COVERAL 11 to the metal surface, rabble gently and skim off cleanly.

Recommended pouring temperatures are:
Light castings, up to 13 mm section   730°C.
Medium castings, 13–38 mm section   710°C.
Heavy castings, over 38 mm section   690°C.

Metal Treatment for Gravity and Pressure Diecastings

A full metal treatment as outlined above is not usually necessary for gravity and pressure diecastings. What is very necessary, however, is that the holding bale out furnaces are regularly drossed off in order to circumvent the possible formation of hard spot inclusions in the metal. Gravity diecasters will normally hold metal at average temperatures, i.e. around 700–740°C, and can therefore use COVERAL 11 in the amounts and manner already described for drossing off.

Additionally, some gravity diecastings subject to shrinkage unsoundness, etc., benefit greatly from the deliberate introduction of a controlled amount of gas into the metal. Ejection of this gas from solution in the form of widely dispersed pinholes helps to offset possible shrinkage unsoundness during solidification. The gas (hydrogen) may be added by plunging DYCASTAL tablets into the melt (0.15% by weight) or a pinch of DYCASTAL powder may be added to the ladle before baling out.

Pressure diecasters, because of the different requirements of the process, often hold metal in the bale-out furnaces at temperatures only just above the liquidus, i.e. around 610–640°C. At these low temperatures normal drossing fluxes work too slowly if at all, and it becomes necessary to use COVERAL 73 which was specially designed for the purpose. About 0.5% by weight is scattered on the metal surface and lightly rabbled until the reaction is complete. Skim the dry dross cleanly.

Grain Refinement

The grain refinement of commercial die-casting alloy improves feeding characteristics. This treatment reduces the incidence of shrinkage unsoundness in heavy cast sections and thin sections show less tendency to form laps and misruns. NUCLEANT 2 tabletted grain refiner is recommended and this should be plunged into the melt at the rate of 0.25% by weight, prior to drossing off the melt with COVERAL 11.

ALUMINIUM SILICON ALLOYS

The Medium Silicon Group—3-6% Silicon

For example Alcan B116; Alcoa 356; the DTD alloys 716A, 722A, 727A, 735A, LM18, LM25, and LM8 (obsolete).

Modification of these alloys is not essential, but it often improves mechanical properties. Grain refinement can also assist response to heat treatment where this is carried out, and will prevent patchiness due to grain-size variation in articles subsequently anodised.

Add 250 g COVERAL 36A per 50 kg melt as an initial cover. When the charge is molten add a further 250 g, wait until the flux liquifies and then work thoroughly into the metal. Plunge 0.1–0.2% NUCLEANT 2 by weight before degassing with 0.25% by weight DEGASER 190 or 0.35% DEGASER 450. Finally, dross off with a small amount of COVERAL 11. The clean, oxide-free modified metal produced by this technique will exhibit enhanced fluidity and improved mass feeding characteristics.

Suggested pouring temperature for sand castings are:
Light castings, under 13 mm section   730°C.
Medium castings, 13–38 mm section   710°C.
Heavy castings over 38 mm section   690°C.

The Eutectic Alloy Group—10-13% Silicon

For example; Aerospace L33, LM6, LM20; also sometimes known as Alpax, Silumin, etc.

General Notes

The BS 1490 specification covering the aluminium–silicon alloys LM6 and LM20 calls for the castings to be supplied in the as-cast condition, i.e. not heat-treated. As such they are suitable for exposed marine castings, motor-car and lorry fittings, and engine parts (manifolds, casings, etc.), switch boxes and generally where corrosion resistance and ease of casting are essential.

The aluminium–silicon alloys generally have excellent casting properties and very good resistance to corrosion; the LM6 type is the best in this latter respect. Although of medium strength with relatively low hardness and elastic limit, they have excellent ductility. Because of their high silicon content, they have a tendency to drag during machining and cause rapid tool wear. Carbide-tipped tools with large rake angles, low cutting speeds and the use of lubricants and coolants, are recommended.

They can be sand cast, die cast—both gravity and pressure—and are very suitable for low pressure casting.

Test Bars

Test bars are generally dry-sand moulded in Mansfield or similar sand or in oil sand. The bar with the 70 mm diameter head is called for under the BS specifications but where there is some choice in the matter many foundrymen prefer a bar with a larger diameter head. The dimensions of the BSS mould and machined bar are shown on page 159.

Bars are generally poured at a temperature of about 680°C; in no circumstances should the metal be overheated. At the start of pouring, the mould should be held at an angle of about 45° from the vertical and then brought gradually to the upright position as the head fills. Pouring should be carried out steadily and over a period of not less than 10 seconds.

Heat Treatment

The alloys LM6 and LM20 are supplied in the non-heat-treated condition. Apart from stress relieving and for improving ductility, heat treatment is of little interest. It is worth noting, however, that the high silicon alloys, Alpax beta and Alpax gamma covered by BS specification 1490LM9 (TE and TF) are subject to a precipitation treatment and a precipitation/solution treatment respectively.

Casting Characteristics

Fluidity:	Excellent—can be cast into thinner and more intricate sections than any other aluminium alloy.
Pressure Tightness:	Very good—especially suitable for castings required to be pressure tight.
Hot-tearing:	Sand and die-cast castings show complete freedom from hot-tearing.
Patternmaker's shrinkage:	1.3%.

Melting Considerations

As with most aluminium alloys, accurate temperature control is essential if best results are to be obtained. Gas pick-up increases with temperature of the melt and the temperature should, therefore, be kept as low as possible consistent with the requirements of correct fluxing treatment. As with both gas and oil-fired furnaces, the temperature continues to rise after heating is stopped, the fuel should be shut off before the required peak temperature is reached. An efficient degassing treatment should always be carried out if the metal is to be used for sand castings.

Modification

Although aluminium–silicon alloys were the first alloys of aluminium to be produced, it was not until 1921, when Dr. Pacz found that their mechanical properties could be considerably improved by the addition of sodium fluoride, that they assumed any industrial importance. The process of adding sodium fluorides or other alkali fluorides to the molten metal to improve the mechanical properties was termed "modification". Later several investigators found that modification could be achieved by the use of other agents, such as alkali metals and alkaline–earth metals. Of these modifying agents, the two most widely used have been sodium metal and alkali fluorides. The use of sodium metal in its usual commercially available condition is hazardous and the use of alkali fluorides is laborious. Both of these methods have now been superseded by the vacuum treated sodium (Navac) process.

In unmodified aluminium–silicon alloys, the aluminium–silicon appears as a coarse needle-like form as shown in Photomicrograph No. 1 (magnification × 50), page 150. Although modification does not actually refine the grain size of the metal, it does break up this needle-like structure within the grains and changes it to the fine dispersed form as shown in Photomicrograph No. 2 (magnification × 50), page 150. At the same time the amount of silicon which can be combined in the aluminium–silicon constituent is increased from 11.7% in the unmodified alloy to about 13% in the fully modified state. Thus, modification can result in the removal of primary (i.e. uncombined) silicon from the alloy.

The effect of modification on all the mechanical properties of the alloy is very marked, but is possibly greatest on the ductility and toughness. The following table gives typical figures before and after modification.

Mechanical properties (Typical figures in Avoirdupois units)	Unmodified alloy Sand cast	Unmodified alloy Chill cast	Modified alloy Sand cast	Modified alloy Chill cast
Ultimate tensile strength (ton/in^2)				
Minimum	6.0	9.0	10.5	13.0
Average	7.0	10.0	11.5	14.5
Elongation (% on 2 in)				
Minimum	1.0	2.0	5.0	7.0
Average	3.0	4.0	8.0	14.0
Impact resistance, Izod (ft/lb)				
Minimum	—	—	4.0	6.0
Average	0.5	1.0	5.0	7.0

Reversion

The modified structure of the aluminium–silicon alloy is unstable and it tends to change back or revert to the unmodified condition. This change is called "reversion" and the rate of change is directly proportional to silicon content, temperature of the melt and time of standing and indirectly proportional to size of melt. Reversion is slow at temperatures below 750°C and does not occur to any considerable extent with a melt of about 50 kg or so during a 10-minute holding period provided it is left undisturbed. Larger melts of 75 kg or more can be left longer without harm, but about 20 minutes should be regarded as a maximum period for all sizes of melt.

Advantages of using Vacuum Treated Sodium (Navac)

Modification may be carried out either by the use of metallic sodium or sodium in the form of salts. Each method has advantages and disadvantages, but both are efficient ways of adding sodium to the alloy. The salts method is clean and safe to handle, but is laborious and the salts have a tendency to pick up moisture. Metallic sodium is cheap but hazardous to handle. To protect it from moisture in the atmosphere it is normally stored under paraffin. Sodium which has been stored in this manner still contains a quantity of paraffin which is a source of hydrogen, and consequently aluminium alloys treated with metallic sodium of this type are usually gassy and sand castings produced from them may be porous. A 1 oz (28 g) cube of sodium can contain as much as 0.2 g of oil which could produce 350 cm^3 of hydrogen gas. Due to reaction with moisture metallic sodium which has been exposed to the air is coated with a thin film of sodium hydroxide, which will decompose when the sodium is plunged into the molten aluminium to form sodium oxide and water vapour, and the latter is a source of hydrogen which is readily absorbed by the metal. There is also a strong possibility that commercially produced sodium contains a quantity of hydrogen in solid solution.

The FOSECO organisation has developed a product which has all the advantages and none of the disadvantages of the two previous methods of modification already discussed; this is NAVAC. NAVAC is metallic sodium specially treated to eliminate completely all traces of hydrogen, paraffin and hydroxide, and is packed in air-tight capsules of pure aluminium to prevent contamination of sodium by moisture in the atmosphere. Packed in this way, sodium can be stored for unlimited periods of time, provided the containers are undamaged.

It is, therefore, strongly recommended that if completely gas free metal is required, as is the case in sand casting, modification should be carried out with the appropriate quantity of NAVAC, which is supplied in accurately weighed units in sealed aluminium containers.

Advantages of Degassing before Modifying

It is sometimes thought that by using modified ingot as supplied by the refiner, subsequent modification processes in the foundry can be cut down or even dispensed with. However, when it is borne in mind that the degree of modification in as-supplied ingot varies considerably and also that the ratio of scrap to ingot in any one charge also varies it will be seen that the only satisfactory method of securing a constant degree of modification is to destroy any existing modification and then remodify to a known and controlled extent.

The main function of degassing is, of course, to remove harmful gases, mainly hydrogen. Where completely gas-free metal is required as in sand casting particularly, a strong degassing treatment with DEGASER 190 should be carried out. However, degassing not only removes harmful gases, it removes sodium as well, and so destroys modification. *It is essential, therefore, that degassing should precede modification and never vice versa.*

Melting Procedure (NAVAC Process)

1. Heat up crucible.
2. Charge in ingot metal and scrap (runners, risers, etc.).
3. As soon as the first part of the charge reaches a pasty condition, sprinkle over its surface sufficient COVERAL 11 to form a cover. Usually about 250 g per 50 kg is required.
4. Add remainder of charge and melt rapidly, keeping the protective flux cover intact as much as possible.
5. Do not permit metal temperature to exceed about 760°C, then stop heating or remove the crucible from the furnace. There should be sufficient heat in the crucible to keep the temperature of the melt from falling, and during this period the melt should be degassed with DEGASER 190 or 301, or the ring-shaped tablet, DEGASER 450.
6. Push the surface flux cover a little to one side and through the cleaned area plunge to the bottom of the melt the DEGASER Tablets, using 0.25% by weight DEGASER 190 or 0.35% by weight DEGASER 450 and hold down until the bubbling action has ceased—usually 2 to 3 minutes. DEGASER 190 is a solid tablet and is plunged by means of a perforated bell plunger. DEGASER 450 are pierced tablets and these are simply placed on a hooked steel rod plunger. Coat the plungers well with FIRIT.
7. With the metal temperature between 720°C and 740°C (1328–1364°F), plunge a NAVAC unit to the bottom of the melt, using a perforated bell plunger. When the reaction has subsided the plunger is raised and

lowered two or three times to stir the metal without breaking the surface, and it is then withdrawn. A NAVAC 25 unit is suitable for a melt of 50 kg. For larger or smaller melts see the table on the NAVAC Leaflet No. B9.
8. Sprinkle a further quantity of COVERAL 11 on the surface of the melt, once again using 250 g per 50 kg and allow the melt to stand quietly for about 5 to 10 minutes. The standing time will depend on the size of the melt and should not exceed 5 minutes for melts of less than 50 kg.
9. The flux layer should now be rabbled into the immediate surface of the melt until a red glowing dross is obtained. Resulting from this exothermic action, a fine powdery dross exceptionally free from metal will be obtained. Skim cleanly.
10. Pour without delay.

N.B.—It is emphasised that hard and fast rules on modification technique cannot be laid down; the above procedure is given only as a guide. It is only by trial and error and exercising a great deal of intelligence that optimum results will be obtained.

Melting Procedure (Salts Methods)

1. Heat up crucible.
2. Charge in ingot metal and scrap (runners, risers, etc.).
3. As soon as the first part of the charge reaches a pasty condition sprinkle over its surface sufficient COVERAL 11 to form a cover. Usually about 250 g per 50 kg is required.
4. Add remainder of charge and melt rapidly, keeping the protective flux cover intact as much as possible.
5. Do not permit metal temperature to exceed about 760°C then remove the crucible from the furnace. There should be sufficient heat in the crucible to keep the temperature of the melt rising. After the maximum temperature has been reached, the melt should be degassed and grain refined on a falling temperature with DEGASER 190 or the ring-shaped DEGASER 450.
6. Push the surface flux cover a little to one side and through the cleaned area plunge to the bottom of the melt the DEGASER Tablets, using 0.25% by weight DEGASER 190 or 0.35% by weight DEGASER 450 and hold down until the bubbling action has ceased—usually 2 to 3 minutes. DEGASER 190 is a solid tablet and is plunged by means of a perforated bell plunger. DEGASER 450 are pierced tablets and these are simply placed on a hooked steel rod plunger. Coat the plungers well with FIRIT.

LIGHT CASTING ALLOYS

7. Rabble the surface flux cover into the surface of the melt and skim cleanly. When the melt has reached a temperature of at least 790°C—and not before—spread over the cleaned surface, a layer of COVERAL 29A, using 500 g per 50 kg.
(Alternatively, COVERAL 36A can be used, in which case the temperature of the melt should be 750°C and the quantity used 1 kg per 50 kg.)
When the flux layer has liquefied—and not before—it should be well worked into the melt for about 3 or 4 minutes, using a plunger in a stirring action.
8. Sprinkle a further quantity of COVERAL 11 on the surface of the melt, using 250 g per 50 kg and, depending on the size of the melt, allow to stand quietly for about 5 to 10 minutes—e.g. 5 minutes for a 25 kg and 10 minutes for a 50 kg melt—or until the correct pouring temperature is reached.
9. The flux layer should now be rabbled into the immediate surface of the melt until a red glowing dross is obtained. Resulting from this exothermic action, a fine powdery dross exceptionally free from metal will be obtained. Skim cleanly.
10. Pour without delay.

N.B.—It is emphasised that hard and fast rules on modification technique cannot be laid down; the above procedure is given only as a guide. A certain degree of skill is required to ensure that the optimum results are obtained.

Grain Refinement

As the maximum mechanical properties of the aluminium–silicon alloys are developed through modification as distinct from grain refinement, the latter process is generally not considered of particular importance. Nevertheless, heavy section sand castings can benefit quite considerably if the metal is given a strong grain refining treatment in addition to that produced during the degassing process with DEGASER 190 or 450. This can be carried out using NUCLEANT 2 tablets at the rate of 125 g per 50 kg. The tablets are plunged either before or simultaneously with the degassing treatment.

Casting Temperatures

Casting temperatures vary with section thickness but generally fall within the following limits:

Light castings: under 13 mm section   730°C.
Medium castings: 13 to 38 mm section   710°C.
Heavy castings: over 38 mm section   690°C.

Moulding Technique

Aluminium–silicon alloys can be cast satisfactorily in green or dry sand moulds, in either natural or synthetic moulding sand, the main requirements of which should be high permeability and fairly low green strength. Suitable properties of both types of sand are as follows:

	Natural Sand	Synthetic Sand
Grain fineness—AFS No.	120	80–100
Clay content	10%	4% (bentonite or similar clay)
Moisture	6–8%	3–4%
Permeability—AFS No.	20–30	40–50
Green compression strength	7lb in^2	5lb in^2

Green sand moulds should be dusted with TERRACOTE 20 and sleeked to improve casting finish. Alternatively, an excellent finish will be obtained by applying MOLDCOTE 9. This spirit-based dressing is ignited and burns off from the treated surface to leave a firm, closely adhering coating and gives the effect of skin-drying or torching.

Dry sand moulds and oil sand cores should be dressed with a water suspension of TERRACOTE 14 for best quality work, or with TERRACOTE 20 for ordinary commercial work. Both moulds and cores should, of course, be thoroughly dried before casting.

Running and feeding methods should be as for a high shrinkage alloy, i.e., large diameter runners and large feeding heads are necessary. Chills may have to be used also on heavy sections.

Castings should be gated in such a way as to promote directional solidification, i.e., the sections furthest away from the gates and heads should solidify first. With complex castings, the use of multiple ingates is recommended as this method, in conjunction with as low a pouring temperature as is possible, tends to minimise localised overheating of the mould face and helps to promote rapid and even solidification throughout the casting. Runners and gates should be cut in such a way as to allow the metal to enter the mould fairly rapidly but at the same time with minimum turbulence so as to avoid the entrapment of air and the formation of oxides.

Feeder heads must be of such size as to ensure that the feed metal remains liquid and available to the casting until the casting itself has solidified. If plain sand heads are used this means that the risers must often be of substantial size thus decreasing yield very considerably. Efficiency can be significantly improved by lining the riser cavity with a highly insulating resilient refractory sleeve of KALMIN 33. These are available in a range of ready made sizes and shapes but may be cut and shaped easily with a sharp knife to fit an unusual location. All riser heads should be covered with a layer of FEEDOL 9 immediately they are full in order to prevent heat loss to atmosphere.

Gravity Diecasting

In view of the effect of shock cooling on the metal it is sometimes considered unnecessary to modify aluminium–silicon alloy which has to be gravity diecast. Generally, however, metal for casting sections 5 mm and over should be modified for best results.

Since, in gravity diecasting, a small amount of gas in the metal is beneficial in countering shrinkage, modification can be carried out using NAMETAL. The use also of DYCASTAL which induces limited and controlled gassing of the melt, should be considered. Neither of these recommendations apply, however, to castings which have to pass radiological examination.

NAMETAL is used in the same way as NAVAC but at the rate of about 45 g per 50 kg and plunged at a temperature of 750°C. Further regular additions of about 30 g per 50 kg of melt should be made every ½ or 1 hour depending on rate at which more metal is added. (Care must be taken not to add too much sodium metal as excess tends to make the metal sluggish and die dressings will be blackened.)

Pressure Die Casting

It is generally agreed that aluminium–silicon alloy for pressure die casting does not require to be modified. The temperature of the metal as injected is very low and this, coupled with shock cooling, combine to produce a spontaneous modification. Nevertheless, where there are any somewhat heavy sections the metal undoubtedly benefits from a mild modification. In such cases 20 g of NAMETAL per 50 kg of metal should be used.

Fig. 4.1. Photomicrograph No. 1×50 magnifications. Showing unmodified structure.

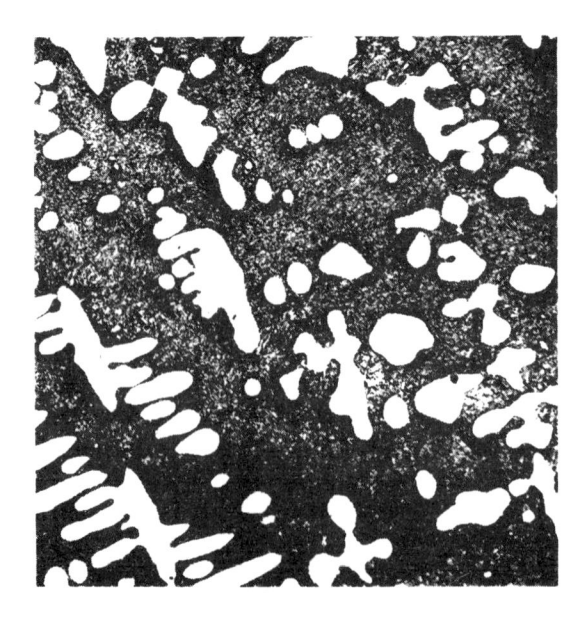

Fig. 4.2. Photomicrograph No. 2×50 magnifications. Showing modified structure.

ALTERNATIVE METHODS OF REFINING THE STRUCTURE OF THE ALUMINIUM–SILICON ALLOYS

The method by which sodium affects or "modifies" the microstructure of the eutectic (and to a smaller extent some of the hypo-eutectic) alloys has been dealt with in some detail. Other elements are also capable of exerting a similar effect chief among them being strontium, lithium, calcium, potassium and antimony. Commercially only strontium is of any real importance and indeed strontium modified material is available from the ingot suppliers. Its main attraction is to the gravity diecaster in that the level of modification due to strontium is very much longer lasting and is capable of surviving several remeltings. The effect on the microstructure is not quite so dramatic or quite so complete as that of sodium and it may be difficult consistently to reach required mechanical properties if it is used as the sole modifier. Strontium modified metal cannot be treated conventionally as all fluxes and tabletted degassing/grain refining agents contain halides or release chlorine all which are effective scavengers for strontium and remove it rapidly from the melt. Only a degassing treatment with nitrogen gas is possible on this basis and it may well be that this alone would provide insufficient cleaning effect especially for re-circulated metal or melts containing returned runners/risers/swarf, etc, that would be expected to contain fairly high proportions of non-metallics. Strontium and sodium are however compatible so some well judged restoration of properties may be possible for affected metal. Unlike sodium, where the founder has a choice of how to treat his metal, strontium can only be added as a hardener alloy not generally available to the foundry and then extremely expensive.

A further method of controlling the microstructure of the peri-eutectic aluminium/silicon alloys exists through the effect of phosphorus. Its effect on the hypereutectic alloys in respect of the very considerable refinement of the primary silicon phase is well recognised and is essential to the satisfactory production and engineering use of this group of high silicon compositions. When used on the 10–13% silicon eutectic compositions, where there are no primary silicon cuboids, the result is to stabilise and slightly exaggerate a completely "unmodified" (in terms of sodium modification) structure. This type of microstructure has low expansion thermal characteristics and some frictional properties that make it attractive, for example, to piston manufacturers. Again, phosphorus treated metal (the term "modification" is possibly best reserved for the effect produced by sodium or strontium: the term "refinement" is perhaps best employed to distinguish the effect of phosphorus) can be obtained from the suppliers but the effect of the addition will fade slowly with time, agitation, temperature etc. To some extent it can be regenerated by bubbling chlorine through the melt so a conventional degassing process, using a chlorine or freon containing gas, or hexachloroethane-based tablets can have a beneficial result.

When a re-treatment does become necessary NUCLEANT 120 tablets can be used. These are self sinking tablets containing a phosphorus addition together with grain refining agents and are used at the rate of about 0.4% of the metal weight. Alloys to be treated should be tolerant of a small increment in the copper content.

As might be deduced from the above, sodium and strontium have an opposing effect to that of phosphorus and vice versa. Indeed either can be used to remove the other(s) from an alloy and obviously they should never be added together to the same melt. Where both processes are used in the same foundry care must be exercised in clearly segregating and identifying returned scraps, off-cuts, etc.

THE HYPER-EUTECTIC ALUMINIUM–SILICON ALLOYS

Hyper-eutectic aluminium–silicon alloys are those aluminium alloys which contain more than the eutectic quantity, i.e. 11.6% of silicon, but more specifically 18–26% of silicon. These alloys are used principally in the production of pistons for internal combustion engines. They are not new; as far back as the 1920s various patents were granted for alloys containing from 15 to 50% silicon and later, in the 1930s, interest was again aroused in pistons produced in alloys containing more than 20% silicon and up to 3% nickel because of their very low coefficient of thermal expansion. However, the advantage of the very low thermal expansion was largely offset by poor foundry characteristics and machining properties of these alloys. The conclusion of these early investigations was the temporary abandonment of the hyper-eutectic alloys and acceptance of Lo-Ex (Alcoa 32, Alcan 162, LM.13) as the most satisfactory piston alloy.

However, in spite of these set-backs, the development of hyper-eutectic aluminium–silicon alloys continued in Europe and in the last decade the production of pistons in such alloys has been quite considerable. More recently, interest has been stimulated by information regarding the use of these alloys for engine cylinder blocks and heads. The development of all aluminium automotive and industrial engines is being rapidly pursued, principally in the United States, and consequently these alloys of low thermal expansion properties are being thoroughly investigated.

Probably the most important single advantage that has made it possible to use the hyper-eutectic alloys for pistons has been the development of a satisfactory technique for producing a fine homogeneous alloy structure. At room temperatures silicon is almost insoluble in aluminium. On solidification of the untreated alloy, silicon is deposited either in the form of a eutectic with aluminium, or as a primary constituent together with the aluminium–silicon eutectic, which characterises alloys of the hyper-eutectic series. In either case silicon takes the form of coarse plates and needles and it is this inherently weak type of structure which gives the untreated alloy its poor mechanical properties.

The process of modification, that is the treatment of the molten eutectic alloy with metallic sodium or salt mixtures based on sodium fluoride, markedly refines the structure of the aluminium–silicon eutectic. In addition, modification increases the silicon content of the eutectic from 11.7% to about 13%. Therefore, alloys containing less than 13% silicon after modification are characterised by a fine, globular, eutectiferous structure containing no primary silicon. Modification improves the mechanical properties of the eutectic and the hypo-eutectic alloys. Alloys containing more than 13% silicon, namely the hyper-eutectic type, still exhibit coarse, primary plates of silicon even after attempts to modify with sodium. These alloys are generally cast in metal moulds and consequently, due to the rapid rate of cooling, the unmodified alloy already exhibits a fine eutectic structure and this is to some extent further refined by modification treatment with sodium. The primary phase, however, remains relatively unaffected irrespective of whether sodium is added or not. It is the coarse unrefined primary phase which gave the original hyper-eutectic alloys such poor machining and physical properties.

Search for Refining Agent

Research work carried out by numerous investigators indicated that phosphorus, added to a hyper-eutectic aluminium–silicon melt, produced an outstanding refinement of the primary silicon. It was shown that phosphorus and aluminium formed nuclei of aluminium–phosphide and that this compound possesses all the qualities which favour nucleation of silicon, such as an identical lattice structure and very similar lattice parameters. It is considered that it is the introduction of nuclei of aluminium–phosphide which causes the refinement of the primary silicon phase and retards the modification of the eutectic when this alloy is poured into a chill mould.

Numerous preparations containing phosphorus were introduced on to the market after these investigators had published the results of their work but many of these products were undesirable in practice, either because of their hygroscopic nature or because they were difficult to introduce into molten alloys. Figure 1 shows an untreated British alloy of this type. Figure 2 shows the effect of a 0.4% addition of NUCLEANT 120. The very striking effect of these powerful refining additions is most obvious. The primary silicon is uniformly distributed throughout the metal section and no longer exists as coarse crystals. Metallurgical techniques using these grain refining agents are now an accepted part of the production of hyper-eutectic aluminium–silicon pistons to produce homogeneous structures of extreme fineness. Not only is the fine structure essential to achieve the required physical properties, but the uniform distribution of fine silicon crystals appreciably enhances the machining properties of the alloy. Coarse primary silicon unevenly distributed, as shown in Fig. 4.3(a), tends to be torn from the soft matrix during the machining operation, giving a rough surface even after careful diamond turning. A fine homogeneous structure, however, is capable of producing a very much smoother surface with consequent improved piston running properties and a reduction in oil consumption.

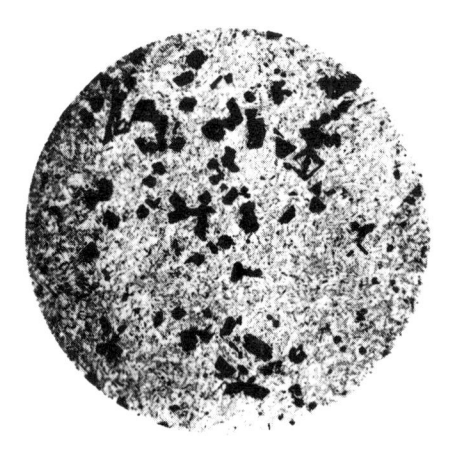

Fig. 4.3(a). Photomicrograph of an untreated 19% Si alloy (×50).

Fig. 4.3(b). Photomicrograph of same alloy treated with 0.3% of NUCLEANT 10, (×50).

Application of Hyper-eutectic Alloys

Many aluminium–silicon alloys of this type have been, or are being, investigated principally by the automotive industry for such parts as pistons, cylinder liners, rocker arms, cylinder blocks, cylinder heads, and brake drums. Generally speaking, these alloys are cast by permanent mould or gravity diecasting methods but the pressure casting of entire cylinder blocks or cylinder heads is being investigated.

Some Typical Hyper-eutectic Alloys

The table on page 157 gives details of some typical hyper-eutectic alloys at present being used in the internal combustion engine industries in Europe, America and Asia.

Heat Treatment

These alloys respond well to heat treatment and improved tensile and hardness properties are obtained after solution and stabilisation treatments; a typical example is to solution heat treat at a temperature of approximately 510°C (950°F) for 6 hours, to quench in water at 85°C (185°F), then to age for 12 hours at 227°C (440°F) and finally to air cool.

Foundry Characteristics

The casting characteristics of these alloys make them rather difficult to handle in the foundry. Carefully controlled foundry techniques are required with special attention to the grain refinement treatment. Most of these alloys have excellent fluidity but are subject to serious shrinkage defects if castings are not fed with adequate risers. They are rarely used for producing sand castings but satisfactory castings have been produced using the moulding and running techniques which apply for the more common aluminium–12% silicon alloy.

Casting temperatures for the hyper-eutectic aluminium silicon alloys are significantly higher than for other aluminium alloys. Extra care must therefore be taken to avoid undue oxidation and gas pick-up. Because the alloys mostly contain some magnesium and the temperatures are high, extensive degassing may be necessary to produce a satisfactory level of gas in the casting.

Melting Procedure

Most of these alloys contain between 0.5 and 2.0% magnesium and they therefore exhibit a strong tendency to form oxide inclusions. These inclusions are often detected during the machining process and are very easily identified as black "hard spots". Therefore, it is essential that these alloys should be treated with a fluxing technique which will reduce the tendency to form oxide inclusions and also wash out any which do happen to be produced during the melting process.

It is also most essential to use fluxing compounds which do not contain sodium because the presence of this element causes the formation of sodium phosphide which rises as a slag, thereby preventing the formation of aluminium phosphide nuclei.

The recommended melting procedure is as follows:
1. Charge in ingot and returns or returned scrap (runners, risers, etc.). It is recommended that for high quality work the amount of virgin ingot should be at least 50%; this will ensure that an excessive build-up of phosphorus (from previous treatments) does not occur and it also ensures that the alloy is kept within specification.
2. As soon as the first part of the charge reaches a pasty condition, sprinkle over its surface sufficient COVERAL 65 to form a cover. Usually about 250 g per 50 kg of charge is necessary (8 oz per 100 lb).
3. Add the remainder of the charge and melt rapidly, keeping the protective flux cover intact as much as possible.
4. When a temperature of about 780°C (1436°F) has been reached stop heating or remove the crucible from the furnace. There should be sufficient heat in the crucible and furnace walls to keep the temperature of the melt constant.
5. With the temperature at approximately 780°C (1436°F)* add NU-CLEANT 120 at the rate of at least 0.4%. Allow a few moments for the tablet to dissolve.
6. By this time the metal temperature will be falling slightly and the melt should be degassed by plunging to the bottom of the melt DEGASER 450 Tablets using 6 oz per 100 lb of metal (150 g per 50 kg).
7. Skim the melt cleanly and apply another cover of COVERAL 65. Wash the COVERAL 65 thoroughly into the melt by stirring or plunging.
8. Skim the melt cleanly and pour without delay.

*These alloys, unlike most other aluminium alloys, benefit from high temperature treatment and high pouring temperatures. Grain refinement treatment at temperatures well in excess of 800°C (1475°F) is not detrimental.

Generally speaking, the melting and pouring temperatures for all the hyper-eutectic aluminium–silicon alloys are higher than for conventional aluminium casting alloys.

LIGHT CASTING ALLOYS

Some Typical Hyper-eutectic Al-Si Alloys

	Si	Cu	Mg	Fe	Mn	Ni	Zn	Cr	Ti	Co
American	20.0	2.0	1.0	—	0.5	—	—	—	—	—
American	21.0	1.6	0.7	—	0.5	0.4	—	—	—	—
American	21.3	1.37	0.38	1.28	0.03	0.01	0.09	—	0.03	—
American	21.4	1.32	1.9	0.3	0.3	0.3	—	—	—	—
French (A-S20-U)	20.0	2.0	0.1	0.8	0.1	0.1	0.05	—	0.1	—
German (Nural 1761)	17.0	1.0	1.0	0.8	0.5	0.5	0.2	0.5	0.2	—
German (Nural 2361)	23.5	1.0	1.0	0.8	0.2	0.9	0.2	0.5	0.2	—
German (KS.280)	21.0	1.5	0.5	0.7	0.7	1.5	0.2	0.8	0.2	—
German (KS.281)	20.0	1.5	1.5	0.7	0.5	1.5	0.2	—	—	—
German (KS.245)	14.0	1.5	0.7	—	1.0	0.5	—	—	—	—
German (KS. Alusil)	20.0	1.5	—	—	—	0.5	—	—	—	—
Japanese	22.0	2.8	0.6	—	—	0.9	—	—	—	—
British LM.28	17–20	1.3–1.8	0.8–1.5	0.7	0.6	0.8–1.5	0.2	1.1	0.2	0.5
British LM.29	22–25	0.8–1.3	0.8–1.3	0.7	0.6	0.8–1.3	0.2	0.6	0.2	0.5
British LM.30	16–18	4.0–5.0	0.4–0.7	1.1	0.3	0.1	0.2	—	0.2	—

Minimum Mechanical Properties of the LM Series

BS.1490	Sand Cast			Chill Cast			
	0.2% proof stress N/mm²	Tensile stress N/mm²	% Elongation	0.2% proof stress N/mm²	Tensile stress N/mm²	% Elongation	Brinell
LM28–TE	—	—	—	170	170	—	90–130
LM28–TF	—	120	—	160	190	—	100–140
LM29–TE	120	120	—	170	190	—	100–140
LM29–TF	120	120	—	170	190	—	100–140
LM30–M	—	—	—	150	150	—	—
LM30–TS	—	—	—	160	160	—	—

M—As cast
TE—Precipitation treated
TF—Solution and precipitation treated
TS—Stress relieved

STANDARD TEST BAR FOR ALUMINIUM ALLOYS SAND CAST

MACHINED TENSILE TEST PIECE

The ends to be shaped to fit the axial loading shackles of the testing machine.

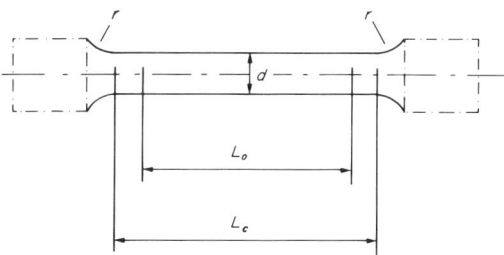

Fig. 4.4.

Cross-sectional area A_0 mm²	Diameter D mm	Gauge length L_0 mm	Minimum parallel length P mm	Minimum radius at shoulder R mm
150	13.82	69	76	26

PROPORTIONAL TEST PIECE

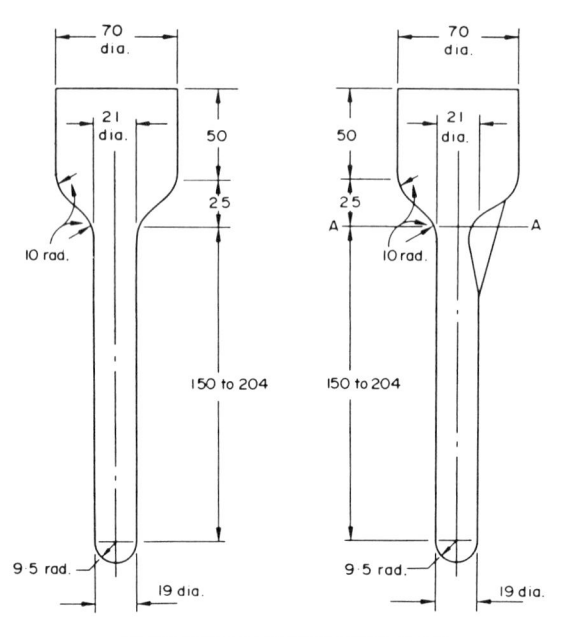

All dimensions are nominal and are in mm.
Fig. 4.5.

ALUMINIUM-MAGNESIUM CASTING ALLOY BS 1490—LM5

The melting, fluxing and degassing technique described for LM5 applies equally to LM14 and also to the L52 alloy.

Related Specifications are:
ALCAN GB.B320   ALCOA 214

Test Bars

Sand-cast test bars should conform to the drawings given above. The dimensions given are those stipulated in BS 1490.

Test bars are generally moulded in dry sand or oil sand, as the specification stipulates a dried mould: this eliminates the possibility of variable chilling due to an uneven moisture content. The steel tube for the mould should be perforated so as to form vents for the escape of mould gas. It should be not less than 80 mm in diameter.

LIGHT CASTING ALLOYS 161

The bars should be poured at a temperature of 690–700°C (1274–1292°F) and it is important if maximum physical properties are to be obtained that the mould be at atmospheric temperature. In no circumstances should the alloy be overheated. At the start of pouring, the mould is held at an angle of about 45° from the vertical and then brought gradually to the upright position as the head fills. Pouring is carried out steadily and should occupy a period of not less than ten seconds: turbulence within the test bar mould should be avoided at all costs.

Heat Treatment

None is required. The BS 1490 Specification calls for the alloy LM5 to be supplied in the as-cast (M) condition.

Physical Properties

Density	2.66
Electrical conductivity (%IACS at 20°C)	29 approximately
Brinell Hardness (sand cast)	60.

Founding Characteristics

Fluidity	Fair.
Hot tearing	Fair resistance.
Pressure tightness	Poor.
Machinability	Good.
Resistance to corrosion	Excellent.

Melting Hazards (Gas absorption)

Aluminium–magnesium alloys of the LM5 type tend to oxidise and pick up hydrogen very readily during melting. Thus every precaution should be taken to protect the melt and the use of a low melting point flux with a strong cleansing action should be employed. The flux must also have the property of dissolving and coalescing oxides, particularly magnesium oxide. Such a flux is COVERAL 65.

As LM5 is susceptible to grain growth, a grain refining treatment is necessary. This is combined in the degassing action of DEGASER 190, which has the additional advantage that it forms magnesium chloride and so is beneficial in helping to free the melt from entrained oxides and non-metallics.

Melting Procedure

1. Heat up the crucible.
2. Charge ingot and scrap (runners, risers, etc.).
3. Dust the solid metal with a generous quantity of COVERAL 65, using about 250 g per 50 kg of melt.
4. Add the remainder of the charge and melt rapidly, keeping the fluid protective cover intact as much as possible.
5. When the melt has reached the plastic condition, i.e. at a temperature of about 600°C (1110°F), add a further quantity of COVERAL 65, spreading it evenly over the surface of the melt. A quantity of at least 1 kg per 50 kg is required.
6. On attaining a temperature of about 700°C (1290°F), stop heating or remove the crucible from the furnace. There should be sufficient residual heat in the refractory crucible and furnace lining to bring the temperature of the charge up to, but not exceeding, 750°C (1380°F). Careful control is necessary at this stage, as the temperature must not be allowed to exceed 750°C.
7. The flux cover, which will now be in a fluid condition, is thoroughly "washed" into the molten alloy, using a perforated plunger in a stirring action, alternatively, a skimmer can be used, stirring in such a way as to bring the fluid flux into contact as much as possible with the molten metal.
8. Push the slag on the surface of the charge, a little to one side and plunge to the bottom, through the cleaned area, DEGASER tablets 190. Use at least 250 g per 50 kg. In the case of large melts, it is recommended that the DEGASER be plunged in at least two stages, using a proportion of the total quantity necessary at each stage. Hold the DEGASER tablets at the bottom of the melt until the bubbling action has ceased.
9. Sprinkle on the surface of the melt a quantity of COVERAL 66, using about 250 g per 50 kg and allow to stand undisturbed for about 5 minutes.
10. The flux is now rabbled into the *immediate surface only* until a glowing red dross is formed. As a result of this exothermic reaction, a fine powdery dross, exceptionally free from entrained metal, is obtained on skimming. Check the temperature and pour without any further delay.

To prevent iron pick-up, plungers and all other iron furnace tools are kept coated with a refractory wash of FIRIT. They must be dried and thoroughly heated before use.

The risk of iron pick-up is minimised and degassing activity enhanced by using DEGASER 450 as an alternative. These tablets are provided with a central hole so that plunging can be carried out by means of a simple hooked rod, eliminating bell plungers which trap and release the scavenging gas as large bubbles.

DEGASER 450 is plunged at the rate of 200 g per 50 kg of metal.

Casting Temperatures

These vary according to thickness of casting section, but generally fall within the following limits:

Light castings—
under 13 mm section   720°C (1330°F)

Medium castings—
13 to 38 mm section   700°C (1290°F)

Heavy castings—
over 38 mm section   680°C (1260°F)

Sand Properties

Both naturally bonded and synthetic sands may be used, as well as silicate bonded sand and those containing synthetic resins as used for shell moulding. In Great Britain natural quarried sands such as Bromsgrove and Mansfield red, are quite suitable. Typical properties are indicated in the following table:

	Natural sand	Synthetic sand
Grain fineness—AFS No.	120	80–100
Clay content (%)	10	4
Moisture (%, green)	6	4.5–5.0
Permeability—AFS No.	50	100–120
Green compression strength (lb/in^2)	7–8	5–6

Metal–Mould Reaction

Although metal–mould reaction is greatest in magnesium bearing alloys of the LM10 type, which contain 10% of magnesium, it is also very pronounced in the LM5 alloys. Its occurrence causes the casting surface to show patches of oxidation which may be severe and deep seated in thick sections. Increase in magnesium content and the length of time taken for the casting to solidify magnifies the trouble, which is not avoided by the use of dried sand moulds and cores.

This metal–mould reaction results from the action of the magnesium present in splitting up the moisture in the sand into its constituents hydrogen and oxygen. The latter forms oxide with the alloy and the hydrogen, in the atomic state, is released into the molten metal. As the casting solidifies the hydrogen is expelled from solution and porosity is the result. This porosity is concentrated in the outer layers of the casting section but diminishes towards its centre. Even with dried sand moulds and cores there is sufficient water combined with the clay bond to react with the magnesium in this way.

J. Lund has shown that mould reaction in the aluminium 5% magnesium alloys is reduced by hard ramming and by increasing the moisture content of green sands. This, in effect, means that influences which increase the rate of freezing of the casting lessen the severity of the mould reaction. It follows, therefore, that increase of pouring temperature will cause a higher degree of porosity.

A particularly efficient mould or core dressing designed to reduce and often eliminate metal–mould reaction is MOLDCOTE 41, a self-drying type of dressing which contains inhibitors.

Facing Sand Inhibitors

Metal–mould reaction can largely be prevented in the aluminium alloys containing magnesium by making additions of chemicals known as inhibitors to the facing sand. The inhibitor most used is boric acid and it is added at the rate of about 2%, although more may be needed in the case of very thick sectioned castings. Alternatively chills may be employed in such cases to increase the rate of cooling.

Ammonium bifluoride, added to the facing sand in the same ratio, may be used in place of boric acid. However, the latter is usually preferred because it does not produce objectionable fumes when the metal is poured but it has the disadvantage that it tends to adhere to the casting and causes a hardening of the sand which hinders knocking-out.

Silicate Bonded Sand

Neither boric acid nor ammonium bifluoride should be added to sand used in the carbon-dioxide process as they exert a deleterious effect. Fortunately, when employing silicated bonded mould and cores, the use of these inhibitors appears to be unnecessary provided that a suitable dressing or coating is used.

Shell Moulding

If urea formaldehyde resins are used, this gives a natural inhibiting effect. Boric acid can be used with phenol formaldehyde resin and has an additional advantage in helping the curing process.

Segregation of Inhibited Sand

It is most important to note that facing sands containing an inhibitor such as boric acid, give rise to pronounced metal–mould reaction if employed in moulds for modified aluminium–silicon alloys. Such sand must be kept away from that used for these alloys and for those not containing magnesium, especially the high silicon alloys like LM6.

Mould Dressings

Green sand moulds should be dusted with TERRACOTE 20 and sleeked to improve coating finish. Alternatively an excellent finish will be obtained by applying MOLDCOTE 6 or 9. These spirit-based dressings are ignited and burn off from the treated surface to leave a firm, closely adhering coating, and give the effect of skin drying or torching.

Dry sand moulds and oil sand cores should be treated with a water suspension of TERRACOTE 14 for best quality work, or with TERRACOTE 20 for ordinary commercial work. Both moulds and cores must be thoroughly dried after applying these dressings.

Running and Gating Methods

Bearing in mind that aluminium–magnesium alloys tend to oxidise easily and form heavy drosses, runners and gates should be formed in such a way as to allow the metal to enter the mould fairly rapidly but with a minimum of turbulence. This will help to avoid entrapment of air and the formation of oxides. For large castings a ring runner with multiple ingates gives good results. Bottom pouring or gating near the base of the mould is the general rule. Also perforated thin steel plates in the runner systems assist the production of castings free from inclusions. They help, too, in choking the downgate and so are effective in keeping up the head of metal in the runner bush which guards against dross being carried into the mould.

Feeding Methods

Generous feeding heads are necessary. On heavy sections and when casting design prevents the placing of adequate feeding heads, denseners or chills should be used. They must be clean and preferably coated with a dressing such as CHILCOTE 7 and dried thoroughly before use.

Although chills and denseners are most often made of cast iron or steel, there is something to be said for producing them in aluminium alloy so far as light alloy founding is concerned. One obvious advantage is that they can be cast from the molten metal handled in the shop. Other advantages are their high heat conductivity, specific heat and latent heat of fusion. Also they are easier to machine or file and, being light, are much easier to hold in position in a green sand core or mould. For best results, the face of the chill should be grooved to allow for escape of air at the interface. A disadvantage of the use of light alloy chills, particularly in foundries with a mechanised sand system, is that they do not respond to magnetic separation from the knock-out sand.

As with all castings some thought should be given to obtaining as much directional solidification as possible. Those parts of the casting furthest from the ingates should solidify first and feeding should be progressive back to the risers. The feeder heads must remain molten longer than the casting or feeding cannot take place. It is important that risers are not over-large so reducing yield and also important that heat loss from the riser mass is retarded as far as possible.

For this purpose, feeder sleeves for lining the riser cavity can be employed. These sleeves can be obtained in FEEDEX exothermic material or the newer and more effective lightweight, highly insulating KALMIN 33 shapes can be used. Insulating sleeves have low specific heat, extremely good long-term properties, and are completely inert to all aluminium alloys. (They may also be used satisfactorily on magnesium base alloys.)

Whichever form of open-top riser is used, it is important to reduce heat loss to atmosphere as far as possible. This is increasingly important with KALMIN 33 sleeves where loss to atmosphere is the major route for heat loss. All risers should therefore be covered with a layer of FEEDOL 9 anti-piping compound immediately on completion of pouring. The layer should be about 15 mm thick or as thick as the riser sleeve material. Using these techniques, it should be possible to eliminate rod feeding, with all its objections, completely.

Gravity Die Casting

Dies should be designed with the runner-riser system cut on generous lines similar to the practice used for LM14 (Y alloy) and with a view to creating the minimum of turbulence as the stream of metal enters. In some cases slightly thicker sections than usual may be required to obtain the best results in castings.

Melt down the ingot and process scrap under a layer of COVERAL 65. Superheat to a maximum temperature of 750°C and skim. Add a further 2% of the charge weight of COVERAL 65. Allow the flux to fuse on the surface, followed by a thorough puddling of the "wet" slag for 3–4 minutes, employing for the purpose a skimmer tool, making a rotary motion. Any remaining "islands" of slag or dross should be skimmed off and the sides of the crucible, at the metal level, gently scraped. Ladling out then proceeds in the usual manner.

In the case of a bale-out furnace working continuously with more ingot metal and scrap being fed from time to time, the routine outlined is repeated at intervals dependent upon the rate at which the metal is consumed.

These alloys are susceptible to grain growth, even when cooled relatively quickly as in gravity die casting. To ensure the production of a fine grain size, NUCLEANT 2 tablets are plunged on a basis of 0.25% by weight.

Every endeavour should be made via die design, casting technique, etc., to obtain sound castings using the standard metal treatment outlined.

Since most LM5 castings are radiographed, anodised, or surface finished in some way, it is normally undesirable that recourse is made to introducing controlled amounts of gas to the melt with DYCASTAL.

An appropriate DYCOTE insulating die dressing should be applied to the die to control solidification and obtain optimum surface finish.

ALUMINIUM-MAGNESIUM CASTING ALLOY BS 1490—LM10

BSS (Aerospace) L53.
BS/STA7: AC10.   Hiduminium 90.
American ASTM; B179-49T-G10A.
American ALCOA 220.   Canadian ALCAN GB350.
French A-G11.
German DIN 1725.   AlMg 7 and GD AlMg 9.

Experience shows that the magnesium content should be maintained at the higher level of the specification as this helps in obtaining the specified mechanical properties. However, it should not exceed 11% in the interests of maximum resistance to corrosion fatigue. Small quantities of beryllium exert a marked effect in reducing metal/mould reaction. Further comments on this point are included later.

Casting Tensile Test Bars

Tensile test pieces of LM10 alloy should be cast to shape in green sand using a multiple test bar mould known as E Bar pattern. The drawing and dimensions of this mould pattern are shown at the end of these notes. The bars should be cut from the running system before placing in the heat treatment furnace: this will prevent distortion. Bars should not be heat treated in the machined state.

Heat Treatment

Specification BS 1490 calls for castings in this alloy to be supplied in the solution heat-treated (TB) condition. It consists of heating for a minimum period of 8 hours at a temperature between 425° and 435°C (797° and 815°F). Eight hours is the minimum holding time for average castings; it may have to be increased in special cases. Following this the castings may be allowed to cool in the furnace to 385° to 395° before quenching which consists of immersion in oil, the oil having a temperature not in excess of 160°C (320°F). The castings may remain in the oil no longer than 1 hour and are finally quenched in water or cooled in air to atmospheric temperature.

Physical Properties

Specific gravity	2.55 (sand cast) 2.60 (die cast)
Weight per cubic inch (lb)	0.092
Specific resistance, microhms/cm³	8.5
Electrical conductivity (%IACS at 20°C)	20 approx.

Founding Characteristics

Fluidity	Can be described as fair only and relatively large diameter runners and adequate ingates are required. It is not very suitable for thin sectioned castings.
Pressure tightness	Poor: it is difficult to make castings which have to be pressure tight.
Hot tearing	Resistance to hot tearing is good.
Patternmakers' shrinkage	1.3%
	Machinability also is good.

Melting Hazards (Gas absorption)

Aluminium–magnesium alloys of the LM10 type tend to oxidise and pick up hydrogen very readily during melting. Thus every precaution must be taken to protect the melt and the use of a flux with a strong cleansing action and able to dissolve and coalesce oxides, particularly magnesium oxide, is essential. Such a flux is COVERAL 33FF which hasthe further advantage of being completely free from sodium. Contamination of the melt with sodium must be avoided because its presence is detrimental to the development of full mechanical properties.

A further factor which influences the development of maximum mechanical properties is the grain size which should not exceed 0.10 to 0.15 mm diameter. Immediately before degassing the melt should be grain refined by making simultaneously small additions of titanium and boron such as are contained in NUCLEANT 2 tablets.

Aluminium–magnesium alloys of LM10 composition are prone to patches of porosity rather similar to those which form in magnesium base alloys, in which it is referred to as "lake" porosity. Porosity of this kind is very harmful to mechanical properties and the strength of test bars so affected may drop by as much as 70%. It is therefore extremely important that a correct metal treatment is strictly adhered to and also that a suitable test bar mould and casting technique is used.

The melt must not be overheated, and a temperature of 720°C (1330°F) should be kept as a maximum. In this way gas pick-up is reduced and large grain growth suppressed. Accurate temperature control is essential to success.

Melting Procedure

1. Heat up crucible and place in the bottom a quantity of COVERAL 33FF (in the form of "chippings") using 500 g per 50 kg. Where tall, narrow crucibles are used, the quantity may be less as the 1% ratio applies to melting in bale-out and reverberatory furnaces of conventional surface area.
2. Charge ingot metal and scrap (runners, risers, etc.).
3. Melt in the usual way, adding more ingot or scrap until the charge is complete.
4. When the melt has just passed the "plastic" stage add a further quantity of COVERAL 33FF, using about 1.5 kg per 50 kg, spreading it evenly over the metal surface.
5. Continue heating until the temperature reaches 680°C (1250°F), then shut off the fuel supply or remove the crucible from the furnace.
6. The residual heat in the crucible or refractories will raise the temperature to about 720°C and the layer of COVERAL 33FF will be in a semi-liquid condition so that it can be well rabbled into the molten charge from 3 to 4 minutes. Use a plunger and stir so as to bring the flux into contact with the molten metal as much as possible.
7. Now grain refine by plunging NUCLEANT 2 tablets to the bottom of the molten charge. Use 125 g per 50 kg and hold submerged until all bubbling ceases.
8. Follow by degassing with DEGASER 200 tablets. These also are plunged as for NUCLEANT but use 250 g per 50 kg. If the melt is large, degassing is best carried out in two or more stages using appropriate proportions of

the total quantity of tablets at each stage. After degassing do not interfere with the melt until it is poured.

Note.—The quantity of COVERAL 33FF mentioned refers to melting in bale-out and reverberatory furnaces. Where tall narrow crucibles are used, the amount may be reduced to some extent. NUCLEANT and DEGASER tablets must be plunged to the bottom of the melt. This can be done most easily and conveniently by means of a special perforated plunger. This tool is also useful for rabbling in the COVERAL 33FF flux. To restrict iron pick-up, plungers and all other iron furnace tools should be coated with a refractory wash of FIRIT and dried thoroughly before use.

Casting Temperatures

These vary with section thickness but fall, generally, within the following limits:

Light castings— under 13 mm section	720°C (1330°F)
Medium castings— 13–38 mm section	700°C (1290°F)
Heavy castings— over 38 mm section	680°C (1260°F)

Metal–Mould Reaction

It has already been pointed out that the inclusion of 0.004% beryllium in the LM10 alloy helps to reduce metal/mould reaction. None of the common impurities, except sodium, interferes with this beneficial action of beryllium.

When metal–mould reaction occurs the surface of the casting shows oxidation which may be severe and deep seated. Hydrogen, too, is absorbed and liberated, as the casting solidifies, to cause porosity. This porosity is concentrated beneath the casting surface, but diminishes towards the centre of the section. It results from the splitting up of the moisture in the sand due to the presence of magnesium. The oxygen forms oxide with the alloy and the hydrogen is released into the molten metal. Increase in magnesium content and length of time taken for the casting to solidify magnifies the trouble, which is not avoided by the use of dried sand moulds and cores.

Apart from the use of beryllium, which is effective in overcoming metal–mould reaction in the thinner sectioned range of castings, inhibiting agents are added to the facing sand. Thus for casting sections above 25 mm or about 13 mm plate thickness, 2% of boric acid or a similar quantity of ammonium bifluoride is milled into the facing sand. Boric acid is usually preferred because it does not produce objectionable fumes when the metal is poured. However, sand containing boric acid, tends to adhere to the casting and the hardening of the sand hinders knocking-out. For very thick sections a higher proportion of inhibitor will be necessary.

M. Whitaker has worked on the problem of metal–mould reaction in the 10% magnesium–aluminium alloy and found that it is influenced by mould variables such as ramming density, moisture content, and coarseness of the sand grains. These factors affect the chilling power of the mould which should be at a maximum to reduce the reaction. The soundest castings are obtained by using medium or fine sand and ramming as hard as it consistent with adequate permeability and other desirable characteristics.

Other important factors are the maintenance of 0.004% beryllium in the alloy, the use of inhibiting agents in the sand and pouring at a temperature as low as possible consistent with good foundry practice.

Silicate Bonded Sand

Neither boric acid nor ammonium bifluoride should be added to sand used in the carbon-dioxide process as they have a deleterious effect. Fortunately, when employing silicate bonded moulds and cores, the use of these inhibitors appears to be unnecessary. In certain cases, particularly for heavy cast sections, a dusting of INERTEX over mould and core surfaces is advised as a safeguard.

Shell Moulding

If urea formaldehyde resin is used, this gives a natural inhibiting effect. Boric acid can be used with phenol formaldehyde resin and has an additional advantage in helping the "curing" process.

Loss of Beryllium on Re-melting

Some loss of this element does occur during remelting the alloy. Therefore, an occasional check should be made to ensure that the optimum content of 0.004% is maintained. The loss is greatest in small melts, and it is suggested that for melts of about 50 kg a check should be made after about eight or so consecutive remeltings. Experience will determine whether more or less frequent checks are necessary.

Mould Dressings

Green sand, dry sand or CO_2 moulds and cores may be coated with MOLDCOTE 41, a self drying type of dressing, to diminish metal–mould reaction. This dressing may be applied by brush, swab or spray gun and takes approximately 10 minutes to air dry.

Running and Gating Methods

Bearing in mind that aluminium–magnesium alloys tend to oxidise easily and form heavy drosses, runners and gates should be formed in such a way as to allow the metal to enter the mould fairly rapidly but with a minimum of turbulence. This will help to avoid entrapment of air and the formation of oxides. For large castings a ring runner with multiple ingates give good results. Bottom pouring or gating near the base of the mould is the general rule. Also perforated thin steel plates in the runner system, assist the production of castings free from inclusions. They help too in choking the downgate and so are effective in keeping up the head of metal in the runner bush which guards against dross being carried into the mould.

Additional points to be observed are:
1. Sprues or down runners should be tapered, and the smallest sprue that will prevent mis-runs used.
2. Sprue base should be enlarged or a well provided to slow down flow of metal entering the runner bar.
3. Runner bar should be cut in drag and the gates in the cope to ensure keeping runners full and to provide skimming action.
4. Total cross sectional area (CSA) of runner bar should be four to six times restricted CSA of Sprue. This ratio slows metal at gates.
5. Total CSA of gates should equal CSA of runner bar.
6. CSA of runner bar should be reduced by CSA of gate as each gate is passed. This helps to keep runner bar and gates full and equalises the flow through each gate.
7. Abrupt changes in the area of flow channel and in direction should be avoided as should sharp corners.
8. Castings should be fed through risers. For deep castings use slot gates entering casting from side risers.

When parallel-sided down runners are used there is a reduced pressure and gases are aspirated into the metal stream during pouring. The tapered runner helps to set up a back pressure and air and gases are not sucked in.

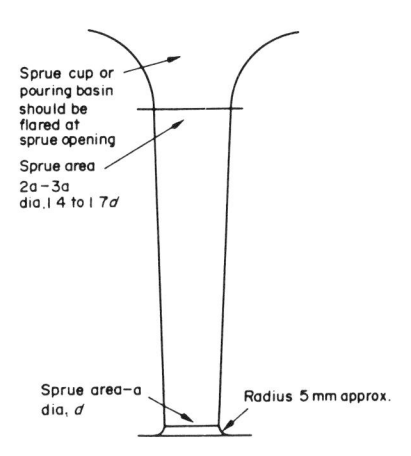

Fig. 4.6. Recommended design for Sprue or down runner pattern for Sand Cast LM10.

Gravity Die Casting

Dies should be designed with the runner-riser system cut on generous lines similar to the practice used for LM14 (Y alloy). In some cases slightly thicker sections than usual may be required to obtain the best results in castings.

Feeding Methods—Sand Castings

Feeder heads of generous proportions are essential. On heavy sections and when casting design prevents the provision of adequate feeding heads, denseners or chills should be used.

They must be clean and preferably coated with a dressing, such as CHILCOTE 7, and dried thoroughly before use.

Satisfactory chills for light alloy castings can be formed from cast iron, steel, brass or gunmetal. However, in a light alloy foundry chills can be made most readily from the alloys being cast. Their advantages are high heat conductivity, specific heat and latent heat of fusion. Also they are easier to machine or file; and being light, are much easier to hold in position in a green sand mould or core. For best results, the face of the chills should be grooved to allow air to escape the metal mould interface. A disadvantage of the use of light alloy chills, particularly in continuous production foundries, is that they do not respond to magnetic separation from the knock-out sand. As in all cases, every effort should be made to promote directional solidification; that is, ensure that both areas remote from the ingates and feeder heads solidify first. Feeder head dimensions should be sufficient to delay solidification until the casting has practically set. Slightly tapering risers, wide end up, are advised; the height to diameter ratio of risers being about 2 to 1.

To improve feeding efficiency, and at the same time reduce riser dimensions, exothermic or insulating sleeves can be of great benefit.

Prefabricated sleeves and shapes can be obtained in either FEEDEX 4 or KALMIN 33. Technically the highly insulating, inert KALMIN 33 has advantages over the exothermic sleeve and for these very reactive alloys increasing use is being made of them. Heat loss to atmosphere from the open top of the riser should be controlled by applying a layer of FEEDOL 9 hot topping compound immediately on completion of casting.

Segregation of Special Sands

Sand which contains any of the inhibiting agents described must be kept apart from other sand mixtures. For example, if inhibiting sand is used for moulds in which silicon bearing alloys are cast, particularly high silicon alloys such as LM6 (Silumin), pronounced metal–mould reaction can cause severe porosity. Likewise, sand containing sulphur, such as is used for magnesium-based alloys, should not be employed for moulds into which LM10 alloy is cast. Even as little as 0.5% sulphur is sufficient to cause metal–mould reaction, despite the presence of beryllium in the metal, or inhibitors in the facing sand.

These precautions are not necessary for silicate-bonded sands whether the hardening agent is carbon dioxide gas or one of the new chemical self-hardening systems.

LIGHT CASTING ALLOYS

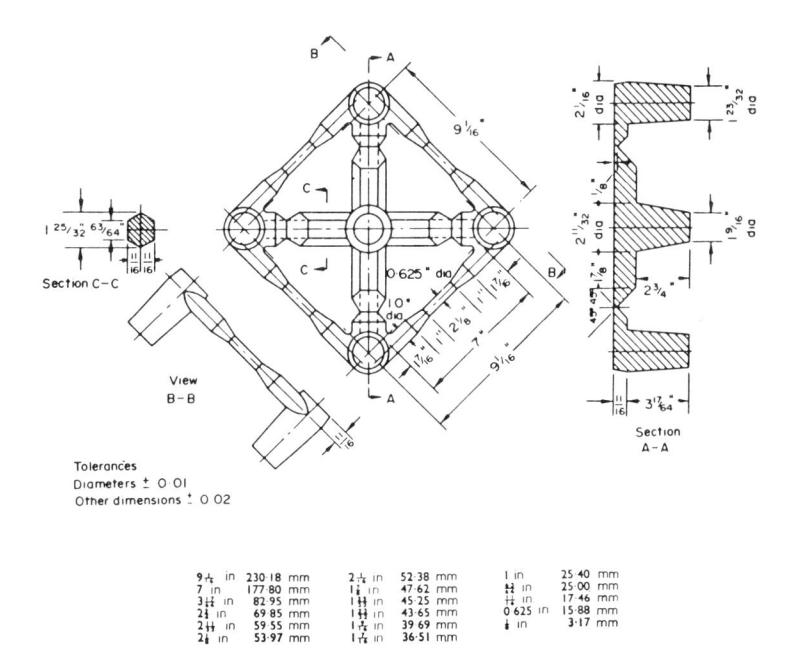

Fig. 4.7. Tensile test sample.

ALUMINIUM-ZINC-MAGNESIUM CASTING ALLOY
FRONTIER 40E ALLOY*

Frontier 40E alloy is an aluminium-zinc-magnesium casting alloy, possessing high strength, pressure tightness and hardness obtained without resorting to heat treatment. It has excellent machinability, exceptional resistance to shock and corrosion, very good anodising properties and weldability. From the point of view of resistance to atmospheric and marine corrosion, it can be compared favourably with the aluminium magnesium alloys.

Related specifications are:
DTD.5008B (British).   Ternalloy 8, Apex Smelting
612A and 612B Alcoa.   Unifont, AIAG.
Tenzaloy, American Smelting   Inafond, Z5F, Montecatini.
  and Refining Co.   French Alliage A/Z 5G.
                                   U.S.A. ASTM B-26-52T.

Chemical Composition of DTD.5008B Alloy

Zinc	4.8–5.7%
Magnesium	0.5–0.75%
Chromium	0.4–0.6%
Titanium	0.25% maximum
Iron	0.5% maximum
Silicon	0.25% maximum
Copper	0.1% maximum
Manganese	0.1% maximum
Lead	0.05% maximum
Tin	0.05% maximum
Aluminium	Remainder

*Registered trade name of the Frontier Bronze Corporation, Niagara Falls, U.S.A.

Influence of Constituent Elements
Zinc and Magnesium

As these two important constituents are increased, so does the tensile strength of age hardened castings increase but at the expense of ductility. Best results appear to be obtained when magnesium is limited to the lower half of the specified range, keeping zinc to the upper half.

Chromium

This element is said to improve stress corrosion resistance but when present to a maximum specified, primary particles may form in the cast structure, having an embrittling effect, so chromium is best kept to the lower range of the specification.

Titanium

Investigations have shown that there is no advantage in exceeding 0.15% titanium. However, adequate grain refinement has been obtained with only 0.03% titanium additions, if boron is also present. Grain refinement with NUCLEANT 2 tablets, plunged into the melt, is therefore recommended, as both titanium and boron are thereby released.

Silicon

This has a harmful effect on tensile properties and should be kept as low as possible.

Required Mechanical Properties

	Sand Cast Test Bar
0.2% proof stress (N/mm^2)	170
Ultimate tensile stress (N/mm^2)	215
Elongation (%)	4

Typical Properties

The outstanding feature of this alloy is its property of age-hardening at room temperature. While artificial ageing at 180°C (356°F) for 10 hours, 48 hours after casting, may be substituted, an initial natural ageing period of 21 days is recommended.

Test Bars

The most suitable form of test bar is the standard E-type bar (as shown on page 175), the moulds for which should be made from an accurate pattern incorporating all the requirements for running and feeding. For diecastings the wedge type metal mould should be used.

Test bars should be cast at 730°C ± 13°C (1346°F ± 24°F).

The following points on the mechanical properties of test bars are given for guidance:
(a) High elongation, low tensile, low yield point—magnesium content to be INCREASED.
(b) Low elongation, high tensile, high yield point—magnesium content to be DECREASED.
(c) Low elongation, low tensile, with or without low yield point, indicates wrong melting or pouring practice. Items to be checked—the test bar pattern and mould, sand moisture, melting and/or pouring temperature, silicon or copper contamination and chromium segregation.
(d) Shiny, star-like fracture of the test bar with low elongation and/or segregation indicates that the metal has not been thoroughly stirred before grain refining, degassing and pouring.
(e) White speckled fracture indicates gassed metal and is due to one, or all, of three things—too slow melting time; excessive temperature during melting; smoky combustion of furnace fuels.
(f) Coarse columnar grain extending inward from outside edge of the bar indicates that the metal has been poured too hot, or need for efficient grain refinement of melt.
(g) An oxide film across the fracture indicates that the sand is either too wet or has been rammed too hard or that the metal has been too cold. If the film across the fracture is heavy enough to be almost a misrun, this indicates cold metal. Turbulent metal flow in the mould is another cause of trapped oxide films.

Melting Technique Using a Dry Flux Cover

1. Heat crucible to red heat.
2. Charge in ingot and scrap material, runners, risers, etc., bearing in mind the resultant alloy may need a further addition of magnesium.
3. Melt as rapidly as possible and when the charge becomes "pasty" sprinkle COVERAL 11 on the surface, using 500 g per 100 kg. The temperature of the metal should not be allowed to exceed 800°C (1472°F).
4. Grain refining and degassing before casting are essential. Plunge to the bottom of the melt NUCLEANT 2 tablets using 250 g per 100 kg when temperature begins to fall. Keep tablets immersed until bubbling ceases. Then plunge the DEGASER 400 in a similar way, using 300 g per 100 kg. Add magnesium, if necessary, by plunging with a pre-warmed hooked steel rod. Plunging tools, skimmers, etc., should be treated with FIRIT 1 to avoid iron pick-up.
5. At appropriate temperature skim off surface dross and pour.
   NB—Always maintain an unbroken flux cover during melting and if the flux thickens or exposes fresh molten metal apply more COVERAL 11.

LIGHT CASTING ALLOYS

NOTE: The foregoing recommendations are based upon the widely accepted practice of using a dry type of covering flux, of which COVERAL 11 has been specified.

Alternatively, a fluid fluxing technique is advised where the magnesium content of the charge metal is already low. A fluid flux cover such as COVERAL 65 prevents the magnesium content becoming too low.

Pouring Temperatures

Under 13 mm section	770°C (1418°F)
13-38 mm section	750°C (1382°F)
Over 38 mm section	730°C (1346°F)

It will be seen that recommended pouring temperatures are about 40°C (104°F) above normal and this is so because the fluidity of Frontier 40E alloy is low at pouring temperatures used for the more common aluminium alloys.

Moulding Techniques

Normal aluminium foundry sands are quite suitable. Moisture in the sand should be around 5-6% maximum, consistent with good moulding practice, as above this figure of moisture there is metal—mould reaction which forms oxide, scum, and other drosses. To obtain excellent casting finish, green sand moulds should be treated with TERRACOTE 20 or with MOLDCOTE 6 or 9 spirit-based types of dressing.

Dry sand moulds and cores should be dressed with a water suspension of TERRACOTE for best quality work or with TERRACOTE 20 for ordinary commercial work.

As Frontier 40E alloy is most susceptible to hot tearing and cracking, moulds should not be so inflexible as to restrain contraction.

Top pouring is very practicable providing it is arranged to minimise turbulence. Isolated heavy sections must be individually risered or chilled but feeding is preferred as the results are more predictable. For feeding extensive heavy sections, the use of excessively large feeders is not advisable because they would take too much room and would retard the cooling of the section with adverse effects. As the shrinkage of Frontier 40E alloy is high compared with the common alloys, ample feeding is necessary particularly during the pasty stage. It is therefore advisable to use FEEDEX or KALMIN sleeves to assist feeding as much as possible.

Diecasting

The fact that Frontier 40E alloy is subject to hot tearing and cracking may have retarded its use in diecasting procedure. Nonetheless quite large and complex diecastings have been produced by using alternative methods of coring such as sand or shell moulded cores. Additionally, a properly designed feeding system can minimise the tendencies to tear and draw.

Owing to the comparatively low fluidity of this particular alloy, it is advisable to cast at a slightly increased temperature, say in the region of 30 Centigrade degrees (54 Fahrenheit degrees) higher than for the more common aluminium alloys.

General Notes

Because of its high strength and shock resistance Frontier 40E alloy is firmly established in the United States of America, notably for the manufacture of services castings and in a wide variety of industrial applications including castings for mining, oil refinery and ventilating equipment, road vehicle parts (e.g. rear axle castings), rail jacks and hand tools.

It is particularly useful as a replacement for heat treated alloys where heat treatment can cause distortion.

MELTING, FLUXING, GRAIN REFINING, ETC., AS APPLIED TO MAGNESIUM BASE ALLOYS

General Notes

The primary advantages of magnesium alloys are their lightness combined with strength (they are about 40% lighter than the average aluminium alloy), resistance to fatigue failure and embrittlement in use, and freedom from gas pin-holing and porosity. They are used extensively for aerospace castings, etc. They are also used, but to a more limited extent, in the vehicle industry. They are fairly widely used in the heavy transport industry.

Aluminium is the main alloying constituent. In increases fluidity and decreases shrinkage, but hot shortness becomes more evident with rising aluminium content. Magnesium alloys, because of their lightness, are easy to set up for machining and have outstandingly good machining properties. However, the cutting edge of the tool must not be allowed to dull, as the heat generated might be sufficient to ignite the chips. The heavier the cut the less chance there is of ignition. Dry machining is recommended, but if cooling is required, a perfectly dry air blast should be used.

They can be sand cast and both gravity and pressure die cast.

Two of the principal casting alloys of this group are L123 and L121. These alloys are also known under the trade names of Elektron AZ91 and Elektron A8. Many other proprietary alloys are also in use.

Chemical Composition

	L123	L121
Aluminium	9.0–10.5%	7.5–9.0%
Zinc	0.3–1.0%	0.3–1.0%
Manganese	0.15–0.4%	0.15–0.4%
Copper	0.15% max.	0.15% max.
Iron	0.05% max.	0.05% max.
Silicon	0.30%	0.30% max.
Tin	0.1% max.	0.1% max.
Nickel	0.01% max.	0.01% max.
Magnesium	Remainder	Remainder
Total impurities	0.40% max.	0.40% max.

Typical Mechanical Properties (as cast)

	L123	L121
0.1% proof stress (tons/in^2)	4.5	4.5
Ultimate tensile strength (tons/in^2)	8.5	10.0
Elongation (% on 2 in)	1.0	3.0
Brinell hardness number	55	55

Test Bars

Test bars are generally dry-sand moulded to the DTD pattern (see page 160).

Bars should be poured at about 760–780°C. At the start of pouring, the mould should be held at an angle of about 45° from the vertical and then brought gradually to the upright position as the head fills. Pouring should be carried out steadily and over a period of not less than 10 seconds.

Heat Treatment

Although many magnesium alloys containing not less than 6% aluminium respond to heat treatment, the DTD specification calls for L123 and L121 to be supplied in the as-cast condition.

Casting Characteristics

Fluidity	With most alloys excellent—increasing with aluminium content.
Pressure tightness	Due to absence of gas porosity pressure tightness is generally very good, but L123 gives particularly good results.
Hot tearing	Hot tearing is not pronounced in most magnesium alloys but may increase with aluminium content.
Patternmakers' shrinkage	1.6%.

Melting Considerations

Molten magnesium alloys attack firebrick and refractory furnace linings resulting in harmful silicon contamination and for this reason steel—pressed or cast—crucibles are used. Iron is also slightly soluble in magnesium alloys, but it has a much less harmful effect than silicon. Scrap should be cleaned and if possible shot-blasted to remove any adhering sand as a further precaution against silicon pick-up. To eliminate ladling, the molten alloy should, if possible, be poured direct from the melting pot.

Melting Procedure

1. Heat up the steel crucible and dust the bottom with a little MAGREX 60.
2. Charge in ingot and clean scrap (runners, risers, etc.).
3. Dust the solid charge with a generous quantity of MAGREX 60 using about 500 g per 50 kg of metal.
4. Add remainder of charge and continue melting rapidly, keeping the fluid protective flux cover intact as much as possible.
5. When the melt has reached a temperature of about 750°C stop heating.
6. Skim off the flux remaining on the top surface of the melt and as each patch of metal is exposed dust in more MAGREX 60 until the whole surface is again covered with a fairly thick layer. About 2 kg per 100 kg should be used at this stage.

7. The flux layer will now have melted and should be thoroughly "washed" into the melt, using a plunger (preferably the solid type) in a stirring motion, to bring the fluid flux into contact as much as possible with the molten metal. The plunger should be well coated with a refractory wash of FIRIT and dried thoroughly before use.
8. Stirring should be continued until the metal takes on a bright clean appearance, free from the "frothy" look of partially cleansed metal. As the flux dissolves and entraps oxides it increases in weight and finally sinks to the bottom of the melt, forming a sludge. If the metal is very dirty a further quantity of MAGREX 60 should be stirred in until the metal takes on a bright lustrous appearance.
9. The cleaned surface should now be dusted with a generous layer of MAGREX 60 so as to curtail burning, which, however, should not be excessive if the temperature is maintained at about 750°C.
10. The sludge may be removed at this stage, but it is recommended that it be left in the bottom of the pot and the metal decanted from it in due course.
11. With the temperature of the metal still around 750°C, grain refinement should now be carried out, using NUCLEANT 200. Using 125 g per 50 kg the required amount is thrown on to the surface of the melt. Plunging is not usually necessary as the tablets are sufficiently heavy to sink to the bottom of the melt, where they decompose and exercise a powerful degassing as well as a grain refining action. In some alloys containing large quantities of other elements such as aluminium, plunging may be necessary, when this degassing action will be most beneficial in reducing harmful micro porosity.

This method of grain refining is based on the theory of carbon nucleation; NUCLEANT 200 decomposes to form throughout the melt small "specks" of carbon which act as nuclei and around which a very large number of small grains form instead of a lesser number of larger grains, as would occur in a melt perfectly free of such carbon particles or other materials capable of acting as nuclei. It replaces the old method of grain refining by superheating the melt to a temperature of about 850–900°C, and is much more satisfactory. It also shows a marked economy in fuel consumption.
12. During the time the melt is cooling to the correct temperature it should be skimmed and the cleaned area immediately dusted with INERTEX to prevent burning.
13. Pour carefully, dusting the metal stream as it flows into the mould with sulphur. Care must be taken to keep back any slag and particularly when nearing the end of the pour, to prevent any sludge entering the mould.
14. Remove the sludge from the bottom of the pot and thoroughly scrape the sides and bottom before returning it to the furnace for recharging.

Casting Temperatures

Casting temperatures vary with section thickness but generally fall within the following limits:

Light castings— under 13 mm section	780–810°C
Medium castings— 13–38 mm section	760°C
Heavy castings— over 38 mm section	730°C

Use of Inhibitors

With natural or synthetic sand when used green, it is most important that a chemical inhibitor should be added to the facing sand to prevent the reaction which would occur between the molten metal and the moisture of the sand. Several materials give satisfactory results, but the more generally used are boric acid or ammonium bi-fluoride when used singly, or boric acid with additions of sulphur. Sulphur can also be used in conjunction with ammonium bi-fluoride.

The quantity required in each case depends upon the moisture content of the sand and section thickness of the casting, an increase in either calling for an increase in the amount of inhibitor used. Generally, however, 4–6% sulphur with 0.5% boric acid is used or up to 2% of ammonium bi-fluoride alone.

It is recommended that new facing sand should always be used and that all sand used for such magnesium alloys should be segregated from that used for other light alloy work. This is especially important with aluminium alloys containing silicon as a vigorous metal/mould reaction can ensue.

Moulding Technique

Magnesium alloys can be satisfactorily green or dry sand moulded in either natural or synthetic moulding sand, the requirements being high permeability and fairly low green and dry strength, but it is more normal to use synthetic sands. Magnesium alloys can be cast easily into moulds produced by the carbon dioxide process and by shell moulding processes.

Sand Properties

Suitable properties of both types of sand are as follows:

	Natural sand	Synthetic sand
Grain fineness (AFA No.)	120	80–100
Clay content (%)	10	4
Moisture (%)	5–6	2.5–3.5
Permeability AFA No. (green sand)	30–40	80–100
Green compression strength (lb/in^2)	10–13	11–15

Cores should be made of fairly coarse silica sand bonded with an oil or synthetic resin binder and some agent to give the necessary green bond. The binder should evolve a minimum of gas and absorb a minimum of moisture. Magnesium alloys have appreciable shrinkage and tend to be hot short, thus hard cores must be avoided. It is recommended, therefore, that cores should be well-baked and preferably over-baked rather than under-baked.

Mould Dressings

Green sand moulds should be dusted with TERRACOTE 14 or 20 and sleeked to improve casting finish. Alternatively, an excellent finish will be obtained by applying MOLDCOTE 9. This spirit-based dressing is ignited and burns off from the treated surface to leave a firm, closely-adhering coating and gives the effect of skin-drying or torching.

Dry sand moulds and oil sand cores should be dressed with a water suspension of TERRACOTE 14 for best quality work, or with TERRACOTE 20 for ordinary commercial work. Both moulds and cores should, of course, be thoroughly dried before casting.

Running and Gating Methods

As magnesium alloys oxidise rapidly, every effort should be made to ensure non-turbulent pouring and to prevent dross being carried into the mould cavity. Bottom pouring is generally used, often from a ring runner. A shrink bob is frequently incorporated at some point near the gate, particularly if there are thin sections followed by thick ones. The downgate is generally cut rectangular to prevent a swirling effect of the metal, while runners of triangular section showing a larger surface area than the usual circular section runners, help to remove dross. Alternatively, if runners of circular section are used, a skimgate either of baked core sand or perforated metal should be incorporated. The runner should be enlarged slightly where the gate is situated to ensure it will not choke the flow of metal. Ceramic filters are also an extremely useful method of ensuring clean castings—see separate section on metal filtration.

As magnesium is so light, a pouring basin should be built up above the top surface of the mould to provide greater hydrostatic pressure. The basin should be roughly oblong in shape, with the mouth of the downgate at one end and a shallow depression at the other, into which the metal is poured. This prevents splashing down the gate into the mould and helps also to trap any dross.

Feeding Methods

Running and feeding methods should be as for a moderately high shrinkage alloy, e.g. fairly large diameter runners and large feeding heads should be used. Running methods should promote directional solidification, i.e. the sections furthest away from the gates and heads should solidify first. Heads should not be "necked"; they should be of the full section they have to feed and should be of such a size as to ensure that they remain liquid until the casting itself has solidified. On heavy sections where casting design prevents the placing of adequate heads, chills should be used. They should be perfectly clean and preferably coated with a dressing such as MOLD-COTE 9 and dried thoroughly before use. Chills made of cast iron and perforated or grooved for venting purposes will be found effective.

Change in section of the castings should be gradual, from heavy to light, and generous fillets of a radius not less than 13 mm should be used.

The necessity for relatively large feeder heads has already been touched on. Quite considerable economies in yield can, however, be obtained from the use of KALMIN 33 or KALMIN 4170 insulating riser sleeves. These resilient, highly permeable, high insulation prefabricated sleeves are inert to the magnesium base alloys, and their use can permit the use of risers smaller in mass than would otherwise be possible. Again heat loss to atmosphere becomes increasingly important as insulation of the riser walls improves and a layer of FEEDOL 17 should be applied to the open surface immediately the riser is full.

Gravity Die Casting

Dies should always be designed with ample venting particularly to "blind" areas. Magnesium alloys are so light in weight that little metallostatic pressure is available to displace mould air and the possibility of short runs or cold shuts is enhanced. Downsprues, for the same reason, are large in cross-sectional area relative to aluminium practice and choke is provided by narrow flat ingates. Feeding heads, as already mentioned, should be of large relative mass. A minimum casting wall thickness will be about 5 mm. Die dressings may be the same as used for aluminium alloys.

The following observations will apply to bale-out furnaces fitted with steel crucibles but where no other special adaptation such as hooding or the provision of a sulphur dioxide atmosphere is afforded. Under these open conditions it is essential that the surface of the metal is protected in order to prevent burning. At the same time the introduction of fresh scrap, ingot, etc. means that oxides can build up and a suitable flux cover should be capable of carrying out both tasks.

Three products are commonly utilised for this purpose, viz:

MAGREX 36 FF.—A fused compound based upon magnesium chloride and supplied in the form of "chippings". At the same time, a heavy ingredient is used as part of the mixture, with the result the slag tends to descend of its own volition through the charge as a whole. The product is therefore what might be termed self-cleansing.

NUCLEANT 200 TABLETS. The term "degasser" is used in the sense of expelling hydrogen, although a major influence of the decomposition products is to nucleate the molten metal with carbon, thus effecting considerable grain refinement without recourse to superheating. Furthermore, as nascent magnesium chloride is also formed within the charge, a natural cleansing influence follows as a matter of course.

INERTEX.—A "dusting" compound which is far more efficient than ordinary flowers of sulphur. The compound is partly volatile and partly "static", the latter feature manifesting itself in the form of a thin but tenacious film which is not easily ruptured. This film is far more readily controlled than that applying to a flux slag, and when baling-out, it is a comparatively simple operation to push back the fused Inertex with the hand ladle, thus allowing the latter to fill with clean magnesium.

Set out below is a recommended melting sequence, using the appropriate products.

1. Heat up the steel crucible.
2. Introduce MAGREX 36 FF "lumps" on a basis of about 3% by weight of the eventual charge capacity.
3. Add pre-heated magnesium ingot and/or scrap in the form of runners, risers, etc. This should be done in stages until the melt is complete. Care should be taken to retain the flux "cover", any deficiency in this respect being made good by a further introduction of MAGREX 36 FF. Temperature of the charge should be closely controlled with the objective in view of a working range of 730–750°C.

4. The refining stage has now been reached, the MAGREX 36 FF "wet" slag being thoroughly but gently worked into the bulk of the metal over a period of some 3-4 minutes. This should be carried out until such time as the magnesium assumes a bright lustre. If, after refining, the metal still looks "frothy" then this is an indication it is still in a comparatively dirty condition and requires more refining with a fresh introduction of fluxing compound.

   It will be appreciated that one of the major functions of the fluxing slag is to chemically dissolve and mechanically entrain oxides, plus other non-metallics. This "loaded" slag will therefore tend to cling to the sides of the melting pot, although the bulk of it gradually descends to the bottom. Consequently, atmospheric action upon the relatively bare surface of the melt should be curtailed by adding a little fresh MAGREX 36 FF, thus forming an adequate seal.

5. Grain refinement should now be practised by introducing NUCLEANT 200 tablets at the rate of approximately 0.25% of magnesium base alloy. As the specific gravity on this tabletted compound is of the order of 2.1 they sometimes sink of their own accord dependent upon the density of the particular type of magnesium alloy. However, as plunging is necessary in certain cases, a well-drilled plunging tool should be employed. This tool, of course, should be coated with a refractory wash and FIRIT has proved very suitable for the purpose. It is important also to make sure the plunger is *thoroughly* pre-heated prior to each application bearing in mind the tool inevitably comes into contact with the "wet" fluxing slag, the latter being very hygroscopic when cold.

   The covering slag in the centre of the charge is pushed aside with a skimmer or the plunger itself, and the requisite number of tablets placed on the freshly exposed area. The reaction is by no means violent, decomposition taking place steadily over a period of 1 to 2 minutes.

6. Metal should now be allowed to "settle" for 10 minutes or so (under protective cover of flux).

7. Skim off the surface slag in stages, using a spade type tool for the purpose. As unprotected metal is exposed, the area should be immediately dusted with INERTEX 1, thus preventing any tendency to burn in contact with the atmosphere. We are now left with a thin film of this product and ladling-out can commence. An average pouring temperature is about 750°C.

8. Although the INERTEX film is to a large extent self-healing, it is desirable to keep a quantity of the powder in a drilled canister, thus allowing frequent applications to cover the exposed surfaces.

   As suggested earlier, the diecaster merely pushes back part of the surface skin with the bowl of his ladle, so allowing clean magnesium to

enter. The hand ladle itself should preferably be of a covered or hooded type, and incorporate a "teapot" spout. This gives a controlled pour into the die cavity and, furthermore, the hood does help to prevent surface burning during the short period involved in transferring metal from bale-out furnace to die.
9. The die cavity itself should, of course, be flushed with sulphur dioxide delivered under pressure from a cylinder prior to each pour.
10. The majority of magnesium gravity diecasting exponents employ their bale-out furnaces on a continuous rather than a batch basis. This means that fresh ingot metal and/or scrap is added from time to time to maintain reasonable metal capacity. Such additions, of course, should not be made too frequently, as the general disturbance could lead to the presence of flux inclusions. Consequently, wherever conditions permit, topping-up should be done at prescribed intervals, refining and grain refining following in sequence as already laid down.
11. With regard to the spent flux in the form of "sludge" which accumulates around the sides and at the bottom of the steel crucible, it is customary to remove this on, say, two occasions in an 8 hour shift. This involves thoroughly scraping the walls of the pot with a skimming tool, followed by a suitably shaped tool for removing sludge from the bottom. A lot depends upon the amount of metal which has melted in a certain period, some firms run through a complete shift leaving sludge removal as the last job of the day.

To prevent sticking, to give improved surface finish and—where necessary—to give control over the rate of cooling, the use of DYCOTE is essential. DYCOTE 140 is particularly suitable for use with magnesium alloys. It has especially good insulating properties to prevent too severe a chilling effect of the die and gives a first-class casting finish.

Pressure Die Casting

Because of the choking of the nozzle due to the disturbance of the flux cover with consequent dross formation, the conventional "gooseneck" type of machine cannot be used. The cold chamber machine, however, gives completely satisfactory results. With the cold chamber machine a pressure of about 4000 lb/in^2 forces the metal at a little above its freezing temperature into the die cavity. Both these factors help the production of strong, sound castings free from porosity and with an excellent finish.

Pressure die castings can be made with a wall thickness of less than 3 mm. A small addition of zinc improves fluidity and surface finish but as it tends also to increase hot shortness it should not exceed about 0.6%.

Melting Magnesium Scrap, Swarf, Turnings, etc.

A low fusion-point flux is necessary in order to protect the large surface area of the charge against oxidation and also to absorb dirt and non-metallics. The appropriate flux in this case will be MAGREX 60.

The charge should be commenced by pre-fusing MAGREX 60 to the extent of 5-10% by weight of the charge—the precise quantity must depend on the condition of the scrap and swarf.

The pre-heated scrap or dried swarf should be introduced into this flux bath in relatively small quantities.

The aim at all times should be to preserve a complete layer of molten flux, otherwise the losses by oxidation will be extremely high.

When the charge is complete and at a temperature of approximately 700-750°C add fresh MAGREX 60 to the extent of 2% by weight of the molten metal. The fused flux should then be thoroughly stirred into the melt for a period of from 3 to 5 minutes, using a coated iron bar for this purpose—the bar is suitably bent to enable the operator to "whisk" the metal and flux, thus bringing the two into as intimate contact as possible. During the operation, most of the flux will disappear and form a "sludge" on the bottom of the crucible.

This sludge may be removed immediately by dredging with a suitable tool or the metal may be decanted.

The first method should be carried out as follows:

The sides of the crucible should be well scraped and the sludge finally dredged out, using special "spoon" tools for the purpose. The metal should then be ladled into the ingot moulds, and the necessary protection for the metal remaining in the pot can be maintained with INERTEX. A dusting of INERTEX on the solidifying magnesium ingots will curtail undue burning.

If it is found more convenient to remove the residual flux after pouring, it is necessary to carefully watch that a few inches of magnesium are left in the crucible. This residue of metal and sludge should be poured into separate moulds followed by final and thorough cleaning of the crucible.

MELTING AND RECOVERY OF ALUMINIUM SWARF AND SMALL SCRAP SUCH AS SHEET CLIPPINGS, MILK BOTTLE TOPS, etc.

Rotary type furnaces are favoured. The flame from the burner strikes the furnace lining and the exhaust spreads around the furnace walls which absorb heat. Then as the furnace rotates this heat is taken up by the metal. Melting must take place under a cover of molten flux, otherwise heavy oxidation results, as each piece of swarf, as it melts, forms a globule of liquid surrounded by a shell of oxide. The skin tension of this oxide prevents coalescence and in consequence of the large surface area presented by the mass of globules, increase of oxide formation and loss of yield is bound to take place. Therefore, the skin of oxide has to be ruptured in order to allow coalescence. The molten flux encourages this by chemical action, while the rotation of the furnace provides a mechanical action. COVERAL 48 or 57 are appropriate fluxes melting at 605 and 590°C respectively.

Procedure for 500 kg Furnace

1. Charge 100 kg of flux before introducing the turnings, and heat to 850–900°C.
2. This takes 15–20 minutes, after which the furnace is turned over to the charging position and 30 kg of swarf or smaller scrap fed in.
3. Rotation of the furnace turns the swarf in under the flux. The heat from the flux is rapidly transferred to the metal until a state of balance is reached.
4. Consequent cooling of the flux causes it to become pasty, thus affording a complete cover for the melt.
5. After adding the 30 kg of swarf, the furnace is returned to the firing position and when the flame, caused by oil burning from the swarf, dies the burner is re-started and in from 4 to 5 minutes the temperature is back to 750°C. This temperature should not be greatly exceeded in order to keep gas absorption to a minimum.
6. Repeat the process until the balance of the 500 kg of swarf is fed into the furnace. During the latter stages an increased rate of charge of about 45 kg a time can be used. The complete melt takes about $2\frac{1}{4}$ hours from start to finish.

After the whole 500 kg has melted the flux becomes more viscous through taking up oxides and hydrocarbons from the burnt oil and other impurities. This enables the molten metal to be poured from underneath the flux.

In order to bring back the fluidity of the flux, about 25 kg is removed from the furnace and an equal quantity of fresh flux added. In this way five charges can be completed before a complete change of flux is needed. This brings the flux/metal ratio to 10%. Average reclamation yield is approximately 90%. The recovered metal is degassed and drossed off with COVERAL 11. If magnesium has to be removed, the melt is treated with LOMAG 6. 0.5 kg of LOMAG removes approximately 0.1% of magnesium from a 50 kg melt.

GRAIN REFINEMENT OF ALUMINIUM ALLOYS WITH TITANIUM AND BORON

Both titanium and boron will grain refine or nucleate all the aluminium alloys. They may be added either as the element or as hardener alloys or by direct reduction of titanium and boron containing salts. Hardener alloys have some disadvantages in that considerable amounts of oxide are added simultaneously and slow solution rates often mean areas of high concentration leading to grain boundary borides, etc. The salts method gives good dispersion, but efficiency of pick-up may sometimes be disappointing if conditions are not exactly right.

Considerable experimental work on the individual elements and on combinations has produced results that are often indecisive and inconclusive. On the one hand, it is considered that with the exception of boron, which is assumed to form aluminium boride, the other well-known nucleating agents, e.g. titanium, columbium, cerium, zirconium, etc., all react with any available carbon to form the appropriate carbide. This is said to be the important refining agent as distinct from the metal itself. On the other hand, the view is also held that grain refinement consequent upon the addition of titanium and boron is due to the formation of titanium boride as distinct from duplex nucleation from titanium carbide and aluminium boride.

It is now known that single introduction of titanium, added either as hardener or salt to an aluminium alloy, grain refines due to the formation of titanium carbide. An addition of boron results in the formation of nuclei of aluminium boride which grain refines in a similar way. A combined introduction of titanium and boron results additionally in the formation of nuclei of titanium boride and it is generally conceded that boron and titanium added together have a more marked refining effect than when either is added alone.

From all this background it is clear that a combined addition of titanium and boron is desirable and other considerations indicate that a reduction from the appropriate salt would be the best method. These factors led to the development of NUCLEANT 2 tablets the use of which is now standard practice in many countries where consistent quality in the aluminium produced is of prime importance. In appropriate quantities the product is used for sand and gravity diecastings, forging and extrusion billets, slabs, etc., produced by all conventional processes including semi- and fully continuous methods.

Microspecimens invariably show small and uniform grain structures with columnar tendencies, so often associated with large equi-axial "banding", completely suppressed. At the same time local concentrations of boron, giving rise to grain boundary constituents are avoided. NUCLEANT 2

LIGHT CASTING ALLOYS

tablets utilise a titanium: boron ratio of 6:1, which is generally accepted as being the most effective and the standard recommendation is 0.25% of product by weight with metal temperature approximately 750°C. With large capacity melting and holding furnaces where application efficiency can be improved, quite considerable economies in product application rate can be effected.

While NUCLEANT 2 has been developed to meet the major requirement, for other purposes it is possible that nucleating agents based on either titanium or boron alone may be required. These can be met with NUCLEANT 6 (titanium only) or NUCLEANT 186A (boron only).

For the very large user where individual furnaces often contain many tonnes of aluminium alloy there is a further development known as TILITE 101. This is a lustrous grey coloured block weighing 0.5 kg and containing 40% titanium and 4% boron. The highly compressed block is self-sinking in molten aluminium and application is therefore simply a matter of throwing the requisite number of blocks into the cleaned melt. The blocks will dissolve rapidly in a matter of 15–30 seconds depending on metal temperature—and a gentle stir is all that is required to ensure good dispersion of the nuclei.

Depending therefore on melt size and on type of nucleating agent it is possible to select a suitable grain refiner to give the degree of refinement required in the macrostructure.

MODERN METHODS OF FLUXING MOLTEN ALUMINIUM

It is accepted practice in the aluminium industry that because of the reactive nature of the metal itself a comprehensive cleaning and degassing treatment is necessary for all metal melted if clean, sound castings and sections are required. Conventionally, these treatments have been carried out whatever the type of furnace by surface application of either exothermic or "wet" cleaning fluxes and by plunging degassing tablets/grain refiners or by bubbling inert or active gas, or a mixture, through the melt. Necessarily such methods of application are only as good as the manner in which they are applied and in general they are somewhat prodigal in terms of quantities of flux, etc. used. These standard methods of treatment are still used extensively in the average foundry and are dealt with elsewhere in this handbook.

Flux Washing and F.I.L.D. Process

Where large amounts—say above around 10 tonnes—of metal had to be treated to a high quality, a method of passing the molten metal through a bath of molten flux was developed. Figure 4.8 diagrammatically illustrates the principles involved. Although the method, along with several variations of the main idea, ensured a very considerable improvement in metal cleanliness it did require a fairly substantial and separately heated furnace arrangement with sufficient fall in metal levels to allow insertion of the unit between melting furnace and casting table. For many years this process, or one of its variants, has been the standard method of treatment in most of the large aluminium smelters and wrought plants as well as a number of the more substantial secondary refiners.

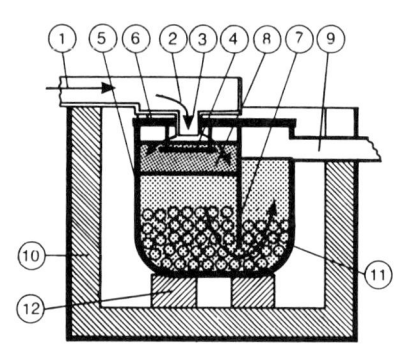

1 Entry Launder	7 Plumbago Baffle
2 Tundish	8 Fluid Flux Layer
3 Downspout	9 Exit Launder
4 Circular Spreader Plate	10 Refractory Furnace Lining
5 Plumbago Crucible	11 Alumina Balls
6 Lid	12 Refractory Stool

Fig. 4.8. Flux washing tank.

The process works by breaking up the stream of molten aluminium into myriads of droplets each of which is exposed to the scavenging effect of the molten flux. Special flux compositions have been produced to suit the basic process, to help removal of magnesium, etc. Subsequent passage through a bed of refractory particles encourages wetting out of slag, carried down flux, etc. and the emergent metal is significantly cleaner in terms of suspended non-metallic inclusions etc. One of the theories of how dissolved hydrogen is held in a melt of aluminium also requires the presence of non-metallic films and particles to provide foci on which the gas is held so removal of these by flux washing should provide a good degree of simultaneous degassing. Such appears to be the case.

A further improvement of this function was obtained by insertion of an inert gas (nitrogen or argon) bubbler into the reaction side of the baffle as shown in Fig. 4.9. The process then became known as the Fumeless in Line Degassing (FILD) process.

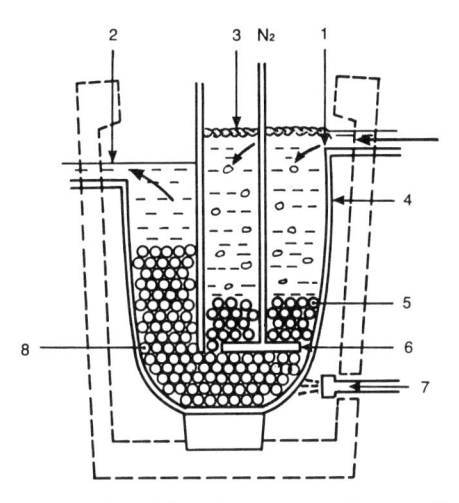

Fig. 4.9. Schematic drawing of the FILD process unit: 1 metal in; 2 metal out; 3 liquid flux; 4 refractory crucible; 5 flux-coated balls; 6 diffuser; 7 gas burner; 8 un-coated balls.

Flux Injection

The idea of increasing flux/metal contact area led to consideration of methods of injecting a stream of powdered flux below the surface of molten aluminium. Such a system would make optimum use of flux as it would be reacting in an enclosed environment and the inert gas (nitrogen or argon) carrier would ensure a simultaneous degree of degassing. After some false starts, mainly due to the aggressive and corrosive nature of the flux mixtures involved, machinery and lances are now available that will produce reliable results in an average foundry background at a very enconomical figure in terms of the consumables—flux and gas—and without the emission of such amounts of fume as to make special individual furnace ventilation necessary. The main advantages of injected flux can be summarised as follows:

1. An enclosed bubble ensures a flux activated inner surface leading to more efficient degassing.
2. Overall melt cleanliness improved by better flux dispersion.
3. Dross contains less metal due to more efficient reaction.
4. Cleaner metal means less rejects.
5. Appreciable process energy savings.
6. Lower direct consumables cost.
7. Improved local working conditions.

8. Reduced die and crucible wear through reduced metal temperatures.
9. Process can be automated or operated by semi-skilled personnel.
10. Adaptable to all types of furnace.

A full range of drossing, modifying and sodium-free fluxes is available to suit this process. A schematic representation of the system is shown in Fig. 4.10.

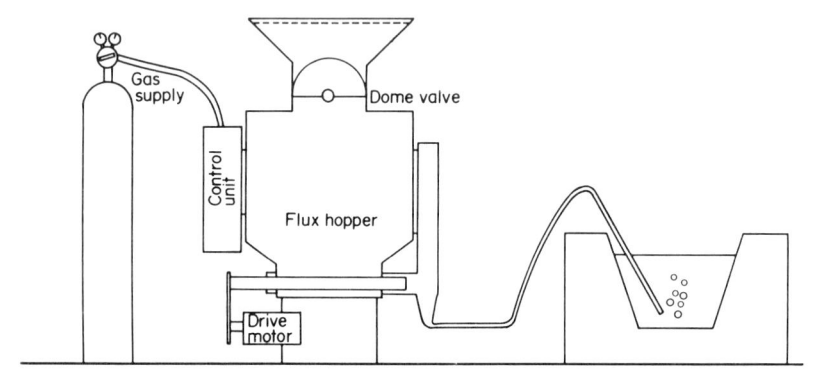

Fig. 4.10. Foseco Macawber Injector-Schematic.

Control of Melt Quality

It is clear from consideration of the previous two sections that very considerable improvements in both the methods used and the efficiency of metal treatment have taken place. If these advances are taken in conjunction with other improvements such as whole melt or individual casting filtration (see separate section on metal filtration) there is obviously enormous potential for a considerable step forward in melt quality to a degree that has never before been available.

Flux washing as such (whether in association with filtration or not) can only be feasible for the bulk melters—say those handling upwards of 10 tonnes of metal—because of the practical difficulties. This will encompass the wrought producers but can also include some of the larger scrap reclaimers and casting ingot producers particularly where quality demands are high and a premium on this basis can be obtained. The advent of flux injection however now extends the routine achievement of very high melt quality standards to a much wider field and to smaller furnaces. Indeed there is probably no lower practical limit to the size of furnace (except when it becomes impractical to insert a lance and blow gas through it) that can be treated. Injected flux ensures optimum economic and technical advantage of formulation together with a strong degassing effect. It is possible to adapt almost any type of flux—cleaning, drossing, degassing, modifying or refining—to this process. The only real disadvantage is the cost of the injector and the carrier gas, currently either dry nitrogen or argon. It is now a practical foundry tool that can be operated by semi-skilled staff.

If these new and highly effective melt quality treatment methods are used in association with strategic in-mould filtration and flow control new standards for individual casting integrity and appearance may well be set.

SECTION V
NON-FERROUS CASTING ALLOYS

THE ORDER OF ALLOYING

The table below shows the order in which various elements are added in the alloying of copper and nickel alloys.

The base metal with which the melt commences is usually marked as "1", whilst the various elements added afterwards are marked under their respective heading with the number giving their stage of addition.
Example: GUNMETAL. 88% copper, 10% tin, 2% zinc.
Charge crucible with "1" copper, then add "2" tin, afterwards "3" zinc.
This table applies to virgin metals. If scrap is used in addition this is usually added together with "1".

	Copper	Tin	Zinc	Lead	Nickel	Phos-phorus	Remarks
Brass	1	—	2	—	—	—	Stir very well after zinc has been added
Naval brass	1	*2	3	—	—	—	Stir very well after zinc has been added
Gunmetal copper, tin, zinc, such as 88/10/2	1	*2	3	—	—	—	Stir well before pouring
Gunmetal with lead such as 85/5/5/5.	1	*2	4	3	—	—	Stir very well before pouring
Bronzes such as 90% copper, 10% tin	1	*2	—	—	—	—	Stir well before pouring
Phosphor bronze	1	3	—	—	—	**2 **4	Stir well
Nickel silver or white metal	***2	3 if any	5	5	***1	—	Stir very well
Cupro nickel or nickel bronzes	***2	—	—	—	***1	—	Stir well

*Deoxidise well with deoxidising tubes DS before "2" is added.
**Phosphorus is best added as phosphor copper. Half the total amount before "3" is added, the other half afterwards.
***Charge the total amount of nickel, then add as much of the copper as crucible will hold. Then commence melting and continue to add copper as melt proceeds in the usual fashion.

BRITISH STANDARD SPECIFICATIONS FOR THE PRINCIPAL NON-FERROUS CASTING ALLOYS

BS 1400.1969 is the first revision of the specification in metric terms. At the same time, the alloys covered have been grouped into three categories according to the approximate degree of usage and availability; There are some additions, some revisions, and some deletions with respect to previous editions of the BS 1400 Specification.

The main alloy groups are as follows:
Group A. Alloys in common use. PB4, LPB1, LB2, LB4, LG2, LG4, SCB1, SCB3, SCB6, DCB1, DCB3, PCB1.
Group B. Special purpose alloys. HCC1, CC1-WP, PB1, PB2, CT1, LB5, LG1, AB1, AB2, CMA1, CMA2, HTB1, HTB3.
Group C. Alloys in limited production. LB1, G1, G3, G3-WP, SCB4.

Tables giving chemical composition, mechanical properties, density, and coefficient of thermal expansion of these alloys are given on the following pages.

CHEMICAL COMPOSITION OF GROUP A CASTINGS

Designation:	PB4		LPB1		LB2		LB4		LG2		LG4	
Material:	Phosphor bronze (copper–tin–phosphorus)		Leaded phosphor bronze		80/10/0/10 leaded bronze		85/10/0/10 leaded bronze		85/5/5/5 leaded gunmetal		87/7/3/3 leaded gunmetal	
Elements	min. (%)	max. (%)	min. (%)	max. (%)	min. (%)	max. (%)	min. (%)	max. (%)	min. (%)	max. (%)	min. (%)	max. (%)
Copper	REM		REM		REM		REM		REM		REM	
Tin	9.5	—	6.5	8.5	9.0	11.0	4.0	6.0	4.0	6.0	6.0	8.05
Zinc	—	0.5	—	2.0	—	1.0	—	2.0	4.0	6.0	1.5	3.0
Lead	—	0.75	2.0	5.0	8.5	11.0	8.0	10.0	4.0	6.0	2.5	3.5
Phosphorus	0.4	—	0.3	—	—	0.10	—	0.10	—	—	—	—
Nickel	—	0.5	—	1.0	—	2.0	—	2.0	—	2.0	—	2.0§
Iron	—	—	—	—	—	0.15*	—	—	—	—	—	0.20*
Aluminium	—	—	—	—	—	0.01*	—	—	—	0.01*	—	0.01*
Manganese	—	—	—	—	—	—	—	—	—	—	—	—
Antimony	—	—	—	—	—	0.5	—	0.5	—	—	—	0.25*
Arsenic	—	—	—	—	—	—	—	—	—	—	—	—
Iron + arsenic antimony	—	—	—	—	—	—	—	—	—	0.50*	—	0.40*
Silicon	—	—	—	—	—	0.02*	—	0.02*	—	0.02*	—	0.01*
Bismuth	—	—	—	—	—	—	—	—	—	0.05*	—	0.05*
Total impurities	—	0.50	—	0.50	—	0.50	—	0.50	—	0.80	—	0.70

Total impurities include elements marked*

†DCB1—0.1% lead if required.
‡DCB3—nickel to be counted as copper.
§Tin + ½ nickel content 7.0–8.0%.

CHEMICAL COMPOSITION OF GROUP A CASTINGS

Designation:	SCB1		SCB3		SCB6		DCB1†		DCB3‡		PCB1	
Material:	Brass for sand casting		Brass for sand casting		Brass for brazable casting		Brass for die casting		Brass for die casting		Brass for pressure die casting	
Elements	min. (%)	max. (%)	min. (%)	max. (%)	min. (%)	max. (%)	min. (%)	max. (%)	min. (%)	max. (%)	min. (%)	max. (%)
Copper	70.0	80.0	63.0	70.0	83.0	88.0	59.0	63.0	58.0	63.0	57.0	60.0
Tin	1.0	3.0	—	1.5	—	—	—	—	—	1.0	—	0.5
Zinc	REM		REM		REM		REM		REM		REM	
Lead	2.0	5.0	1.0	3.0	—	0.5	—	0.25	0.5	2.5	0.5	2.5
Phosphorus	—	—	—	—	—	—	—	—	—	—	—	—
Nickel	—	1.0	—	1.0	—	—	—	—	—	1.0	—	—
Iron	—	0.75	—	0.75	—	—	—	—	—	0.8	—	0.3
Aluminium	—	0.01*	—	0.1*	—	—	—	0.5	0.2	0.8	—	0.5
Manganese	—	—	—	—	—	—	—	—	—	0.5	—	—
Antimony	—	—	—	—	—	—	—	—	—	—	—	—
Arsenic	—	—	—	—	0.05	0.20	—	—	—	—	—	—
Iron + arsenic + antimony	—	—	—	—	—	—	—	—	—	—	—	—
Silicon	—	—	—	—	—	—	—	—	—	0.05*	—	—
Bismuth	—	—	—	—	—	—	—	—	—	—	—	—
Total impurities	—	1.0	—	1.0	—	1.0 (incl. Pb)	—	0.75	—	2.0 (excl. Ni+ Pb+ Al)	—	0.5

Total impurities include elements marked*

†DCB1—0.1% lead if required.
‡DCB3—nickel to be counted as copper.
§Tin + ½ nickel content 7.0–8.0%.

Chemical Composition of Group B Castings

Designation	HCC1† High conductivity copper		CC1-WP‡ Copper chromium		PB1 Phosphor bronze (copper-tin-phosphorus)		PB2 Phosphor bronze (copper-tin-phosphorus)		CT1 Copper-tin		LB5 75/5/0/20 leaded bronze	
Material												
Elements	min. (%)	max. (%)	min. (%)	max. (%)	min. (%)	max. (%)	min. (%)	max. (%)	min. (%)	max. (%)	min. (%)	max. (%)
Copper	REM		REM		REM		REM		REM		REM	
Tin					10.0	—	11.0	13.0	9.0	11.0	4.0	6.0
Zinc					—	0.05*	—	0.30	—	0.05*	—	1.0
Lead	Castings shall be made from copper complying with BS 1035–1037. Added elements and impurities not specified		—	—	—	0.25*	—	0.50	—	0.25*	18.0	23.0
Phosphorus			—	—	0.50	—	0.15	—	—	0.15	—	0.10**
Nickel			—	—	—	0.10*	—	0.50	—	0.25*	—	2.0
Iron			—	—	—	0.10*	—	0.15*	—	—	—	—
Aluminium			—	—	—	0.01*	—	0.01*	—	—	—	—
Manganese			—	—	—	—	—	—	—	—	—	—
Antimony			—	—	—	—	—	—	—	—	—	0.50
Arsenic			—	—	—	—	—	—	—	—	—	—
Iron + arsenic + antimony			—	—	—	—	—	—	—	—	—	—
Silicon			—	—	—	0.02*	—	0.02*	—	—	—	0.01**
Bismuth			—	—	—	—	—	—	—	—	—	—
Magnesium			—	—	—	—	—	—	—	—	—	—
Chromium			0.60	1.2	—	—	—	—	—	—	—	—
Total impurities			—	—	—	0.60	—	0.20	—	0.80	—	0.30

†HCC1—maximum resistivity = 0.019 microhm-metre (approximately 90% IACS).
‡CC1-WP—maximum resistivity = 0.022 microhm-metre (approximately 80% IACS).

NOTE. Copper which has a resistivity of 0.017 241 microhm-metre is said to have a conductivity of 100% IACS.

NON-FERROUS CASTING ALLOYS

Chemical Composition of Group A Castings

Designation	LG1		AB1§		AB2§		CMA1		CMA2		HTB ‖		HTB3	
Material	83/3/9/5 leaded gunmetal		Aluminium bronze (copper–aluminium)		Aluminium bronze (copper–aluminium)		Copper–manganese aluminium		Copper–manganese aluminium		46 hbar¶ high tensile brass		74hbar¶ high tensile beta brass	
Elements	min. (%)	max. (%)	min. (%)	max. (%)	min. (%)	max. (%)	min. (%)	max. (%)	min. (%)	max. (%)	min. (%)	max. (%)	min. (%)	max. (%)
Copper	REM		REM		REM		REM		REM		55.0	—	55.0	—
Tin	2.0	3.5	—	0.1**	—	0.1**	—	1.0	—	1.0	—	1.0	—	0.20
Zinc	7.0	9.5	—	0.50	—	0.50	—	0.50	—	0.50	REM		REM	
Lead	4.0	6.0	—	0.05**	—	0.05*	—	0.05*	—	0.05*	—	0.50	—	0.20
Phosphorus	—	—	—	—	—	—	—	0.05*	—	0.05*	—	—	—	—
Nickel	—	2.0	—	1.0	4.0	5.5	1.5	4.5	1.5	4.5	—	1.0	—	1.0
Iron	—	—	1.5	3.5	4.0	5.5	2.0	4.0	2.0	4.0	0.7	2.0	1.5	3.25
Aluminium	—	0.01**	8.5	10.5	8.8	10.0	7.5	8.5	8.5	9.0	0.5	2.5	3.0	6.0
Manganese	—	—	—	1.0	—	1.5	11.0	15.0	11.0	15.0	—	3.0	—	4.0
Antimony	—	—	—	—	—	—	—	—	—	—	—	—	—	—
Arsenic	—	—	—	—	—	—	—	—	—	—	—	—	—	—
*Iron + arsenic + antimony	—	0.75**	—	—	—	—	—	—	—	—	—	—	—	—
Silicon	—	0.02**	—	0.25**	—	0.1*	—	0.15*	—	0.15*	—	0.10*	—	0.10*
Bismuth	—	0.10**	—	—	—	—	—	—	—	—	—	—	—	—
Magnesium	—	—	—	0.05**	—	0.05*	—	—	—	—	—	—	—	—
Chromium	—	—	—	—	—	—	—	—	—	—	—	—	—	—
Total impurities	—	1.0	—	0.30	—	0.30	—	0.30	—	0.30	—	0.2	—	0.2

Analysis must include all elements marked * but not necessarily those marked **.
Total impurities include elements marked * and **.
§AB1 and AB2—when castings are to be welded lead not to exceed 0.01%.
‖ HTB1—subject also to microstructure requirements.
¶Material was designated 30 ton and 48 ton respectively.

Chemical Composition of Group C Castings

Designation	LB1		G1		G3		G3WP		SCB4	
Material	76/9/0/15 leaded bronze		88/10/2 gunmetal		Nickel gunmetal		Nickel gunmetal fully heat treated		Naval brass for sand casting	
Elements	min. (%)	max. (%)	min. (%)	max. (%)	min. (%)	max. (%)	min. (%)	max. (%)	min. (%)	max. (%)
Copper	REM		REM		REM		REM		60.0	63.0
Tin	8.0	10.0	9.5	10.5	6.5	7.5	6.5	7.5	1.0	1.5
Zinc	—	1.0	1.75	2.75	1.5	3.0	1.5	3.0	REM	
Lead	13.0	17.0	—	1.5	0.10	0.50	0.10	0.50	—	0.5
Phosphorus	—	0.1**	—	—	—	0.02	—	0.50	—	—
Nickel	—	2.0	—	1.0	5.25	5.75	5.25	5.75	—	—
Iron	—	—	—	0.15*	—	—	—	—	—	—
Aluminium	—	—	—	0.01*	—	0.01*	—	0.01*	—	0.01**
Manganese	—	—	—	—	—	0.20*	—	0.20*	—	—
Antimony	—	0.5	—	—	—	—	—	—	—	—
Arsenic	—	—	—	—	—	—	—	—	—	—
Iron + arsenic + antimony	—	—	—	0.20*	—	0.20*	—	0.20*	—	—
Silicon	—	0.02**	—	0.02*	—	0.01*	—	0.01*	—	—
Bismuth	—	—	—	0.03*	—	0.02*	—	0.02*	—	—
Total impurities	—	0.30	—	0.50	—	0.50	—	0.50	—	0.75

N.B. Analysis must include all elements marked * but not necessarily those marked **. Total impurities include elements marked * and **

Colour Code for Ingots

Designation	Colour code	Designation	Colour code
Group A		Group B	
PB4	Black/red	PB1	Yellow
LPB1	Black	PB2	Yellow/Red
LB2	White	CT1	Black/aluminium
LB4	White/green	LB5	White/brown
LG2	Blue	LG1	Blue/red
LG4	Blue/brown	AB1	Aluminium
SCB1	Green/blue	AB2	Aluminium/green
SCB3	Green	CMA1	Aluminium/red
SCB6	Green/brown	CMA2	Aluminium/yellow
DCB1	Yellow/blue	HTB1	Brown
DCB3	Yellow/brown	HTB3	Brown/red
PCB1	White/blue		
		Group C	
		LB1	White/black
		G3	Blue/black
		SCB4	Green/yellow
		G1	Red

MECHANICAL PROPERTIES OF TEST SAMPLES FOR CASTINGS

Designation	Tensile strength				0.2% proof stress				Elongation on $5.65\sqrt{S_o}$			
	Sand min. hbar ‖	Chill min. hbar ‖	Continuous min. hbar ‖	Centrifugal§ min. hbar ‖	Sand min. hbar ‖	Chill min. hbar ‖	Continuous min. hbar ‖	Centrifugal§ min. hbar ‖	Sand min. %	Chill min. %	Continuous min. %	Centrifugal§ min. %
Group A alloys												
PB4	19	27	33	28	10*	14*	16*	14*	3	2	7	4
LPB1	19	22	27	23	8*	13*	13*	13*	3	2	5	4
LB2	19	22	28	23	8*	14*	16*	14*	5	3	6	5
LB4	16	20	23	22	6*	8*	13*	8*	7	5	9	6
LG2	20	20	27	22	10*	11*	10*	11*	13	6	13	8
LG4	25	25	30	25	13*	13*	13*	13*	16	5	13	6
Group B alloys												
CC1WP†	—	—	—	—	—	—	—	—	—	—	—	—
PB1	22	31	36	33	13*	17*	17*	17*	3	2	6	4
PB2	22	27	31	28	13*	17*	17*	17*	5	3	5	3
CT1	23	27	31	28	13*	14*	16*	14*	6	5	9	6
LB5	16	17	19	19	6*	8*	10*	8*	5	5	8	7
LG1	18	18	—	—	8*	—	—	—	11	2	—	—
AB1	50	54	—	56	17*	20*	—	20*	18	18	—	20
AB2	64	65	—	67	25	25	—	25	13	13	—	13
CMA1	65	67	—	—	28	31	—	—	18	27	—	—

CMA2	74	—	—	—	38	—	—	—	9	—	—
HTB1	47	50	—	50	17	21	—	21	18	18	20
HTB3	74	—	—	74	40	—	—	40	11	—	13
Group C alloys
LB1	17	20	23	22	8*	13*	13*	13*	4	3	9	4
G1	27	23	30	25	13*	13*	14*	13*	13	3	9	5
G3	28	—	34	—	14*	—	17*	—	16	—	18	—
G3WP‡	43	—	43	—	28*	—	28*	—	3	—	3	—

*For information only.
†The minimum hardness of CC1WP after heat treatment shall be 100 HB
‡The minimum hardness of G3WP after heat treatment shall be 160 HB
§Values apply to samples cut from centrifugal castings made in metallic moulds. Minimum properties of centrifugal castings made in sand moulds are the same as for other sand castings.
∥ 1 hbar = 10 MN/m^2 = 10 N/mm^2.

DENSITY AND COEFFICIENT OF THERMAL EXPANSION

Alloy designation	Density (g/cm³)	Linear coefficient of thermal expansion × 10⁻⁶ 0–250°C (per °C)
Group A		
PB4	8.8	18.3
LPB1	8.8	18.0
LB2	9.0	18.6
LB4	9.0	18.4
LG2	8.8	18.0
LG4	8.8	18.2
SCB1	8.5	19.3
SCB3	8.4	20.2
SCB6	8.6	18.6
DCB1	8.3	20.8
DCB3	8.3	20.8
PCB1	8.3	20.8
Group B		
HCC1	8.9	17.3
CC1	8.85	17.3
PB1	8.8	18.3
PB2	8.8	18.5
CT1	8.8	18.1
LB5	9.2	18.9
LG1	8.8	17.8
AB1	7.6	17.0
AB2	7.6	17.0
CMA1	7.5	18.6
CMA2	7.5	18.6
HTB1	8.3	20.5
HTB3	7.9	20.5
Group C		
LB1	9.1	18.7
G1	8.8	18.3
G3	8.8	18.3
SCB4	8.3	20.8

THE IMPORTANCE OF FURNACE ATMOSPHERE WHEN MELTING COPPER-BASED ALLOYS

In metal melting one of three types of atmosphere may be experienced, depending on the condition of melting. By atmosphere is meant the space immediately above the metal being melted and this, of course, means inside the crucible or furnace being used.

This atmosphere can be either: (*a*) neutral, (*b*) oxidising, or (*c*) reducing, and upon this and the nature of the metal, will depend whether the metal is good, bad, or indifferent.

If the Atmosphere is Neutral

This means that the atmosphere is not reacting either for or against the metal and that the coke or fuel must be burning completely without excess of oxygen. A neutral atmosphere is therefore obviously a rarity.

If the Atmosphere is Reducing

This means that the fuel is burning with insufficient oxygen for complete combustion and is therefore generating gases which will react with additional oxygen and burn to complete combustion. These are known as reducing gases. If they are soluble in the molten metal, trouble can be expected through gassy metal. These gases are chiefly carbon monoxide, hydrogen, and gaseous compounds of carbon with hydrogen.

If the Atmosphere is Oxidising

This means that the fuel is burning with an excess of oxygen and the conditions are said to be oxidising. In the case of complete combustion, the gases, water vapour and carbon dioxide are formed which, with the addition of some excess oxygen, will be free to affect the metal. In this case oxidation will take place and must be removed by subsequent deoxidation.

Gases which may occur during the melting process from the combustion of the fuel are:

| Carbon monoxide | Hydrogen | Water vapour |
| Carbon dioxide | Hydrocarbon | |

Those which are extremely detrimental to the melt are hydrogen, carbon monoxide, and the hydrocarbons. These are all reducing gases and therefore will create a reducing atmosphere. In the case of brasses, bronzes, copper and nickel alloys, etc. these are best avoided by using an oxidising flux, thus creating an oxidising atmosphere around the metal; the excess oxygen is then removed by subsequent deoxidation.

It will be seen that control of furnace atmosphere, by adjustment of fuel:air ratio and by the use of the correct flux in contact with the metal, will give direct control of metal quality in respect of ingress of harmful gases. The importance of correct furnace operation and careful choice of flux cover is clear if reliable, consistent results are required.

MELTING, FLUXING, AND DEGASSING PROCEDURE

BRASSES, BRONZES, AND GUNMETALS

Copper-based alloys which contain relatively high percentages of zinc, such as the brasses, are normally free from porosity trouble caused by dissolved hydrogen. This is because the vapour pressure of the zinc at the melting temperatures involved is effective in restricting hydrogen absorption. Zinc also acts as a deoxidising agent; therefore molten brass does not need the addition of deoxidisers for the removal of oxygen or oxides. However, the introduction of a small proportion of phosphorus is valuable as a means of increasing fluidity.

For straight copper-zinc alloys the CUPRIT range of fluxes is recommended. Where aluminium forms a part of the alloy, as in manganese bronze and aluminium bronze, it is advisable to employ a flux capable of dissolving and removing oxide of aluminium such as ALBRAL. The silicon bronzes also benefit from treatment with ALBRAL flux.

An entirely different set of circumstances are experienced in the melting of bronzes and gunmetals. These alloys contain a high ratio of copper and little or no zinc so that gas porosity due to hydrogen pick-up becomes more of a possibility.

Hydrogen can enter the melt in a variety of ways such as from atmosphere, corrosion products on scrap, wet furnace tools, damp ladle linings, moisture in fluxes, and so on. The complete combustion of 5 litres of fuel oil results in the formation of about 6 kg of water vapour which may come into direct contact with the surface of the melt.

If water vapour remained as such, little difficulty would follow, for it is doubtful if water vapour is absorbed. Unfortunately, it tends to split up into hydrogen and oxygen at molten metal temperatures and the gases so formed are atomic or nascent and very active. Although the reaction is reversible, in practice equilibrium is set up wherein the amount of dissociation balances recombination. At any given temperature a mixture of hydrogen, oxygen and water vapour is present, the proportions depending on the temperature. As the temperature rises more hydrogen and oxygen are formed. The nascent oxygen immediately unites with metallic constituents to form oxides, whilst the hydrogen dissolves in the melt. So both are removed from the atmosphere above the alloy, which results in the dissociation of more water vapour, giving rise to more atomic hydrogen and oxygen, which in turn will be dissolved. Thus the build-up of gas in the metal continues.

In order to protect molten bronze from attack in this way, it is necessary to provide an oxidising atmosphere during melting. This is achieved by the use of CUPREX fluxes in either block or powder form. The flux is placed in the required ratio of 1% of the charge weight into the base of the preheated crucible, followed by the charge. Then whilst melting proceeds oxidising gases are evolved which bubble up through the melt, thereby acting as an effective barrier against hydrogen pick-up. Towards the latter stages of melting the block fuses and rises to the surface to form a protective layer.

Finally, just prior to pouring, the flux layer is skimmed off and any surplus oxygen removed by plunging deoxidising tubes into the melt. This method of treatment is known as the oxidation-deoxidation technique and is widely recognised as the best procedure for bronzes and gunmetals.

Degassing Copper-based Alloys

For important castings and those which must meet stringent inspection by X-rays and other non-destructive methods, a separate degassing treatment may be advisable. This can be accomplished by plunging and thoroughly scrubbing specialised fluxes into the melt. These REGENEX fluxes are classed as "oxidising" for the gunmetal, bronze, and nickel alloy group, and "neutral" for phosphor bronze and other alloys containing easily oxidisable constituents.

Alternatively, LOGAS 50 briquettes may be plunged into the melt with a suitably protected plunging tool. These briquettes evolve inert gases at the temperatures involved, and degassing is accomplished by a process of diffusion of hydrogen from the metal into the rising bubbles. The process may be used with any copper-based alloy including high conductivity copper, does not affect metal composition, and evolves no fume. However, some cooling of the metal is unavoidable, and crucibles cannot be completely filled with metal because of the necessity to accommodate plunger displacement.

High Lead Bronzes

High lead bronzes, for the purpose of these notes, are considered to be those containing 10% or more of lead. Alloys of copper containing up to 35% of lead form a homogeneous liquid melt above 1083°C. During cooling, solidification begins and copper separates out at temperatures which vary between 1080° and 954°C as the lead varies from 0 up to 36%. On further cooling from 954° to 326°C, the copper and lead exist as separate constituents. The size and shape of the copper and lead particles are of great practical importance and are affected, amongst other things, by the rate of cooling. When solidification is very slow, the lead has time to coalesce into larger "lakes", while the matrix is still in the pasty condition. If, however, solidification is very rapid the lead is entrapped in a fine network of primary copper, which therefore controls the size of the particles.

The high lead bronzes with lead above 36% behave differently from those with a lead content below that figure. The copper and lead, though both molten at 954°C, do not form a homogeneous liquid, but exist largely in the form of fine droplets as an emulsion. As the liquid particles of lead are heavier, they tend to sink to the bottom of the melt. The lower the temperature (above the melting point), the greater is the tendency, which is very marked below about 1000°C. However, the melt can be brought into a uniform and fairly stable emulsion by vigorous agitation. On the other hand, the two metals will form a homogeneous liquid if the temperature exceeds 1325°C. If the rate of cooling from this temperature is sufficiently rapid, then the lead is retained in a fine state of division.

High lead bronze should be melted down under PLUMBRAL flux. Use $\frac{1}{2}$% of the charge weight of PLUMBRAL added with the charge materials or when they commence to melt. Superheat rapidly to the required temperature and then some 5 minutes before pouring, slowly plunge another $\frac{1}{2}$% by weight of PLUMBRAL. This operation is best carried out with a graphite tool or if an iron or steel plunger is used it should be coated previously with a dressing of DYCOTE 36, and thoroughly dried before use.

Small percentages of nickel are an aid to preventing segregation of the lead, especially when more than 20% of lead is present. An addition of 1% to the original charge in the crucible prior to melting will be adequate for most purposes. When nickel is added the pouring temperature should be raised by about 20°C. With very high lead contents the introduction of sulphur also exercises an important influence on keeping the lead dispersed in a finely divided form.

Suggested pouring temperatures are:

15% lead	1080–1100°C
20% lead	1050–1070°C
30% lead	1030–1050°C

MELTING AND FLUXING PROCEDURE FOR HIGH CONDUCTIVITY AND COMMERCIAL COPPER CASTINGS

High Conductivity Copper

The production of castings from pure copper presents many difficulties, consequently, even pure copper castings seldom have conductivities higher than 80% as measured electrically in International Annealed Copper Standard units. Copper in the annealed condition, which has a specific electrical resistance of 1.72 microhms per cubic centimetre at 20°C, is said to have 100% electrical conductivity IACS.

It is often more convenient to refer to the relative electrical conductivity of copper rather than to its resistance, and such conductivity is expressed as a percentage of that of annealed copper based on the international standard. This standard does not represent the highest possible conductivity and it is possible to obtain copper having a conductivity of 1% or even up to 2% higher than that taken as the standard.

Thermal conductivity is also an important property which has to receive major consideration for some applications. It is measured in calories per second per square centimetre per degree centigrade. Generally speaking, purely thermal applications are less critical and lower copper contents (99.50% min.) are common. Thus, any copper casting produced with a view to the development of maximum electrical conductivity will automatically satisfy the requirements of thermal conductivity as well.

Many of the impurities likely to be present in copper lower its electrical conductivity seriously. The commonly found impurities are residues from the deoxidation treatment which are ordinarily soluble in copper and mostly have the effect of depressing the conductivity values. Conductivity is decreased by any element which enters into solid solution with copper. Even silver which has a higher conductivity than copper will reduce the overall conductivity slightly when added to it. The accompanying list gives some idea of the influence of a number of elements, which are commonly present as impurities, in lowering the electrical conductivity.

0.1%	Aluminium lowers electrical conductivity to	85%
0.1%	Antimony ,, ,, ,, ,,	90%
0.1%	Arsenic ,, ,, ,, ,,	75%
0.1%	Beryllium ,, ,, ,, ,,	85%
1.0%	Cadmium ,, ,, ,, ,,	90%
0.1%	Calcium ,, ,, ,, ,,	98%
0.1%	Iron ,, ,, ,, ,,	70%
0.1%	Magnesium ,, ,, ,, ,,	94%
0.04%	Manganese ,, ,, ,, ,,	95%

0.1%	Nickel	,,	,,	,,	,,	95%
0.1%	Phosphorus	,,	,,	,,	,,	50%
0.1%	Silicon	,,	,,	,,	,,	65%
1.0%	Silver	,,	,,	,,	,,	97%
1.0%	Tin	,,	,,	,,	,,	55%
1.0%	Zinc	,,	,,	,,	,,	90%

There are many applications which require a relatively high combination of tensile strength or hardness and electrical conductivity. Until recent years, such combinations of properties were not available, and applications requiring high conductivity were met as well as possible by the use of pure copper castings. With the development of the precipitation hardening process, it became possible to obtain high combinations of strength and conductivity together with casting characteristics much superior to those possessed by pure copper. These heat treatments usually involve heating the casting to a temperature in the neighbourhood of 900°C for at least one hour, followed by quenching to room temperature and re-heating for one to 5 hours to about 500°C.

Some High Conductivity Copper Alloys

Silver—Copper (0.03%–0.1% Ag)	Rarely used in cast form. Supplied as wrought material for electrical parts, photogravure plates, etc.
Cadmium—Copper (0.7–1.0% Cd)	Mainly used in form of drawn wire for electrical transmission wires
Chromium—Copper (0.4–0.8% Cr)	Responds to heat-treatment, producing castings of high strength
Tellurium—Copper (0.3–0.7% Te)	Free machining copper alloy for castings or bar stock for repetition machining

Apart from the adverse effects of impurities upon conductivity, soundness and freedom from porosity in the castings must also have serious consideration. Unfortunately, both hydrogen and oxygen have a strong affinity for copper. The former gas dissolves in it, and the latter combines to form cuprous oxide which is soluble in molten copper. When hydrogen and oxygen are both present, gross unsoundness in the form of widespread blowholes is likely to develop from what is commonly referred to as "steam reaction".

As the molten copper cools in the mould its power of holding hydrogen in solution lessens and some hydrogen is expelled. This nascent or atomic hydrogen is very active and at the high temperature involved reacts with any cuprous oxide present to form water vapour. Such an occurrence is often accompanied by visible signs such as the "backing up" of runner and riser heads to form "cauliflowers".

It is apparent, therefore, that during melting, steps must be taken to exclude both hydrogen and oxygen from high conductivity melts. In practice, however, when melting conditions are ideal, it is found that deoxidisation alone is sufficient to prevent the "steam reaction". The need to remove excessive hydrogen from a melt may only arise when the charge consists of cathode copper, which by the nature of its manufacturing process, contains much dissolved hydrogen.

GENERAL MELTING PROCEDURE

Treatment

Melt down under a reducing cover of CUPRIT 8. Deoxidise first with DEOXIDISING TUBES DS containing phosphorus and finally plunge DEOXIDISING TUBES CB containing calcium boride; or DEOXIDISING TUBES L containing lithium.

Quantity Required per 50 kg Melt

0.5 kg of CUPRIT 8.
DEOXIDISING TUBES DS.
DEOXIDISING TUBES CB or L.
   (See sub-section—Deoxidation Techniques.)

Recommended Casting Temperatures

Light castings—under 13 mm section.	1250°C.
Medium castings—13–38 mm section.	1200°C.
Heavy castings—over 38 mm section.	1150°C.

Deoxidation Techniques

With few exceptions, the deoxidation of high conductivity copper alloy is best carried out by, first, phosphorus to remove most of the dissolved oxygen and finally by a deoxidant based on calcium boride or lithium. The latter deoxidants ensure adequate deoxidation while at the same time any small residual quantities will not adversely reduce conductivity.

Method of Determining Quantity of Deoxidant Required

Owing to the varying degrees of oxidation from one melt to another the quantities of deoxidant plunged before pouring will vary accordingly. Since it is only necessary to add sufficient deoxidant to prevent the "steam reaction", precise recommendations cannot be given. Instead, a series of simple tests should be made in the foundry which involve observing the solidification characteristics of a given melt before any of the castings are poured.

In a standard moulding box, prepare some simple open-topped cylindrical mould impressions measuring about 75 mm high by 50 mm diameter. When the melt is ready for pouring, but before deoxidation, a sample of the copper should be ladled into one of the moulds and the test piece allowed to solidify. If the head rises appreciably as shown in Fig. 5.1a, very gassy metal is indicated. DEOXIDISING TUBES DS containing phosphorus must then be plunged and further test castings made. At the point when the quantity of phosphorus added results in a shallow sink in the head as shown in Fig. 5.1b, it can be assumed that the residual phosphorus content of the melt is nil but that approximately 0.008% of oxygen remains to be removed.

Deoxidation should now be completed by plunging DEOXIDISING TUBES CB or L, adding sufficient to produce a test casting which exhibits a head with a deep sink as shown in Fig. 5.1c. At this stage, the melt is in a condition for producing castings free from porosity.

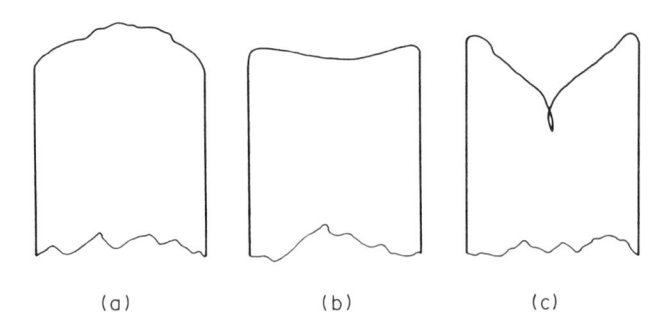

Fig. 5.1

Degassing

If a separate degassing treatment using LOGAS 50 is considered necessary, then the briquettes should be plunged into the melt prior to deoxidation and not afterwards.

Degassing by means of LOGAS 50 is not practical for melts of less than 100 kg capacity owing to the fact that melts of small size would drop appreciably in temperature during the degassing process. For degassing a 100 kg melt, one size 1a or one size 1b LOGAS 50 briquette is sufficient.

Degassing and Deoxidation Techniques for High Conductivity Copper Alloys

Since most high conductivity copper melts are of copper alloy, as distinct from pure copper, it is important to observe degassing and/or deoxidation procedures given in the following notes. This applies particularly when preparing the alloy from virgin materials.

Silver-Copper

Silver additions should be made in the form of copper–silver master alloy and introduced into the melt prior to deoxidation but after a degassing operation if the latter is carried out. The dual deoxidation process is recommended, plunging DEOXIDISING TUBES DS followed by DEOXIDISING TUBES CB or L.

Cadmium-Copper

Degassing, if necessary, and a full deoxidation treatment by the dual process must be completed before making cadmium additions. Cadmium is often introduced by tapping the molten copper directly on to pure cadmium metal, on transferring the melt from the furnace to a pouring vessel. In crucible melting practice the use of copper–cadmium master alloy may be more convenient and result in lower cadmium losses when added. Ventilation of the foundry should be adequate since molten cadmium evolves toxic brown fumes.

Chromium-Copper

Chromium additions in the form of copper–chromium master alloy should be made after a degassing operation, if carried out, but before full deoxidation by the dual process. The chromium alloy should be stirred thoroughly into the melt to ensure a homogeneous solution of chromium. A chromium loss varying between 10–30% should be expected depending on the state of oxidation of the melt and alloying techniques. Normally, the chromium additions and a final deoxidation by calcium boride or lithium will deoxidise adequately. Phosphorus additions should only be made if a test casting shows a rising head on solidification. Phosphorus additions must then be limited to that quantity which produces a level sink; relying upon the chromium additions and CB or L tube to complete deoxidation. An excess of residual phosphorus would upset response of the alloy to heat treatment.

Tellurium-Copper

Degas the melt first if this is considered necessary then introduce tellurium in the form of copper–tellurium–master alloy. Finally deoxidise by the dual process using DEOXIDISING TUBES followed by DEOXIDISING TUBES CB or L.

COMMERCIAL COPPER CASTINGS

There is frequently a demand for copper of commercial quality in which a degree of conductivity is sacrificed in the interests of good casting qualities and for ease of machining. These copper alloys may contain tin and/or zinc up to 2%. For these it is possible to adopt a less exacting fluxing and deoxidising routine in comparison to that necessary for a high conductivity copper alloy.

For commercial quality copper, melting practices as carried out with bronze and gunmetal procedure are usually adopted. That is to say melting is carried out under oxidising conditions and deoxidation performed by means of plunging phosphorus. It is not necessary to employ the duplex deoxidation technique as previously recommended for high conductivity copper. Therefore, prescribed quantities of phosphorus can be recommended for a given melt and, of course, it is unnecessary to check the degree of deoxidation by means of test castings as previously described.

Treatment

Melt down under an oxidising cover of CUPREX blocks. Just before pouring, deoxidise by plunging DEOXIDISING TUBES DS.

Quantity Required per 50 kg Melt

0.5 kg CUPREX 1 block.
Three DEOXIDISING TUBES DS4 (to be plunged in two stages if desired).

Recommended Casting Temperatures

Light castings—under 13 mm section.   1250°C.
Medium castings—13–38 mm section.   1200°C.
Heavy castings—over 38 mm sections. 1150°C.

Degassing

Refer to previous remarks in melting procedure for high conductivity copper.

MOULDING PROCEDURE FOR HIGH CONDUCTIVITY AND COMMERCIAL COPPER CASTINGS

General Recommendations

With the exception of chromium copper, copper–nickel–silicon, and beryllium copper alloys, moulding procedure should follow the practice generally adopted in brass and bronze foundries. A mould permeability of at least 25 AFS is desirable in order to avoid blow holes. Green or dry sand moulds may be used but for large castings, over one inch in section, dry sand moulds are preferable.

Where castings are of complex shape or incorporate severe sectional changes, moulds and cores must not be excessively strong because some of the copper alloys are weak at the point of solidification, and tend to crack.

Gating Practice

Every effort should be directed towards obtaining the minimum turbulence on pouring. The illustrations below give some idea of the essential requirements for running and gating high conductivity or commercial copper castings.

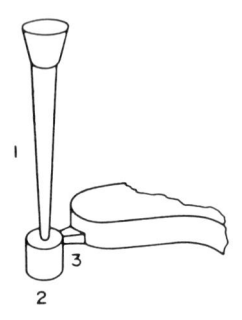

Fig. 5.2. SINGLE INGATE—A
The simplest form of gating for small to medium size castings.
1. Tapered sprue to reduce formation of air bubbles.
2. Deep basin to receive first turbulent impact of metal.
3. Ingate tapering out to reduce metal velocity. Note position of ingate at top of sprue basin.

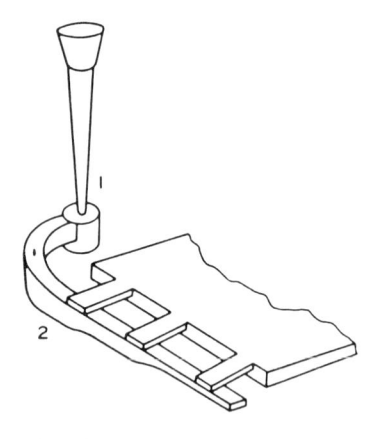

Fig. 5.3. MULTIPLE INGATE—B
For castings of large surface area where more than one ingate helps to fill mould uniformly.
1. Sprue and basin as in Example A.
2. Progressively narrowing runner to reduce metal velocity; this assists in preventing dross formation. Note extension of runner which receives first impact of metal stream, thus acting as dross trap.

COPPER-ZINC ALLOYS—BRASSES

General Notes

The term "brass" should be applied to alloys of copper in which zinc is the other major element. Casting brasses contain copper between the limits 56–76%. Apart from up to 5% in total of other metals, such as tin, lead, iron, and nickel, the remainder is zinc.

Zinc alloyed with copper, within the range indicated increases the tensile strength and hardness but reduces ductility. This is shown in the following table.

Mechanical Properties of Copper-Zinc Alloys—Sand Cast

Zinc %	Alpha %	Tensile tons/in^2	Tensile kg/mm^2	Elongation % on 2 in (50 mm)	Brinell Hardness
33.7	100	15.0	23.6	63	55
35.0	88	19.1	30.1	59	59
38.0	75	21.5	33.9	56	64
40.1	69	23.0	36.2	55	72
41.8	55	24.6	38.7	50	80
43.7	14	27.8	43.8	27	102
45.5	0	30.4	47.9	27	105

Billets for cold rolling have zinc contents between 27% and 38%. Lower zinc values from 5% to 20%, with balance copper, constitute the alloys known as "gilding metals". In these, other constituents must be at a low level for they are virtually simple alloys of zinc with copper.

Some brass compositions have become known by special names such as:

Dutch metal   78/22 Copper–zinc.
Gong metal   76/24 Copper–zinc.
Pinchbeck   93.6/6.4 Copper–zinc.
Tobin bronze 60/38/2 Copper–zinc–tin (average composition).
Tombac bronze 80/17/3 Copper–zinc–tin (average composition).

As is seen from the table referred to previously, with zinc at about 34% the brass has a homogeneous structure made up of a single constituent called alpha. By adding more zinc a second constituent begins to appear when prepared specimens are examined under the microscope. It is called beta and increases in quantity as the zinc is raised up to 46%. At this proportion of zinc all the alpha is replaced by beta.

Therefore, brasses are often classified according to their micro-structure. Those with zinc up to 34% are known as alpha brasses, whilst within the range 34–46% zinc are contained the alloys with a duplex structure referred to as alpha–beta brasses. Beyond 46% zinc the alloys again assume a single-phase solid solution structure and form the series of beta brasses. Actually the range of beta brass is very narrow, about 46–51% zinc, so that beyond this limit a duplex structure returns and a new phase makes its appearance, namely gamma.

The gamma solid solution renders the alloy brittle and these compositions have little engineering use except for the 50/50 copper–zinc alloy which is used for brazing.

Melting Technique (Sand Castings)

The brasses can be melted successfully in any of the conventional type furnaces, crucibles accounting for a high proportion of metal melted in foundries all over the world.

Zinc vapour pressure is sufficient to prevent the ingress of hydrogen into the metal and therefore a neutral or even slightly reducing atmosphere is not deleterious. An oxidising atmosphere should be avoided since this would cause heavy melting losses due to the oxidation of zinc.

The molten alloy is at a temperature much above the boiling point of zinc and unless a suitable covering flux is used zinc will be lost both by direct oxidation and also volatilisation. Additionally, the flux must be capable of absorbing and holding dirt and oxides without becoming so pasty that it will hold heavy quantities of metal.

As in the case of all metals and alloys, the charge should be melted as rapidly as possible and not be allowed to overheat. "Stewing" the melt will result in increased metal loss and poor quality castings.

Deoxidation, in the normally accepted sense of the word, is not applicable to the brasses. The presence of zinc guards against the oxidation of copper and the oxides of zinc and aluminium (if present) are removed by the relevant fluxes.

Nevertheless, the addition of phosphorus by means of DEOXIDISING TUBES DS is a desirable practice in the case of cast brasses, since it will increase the fluidity of the metal, thus allowing the running of thin sections without unduly superheating the charge.

Procedure

1. Heat up the crucible or furnace.
2. Place in the bottom of the empty pre-heated unit a quantity of CUPRIT 1 blocks equal to 1 kg per 100 kg of the metal to be melted.
3. Commence charging the ingots and scrap and melt down as rapidly as possible, maintaining the surface cover of fused flux intact.
   (N.B.—Where CUPRIT 49 powder is used, this should be introduced at the same rate as soon as the first part of the charge reaches a pasty condition.)
4. Bring the metal up to pouring temperature and avoid overheating. When the metal is ready for pouring, the moulds should be ready to receive it.
5. Immediately prior to pouring, the metal should be treated with DEOXIDISING TUBES DS at the rate of one DS2 Tube per 50 kg by plunging the requisite number of tubes deep into the heart of the melt and holding immersed for a few seconds.
   (N.B.—The plunging tool must be pre-heated and should be coated with FIRIT 20 to prevent contamination of the melt and to protect the plunger.)
6. When the metal is at the correct pouring temperature the surface slag should be held back and the metal poured from underneath it. This will curtail "flaring" to the minimum and the unpleasant fuming associated with brass will be almost entirely absent.
7. Recommended pouring temperatures are as follows:

	Under 13 mm section		13–38 mm section		Over 38 mm section	
	(°C)	(°F)	(°C)	(°F)	(°C)	(°F)
Commercial brasses	1150	2102	1100	2012	1070	1958
60/40–65/35 alloys	1100	2012	1050	1922	1020	1868
80/20–70/30 alloys	1150	2102	1100	2012	1070	1958

For charges melted in reverberatory and similar hearth furnaces, CUPRIT 1 blocks are specially recommended. Powder type fluxes are likely to be carried away by the draught and blast from the burners.

Die-castings

The alloys for die-casting are BS 1400:1969, DCB1 and DCB3, both of which contain aluminium. Because of the presence of aluminium a flux

capable of dissolving aluminium oxide gives best results and ALBRAL 3 powder is used as follows:
1. Pre-heat the crucible or furnace.
2. Charge in the ingots and scrap and commence melting.
3. When the first part of the charge reaches a pasty stage sprinkle over its surface a quantity of ALBRAL 3 equal to 500 g per 50 kg of metal to be melted.
4. Continue charging and melt as rapidly as possible under the protective cover.
5. When the metal has been brought up to pouring temperature, take a fresh quantity of ALBRAL 3, 500 g per 50 kg and, with a perforated saucer plunger, plunge the flux slowly to the bottom of the melt and then with a rotary movement of the plunger "wash" the flux in, bringing it into intimate contact with all parts of the melt in order to cleanse it of alumina particles.
N.B.—The plunger should be coated and pre-heated as described earlier.
6. After a "washing" period of 2 or 3 minutes, withdraw the plunger and allow the melt to settle for a minute or so.
7. The metal should now be treated with DEOXIDISING TUBES DS at the rate of one DS2 tube per 50 kg by plunging the tube as described previously.
8. As soon as the metal is at correct working temperature—1100°C (2012°F)—it may be ladled out as required, pushing the surface dross aside to leave a clean working area.
9. From time to time, after scrap and ingots have been added to maintain the metal level, the dross should be skimmed away from the surface and fresh ALBRAL 3 should be washed into the metal as a periodic cleanser. Similarly, DEOXIDISING TUBES DS should be plunged occasionally in order to maintain maximum fluidity. A very considerable improvement in surface finish of gravity die-castings has been reported following the substitution of DS tubes by MG tubes. The addition of a small amount of magnesium to die casting brass apparently improves fluidity and also conditions the surface oxide film to give a much smoother finish to the casting. One MG6 tube should be added to 150 kg approximately of metal, repeating as indicated by results.

Where multiple piece dies are used they are taken apart and dipped in a water suspension of DYCOTE 61 after each pour. In this way the dies are kept cool and the tendency for build up to occur on the face of the dies is reduced. The steel cores are also dipped in the DYCOTE 61 suspension after each cast. Alternative coatings for the die blocks are DYCOTE 36 or 11.

The brass alloy is maintained at 1100°C (2012°F) in the furnace and poured into the dies at about 1050°C (1922°F).

Effect of Added Elements

ALUMINIUM.—Unless added for a definite purpose, as in die-casting brass, it should be absent. Although it gives fluidity and improved definition, it makes founding more difficult in that oxide inclusions and films of alumina tend to cause porosity and unsoundness in castings.

IRON.—In general, small quantities exert a grain refining action. It also has a tendency to increase hardness and tensile strength.

LEAD.—Increases the machinability but must not be allowed to segregate. Lead is insoluble in brass and exists, therefore, as globules. A dispersion of these in as uniform a manner as possible is required.

MANGANESE.—Sometimes used as a deoxidant, its effect is similar to that of iron.

NICKEL.—Improves the mechanical properties and increases resistance to corrosion. Also has a tendency towards grain refinement.

PHOSPHORUS.—Tends to combine with any iron present and increases the hardness. Reduces grain growth and improves fluidity.

SILICON.—Makes founding more difficult but improves corrosion resistance, especially inter-granular attack.

TIN.—Raises the tensile strength and hardness at the expense of ductility but improves corrosion resistance and fluidity.

Other elements sometimes found in small proportions in brass are antimony, arsenic and bismuth. They are regarded as detrimental and should be kept below 0.01% preferably. An exception is made sometimes when arsenic is added up to 0.05% to inhibit dezincification.

Removal of Impurities

1. ALUMINIUM: In sand cast brass, the major impurity encountered is aluminium and this may be removed quite easily by the use of ELIMINAL. It is well-known that in sand castings as little as 0.02% of aluminium is detrimental, since the presence of aluminium oxide films weakens the structure of the metal, causing porosity and leakage under pressure.

Aluminium is fairly easily discernible in molten brass and also in the casting. Even in small quantities it forms a scum on the surface of the molten metal, and in the casting its presence is shown by a silky silvery sheen which usually appears on the extremities.

Up to 0.5% of aluminium may be removed from brass by means of ELIMINAL. If the contamination is heavier than 0.5%, the amount of aluminium should first be reduced by adding metal known to be free from the element.

2. SILICON: ELIMINAL will also remove silicon from molten brass and the treatment is precisely the same as for the removal of aluminium.

Moulding Practice

The patternmaker's contraction allowance for brass is 1.3% and the liquid shrinkage may be taken as 6–7% by volume.

Brasses may be cast easily in either green sand or dry sand moulds, but the latter is normally used for only the heaviest castings. No difficulty should be experienced in running thin sections, provided the metal is free from non-metallics and that maximum fluidity has been obtained by correct metal treatment.

Normal moulding sand of the natural clay bonded type is quite suitable. Moulds should have good permeability and should be well vented to allow the escape of steam, air, and mould gases.

Mould Dressings and Sand Additions

Pin-hole porosity is quite prevalent in brass and recent work has shown that this can largely be prevented by the use of suitable mould dressings and by the control of additions to the facing sand. This pin-hole porosity is generally revealed when the surface of the casting is polished prior to plating operations, and it occurs particularly in the sections which require the higher pouring temperatures.

Wide use has been made of wheat flour and wheat flour substitutes but, although these materials provide a smooth surface on the casting, they encourage sub-surface pin-hole porosity due to a metal-mould reaction which occurs with the wheat flour.

0.5% pelletted pitch or 2% TERRADUST 2 added to the facing sand gives an excellent surface to the casting without the danger of pin-hole porosity.

Self-drying or fired dressings on the surface of the mould will also produce an excellent finish free from pin-holes and the use of either MOLDCOTE 6, MOLDCOTE 9, or MOLDCOTE 11 is recommended, particularly for green sand and silicate bonded moulds.

Running, Gating, and Feeding

The running of brass castings does not present any real problem. Excessive turbulence in the mould should be avoided, but there is no need for elaborate runner systems as in the case of high tensile brass.

NON-FERROUS CASTING ALLOYS 229

Generous feeder heads should be provided over heavy sections and the efficiency of these feeder heads may be greatly improved by the use of FEEDEX, FEEDOL, and KALMIN 44.

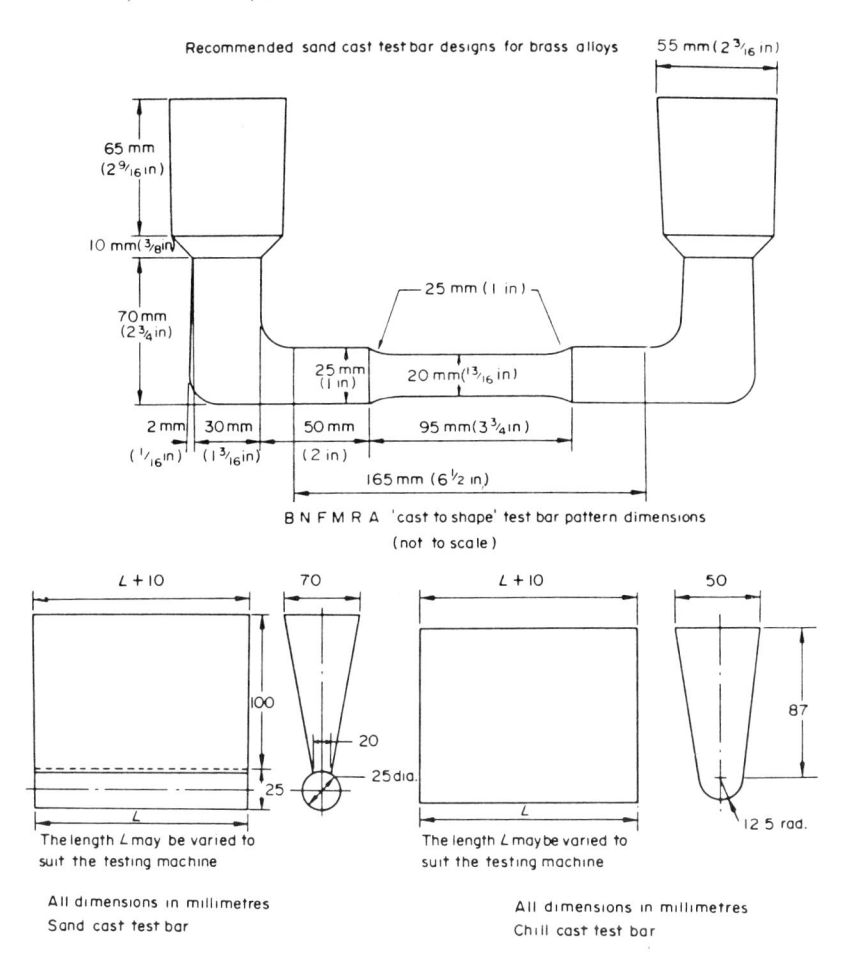

Fig. 5.4.

COPPER ALLOYS—BRONZES AND GUNMETALS

GENERAL NOTES

The original bronze alloys were of copper with tin as the main alloying element. Such a wide variety of bronze and gunmetal alloys now exist, however, that it has become necessary to classify them according to their basic chemical composition. The following examples are given and on pages 201–206 some current British Specifications can be seen.

TIN BRONZES:—Copper-based alloys having tin as the main alloying element. Alloys having between 3% and 7% tin are not normally produced as castings but in billet form for subsequent drawing or rolling operations. Alloys containing between 15% and 20% tin are widely used for the casting of bells. As the tin content reaches this magnitude, the alloy becomes increasingly harder and difficult to machine.

PHOSPHOR BRONZES:—Copper based, having from 5% up to 10% tin as the main alloying constituent together with an intentional phosphorus addition which may vary from 0.5 to 1.0%. These alloys are used to make what are widely known as phosphor–bronze bearings. Tin provides the necessary toughness for the alloy whilst phosphorus additions give castings the necessary hardness and therefore, wear resistance.

GUNMETAL:—Copper based, with tin and zinc as the alloying constituents. Tin usually predominates and an example of this type of alloy is Admiralty gunmetal, 88% copper, 10% tin, 2% zinc. This alloy is used for a variety of purposes, ranging from casting of gears and bearings to hydraulic castings and castings subjected to moderate stresses.

LEADED GUNMETAL:—Copper based with tin, zinc, and lead present in varying proportions. Numerous alloys of this type exist, a widely known example being 85/5/5/5 alloy, in which tin, zinc and led are present in equal proportions. These alloys are extremely popular thoughout the world and are used for a number of purposes, mainly in the casting of hydraulic components, plumbers' fittings, etc. They are also used as bearing alloys operating under moderate loads.

LEADED BRONZE:—Copper–tin, or copper–tin–phosphorus alloys containing from 5–15% lead. These alloys are used almost exclusively for casting heavy duty bearings, in which wear resistance combined with the ability to undergo plastic deformation is required.

BRONZES AND GUNMETALS—MELTING PRACTICE

Effect of Gas Absorption

Bronze and gunmetal alloys are usually melted in coke, gas or oil fired crucible furnaces. Many foundries also employ small oil or gas fired reverberatory furnaces and, in some instances, electric melting by means of an arc or induction furnace is employed.

In the molten condition, these alloys may come into contact with gases which include oxygen, nitrogen, water vapour, hydrogen, carbon monoxide, carbon dioxide, and sulphur dioxide. Carbon dioxide is generally considered to be inert to molten bronze but carbon monoxide and sulphur dioxide have often been reported as minor sources of porosity; the effect of sulphur as metallic sulphides is slight.

Hydrogen is the chief source of trouble for the bronze or gunmetal founder and this may be derived from the products of incomplete combustion during melting, or from water vapour in the atmosphere. Hydrogen is not so soluble in copper alloys as in pure copper, but this gas can cause severe porosity when the gas is ejected during solidification of a casting.

The Steam Reaction

Water vapour reacting with molten copper forms cuprous oxide and hydrogen, both of which are dissolved in the melt. During solidification hydrogen is ejected and reacts immediately with cuprous oxide to form steam. During solidification of a casting the steam evolved causes severe porosity. It is suggested that this reaction is the main cause of porosity in bronze, except that the reacting oxide is oxide of tin.

If this is so then by eliminating both oxygen and hydrogen from a melt the formation of porosity should be prevented. For this reason, it is often advisable to both degas and deoxidise a bronze or gunmetal melt before casting.

THE OXIDATION-DEOXIDATION MELTING PROCESS

If during melting oxidising conditions are maintained, this reduces pick-up of hydrogen gas to a minimum because relatively large volumes of both hydrogen and oxygen cannot exist together as shown in the graph (over page). By maintaining a slightly oxidised melt, therefore, hydrogen pick-up can be prevented and it is only necessary to deoxidise a melt before pouring to ensure freedom from both hydrogen and metallic oxides. This method of melting is known as the oxidation-deoxidation process.

Oxidising conditions are achieved in two ways: by employing an oxidising covering flux such as CUPREX and by adjusting the furnace atmosphere to slightly oxidising burning conditions. Further details are given below.

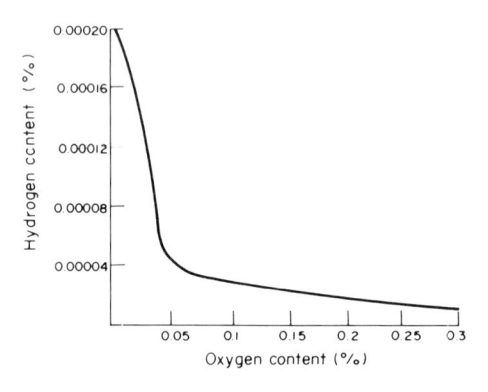

Fig. 5.5.

Degassing

The degassing of bronze or gunmetal melts by means of bubbling inert gases through the melt is widely practised and nitrogen is often used for the purpose. There are disadvantages with nitrogen, however, particularly with respect to the difficulty of handling gas cylinders in a production foundry. Metal tubes for injecting the gas into the metal cannot be used and even the best non-metallic tubes are fragile. Again, in order to overcome the resistance of pressure in deep ladles or crucibles, a fast flow of nitrogen is required. It is generally assumed that nitrogen delivered from a cylinder is dry, provided that pressure does not fall below 20 atmospheres. However, moisture determinations on commercial nitrogen have shown that the moisture in a full cylinder was 0.32 milligrams per litre and 2.7 in a cylinder three-quarters empty. Therefore, 50 litres of nitrogen will contain 80 ml of water in vapour form at a temperature of 1100°C for a full cylinder and approximately 650 ml for a three-quarters empty cylinder.

Alternative Degassing Methods

Alternatively LOGAS 50 briquettes may be plunged into the melt using a suitably protected plunging tool. These briquettes evolve inert gases at the temperatures involved, and degassing—as with nitrogen—is accomplished by a process of diffusion into the rising bubbles. This system is more robust than degassing with nitrogen gas, has none of the limitations, and is equally effective.

A degassing method based on scavenging chemicals is also available. A powdered product known as REGENEX is plunged and well "scrubbed" into the metal with a protected bell-plunger. This method is quick and convenient but may not be the most suitable where very low residual gas levels are required or where close chemical specifications must be held.

Melting Technique-Sand Castings

1. Preheat the crucible or furnace.
2. Place in the bottom of the hot, empty crucible sufficient CUPREX 1 blocks using 1 kg per 100 kg or 1% of the charge weight.
3. Charge in the ingot metal and/or scrap. Alternatively, CUPREX 100 powder can be added with the charge, or at an early stage of melting, again using the 1% ratio.
4. Melt and bring to pouring temperature as rapidly as possible. During melting CUPREX evolves oxidising and scavenging gases which combine with hydrogen and remove it from the melt. An oxidising flux cover is also formed which protects the melt from further hydrogen absorption.
5. Immediately before casting, deoxidise the melt thoroughly by plunging DEOXIDISING TUBES DS using one DS3 tube per 50 kg of melt. The object of this is to introduce enough phosphorus to give a residual phosphorus content of 0.02% which signifies adequate deoxidation and maximum fluidity.
6. If the charge materials contain scrap which is very oily or dirty a large quantity of hydrogen will find its way into the melt. Thus extra degassing may be necessary in such circumstances; also in the case of special castings required to withstand relatively high internal pressure or to be specially sound and free from porosity. Then, before using the DEOXIDISING TUBES, plunge LOGAS 50 briquettes or REGENEX 8 using 100 g to 250 g per 50 kg of melt.
7. Check the correct pouring temperature as given in the accompanying table, skim and cast without delay, taking care to prevent slag entering the mould cavity.
8. Molten bronze should never be held in the furnace waiting for moulds to be ready, the moulds must be prepared in advance to receive the metal as soon as its melting and fluxing treatment is completed.

Reverberatory Furnace Melting

CUPREX 1 blocks are especially suited for fluxing gunmetal and tin-bronze charges melted in open flame reverberatory type furnaces such as the Sklenar. One obvious advantage of the flux in block form is that, unlike powder flux, it is not carried away by the hot air currents and blast pressure from the oil burners.

The CUPREX blocks are placed on the hearth of the furnace in a ratio of 1% of the charge weight and the ingot metal and scrap charged on top of them. Their action is to evolve oxidising and scavenging gases which rise through the melt, cleaning it and removing hydrogen. As already indicated, in the final stages they melt to form a protecting slag cover which guards the molten charge from hydrogen pick-up from the furnace atmosphere. This is an important function considering the relatively large surface area of melt exposed to the furnace gases in shallow hearth furnaces.

Recommended Casting Temperatures

Copper-tin-zinc-lead	Under 13 mm section (°C)	13–38 mm section (°C)	Over 38 mm section (°C)
83/3/9/5	1180	1140	1100
85/5/5/5	1200	1150	1120
86/7/5/2	1200	1160	1120
88/10/2	1200	1170	1130
90/10/(phosphor bronze)	1120	1100	1030

Metal-Mould Reaction

Molten metal poured into a sand mould is immediately confronted with a steam atmosphere which is generally oxidising to the metal. In this way the oxygen becomes fixed and the oxide skin formed is, in most cases, tough and impermeable, thereby stifling further reaction. However, with certain alloys the skin of oxide does not prevent further reaction.

Occurrence of metal-mould reaction depends on the presence of highly reactive elements, the most important of which are magnesium and phosphorus. It occurs also in both aluminium and copper-base alloys which contain small amounts of boron. Substantial metal-mould reaction is produced also in copper-base alloys which contain lead and silicon present together.

The water vapour formed when the molten metal enters the mould, having given up its oxygen to form an oxide skin, releases a quantity of hydrogen. Such "newly born" hydrogen is very active and can enter into solution with the molten metal unless the oxide skin formed on it is very strong and protective.

In copper-base alloys which contain phosphorus the reaction does not form an oxide but probably a phosphate skin which remains liquid at temperatures near the solidification point of the alloy. This liquid does not provide a sufficient barrier on the alloy surface to stifle the reaction. Similarly, in phosphate-free copper-base alloys which contain lead and silicon, the oxidation of these two elements at the metal-mould interface produces lead silicate which acts similarly to phosphate.

With lead or lead-free gunmetal containing zinc, a phosphorus content of 0.03% or above is enough to cause the reaction. A residual phosphorus content of 0.06%–0.08% is usually sufficient to produce an appreciable effect. In general, the degree of gas absorption increases as the pouring temperature is raised and vice versa. For gunmetal alloys and bronzes the reaction can be prevented by coating the moulds and cores with MOLDCOTE 13 which contains a spirit suspension of aluminium.

Metal-mould reaction can sometimes be used to advantage for certain pressure-resisting bronze castings such as taps and valve bodies. The idea is to impart a controlled gassing which serves to disperse concentrated liquid shrinkage and distribute it as uniform microporosity throughout the mass of the casting. To accomplish this effectively the alloy must be gas free as cast and contain sufficient phosphorus to set up a limited metal-mould reaction—this is dependent on the casting section.

Pin-holes

The occurrence of these on the surface of bronze castings which are required to be bright plated is a not uncommon source of trouble. Investigations carried out by the BNFMRA* have shown the cause as a form of metal-mould reaction due to the presence of wheat flour used as an additive to the facing sand or as a mould dressing.

Whilst the defect occurs on thin leaded gunmetal castings, made in such moulds and cast at temperatures above 1160°C, it is much less prevalent in lead-free alloys. Phosphorus contents of the order of 0.08% cause an increase in the amount of pin-holing.

Pin-holing can be reduced by omitting cereal flours from the sand or mould facing and is prevented by spraying with MOLDCOTE 13. These measures may produce castings with a rougher surface than when using flour but this may be improved without risk of pin-holing by adding 0.5% or more of pelletted pitch to the facing sand.

Gravity Die-castings

In melting down the initial bale-out charge, CUPREX 1 blocks should be employed on a basis of 1% of the charge weight. Thereafter melting will be

* Now known as the British Non Ferrous Metals Technology Group.

continuous with fresh ingot metal and returned scrap being added from time to time to maintain the bath of molten alloy. REGENEX should be plunged at appropriate intervals to keep the liquid metal clean and of low gas content.

Following the degassing operation, strong deoxidation is carried out by plunging one DEOXIDISING TUBE DS4 to each 25 kg of melt. This also has the effect of promoting fluidity to the maximum degree. Pouring temperatures range from 1150° to 1200°C and the die should be of heat-resisting cast-iron or steel to stand up successfully.

DYCOTE 36 is recommended for spraying the die blocks, which should be sunk with as much taper as dimensional tolerance will allow, otherwise extraction difficulties will become a major problem.

Where dies are dissembled after each cast and are quenched into a dispersion of lubricant in water, DYCOTE 61, 11, or 38 are the products to use. DYCOTE 11 and 38 are aqueous dispersions of graphite and should be diluted before use. DYCOTE 61 is similar but contains additions of chemicals designed to inhibit zinc oxide build-up on the die face. With all three products, dilution rates of the order of 1 part product to 15 parts of water are common, while for simple castings the proportion can often be increased to 1:30 or even more.

Eliminate any tendency to build up back pressure by cutting venting grooves in the die at as many points as design permits.

Before casting commences, always pre-heat the dies to a working temperature of 350°–400°C with the idea of limiting thermal shock and blowing.

Removal of Aluminium

Aluminium is a common and very deleterious impurity in sand cast gunmetals and bronzes. As little as 0.01% is enough to cause leakage of pressure type castings. This is because the aluminium oxidises and forms tough films and "stringers" which become trapped in the solidifying castings. The presence of the oxide films also interferes with self-feeding properties.

Fortunately, it is not unduly difficult to remove aluminium in quantities up to about 0.5% from molten bronze and gunmetal by means of ELIMINAL.

Founding Properties

Where maximum soundness and pressure-resisting properties are called for the leaded gunmetals are to be preferred. Compositions such as 86/7/5/2 and 85/5/5/5 are best for the production of sound and strong pressure castings for use at atmospheric temperatures. They give maximum pressure tightness coupled with ease of handling in the foundry.

Nickel Additions

Part of the tin content of these alloys can be replaced by nickel when the mechanical properties and corrosion resistance are improved without sacrifice of any of the good founding qualities.

Nickel additions of 1 to 3.5% to pressure bronze castings add measurably to density, tensile yield strength and toughness, especially for tin contents of 2% or more. In addition to grain refining and densening, nickel bolsters the strength of compositions in the low tin range and materially improves their performance at elevated temperatures. When nickel additions to bronzes exceed 4% the alloys are susceptible to age-hardening.

If a nickel addition is made to an alloy, the fluidity at any given temperature is affected and a suitable increase in the pouring temperature should be made.

Running and Feeding

The literature on running and feeding bronzes and gunmetals is extensive. In general those methods best suited to long freezing range alloys should be used, and the most commonly used systems appear to be the non- or only slightly pressurised systems based on a 1:4.1 type of cross-sectional area ratio. Besides giving a good skimming effect, this type of downsprue/runner bar/ingate system can provide a worthwhile source of feed metal to the casting as long as the gate remains unfrozen. Indeed, many thousands of castings (shell mouldings in particular) such as taps, valves, cocks, etc., are made in this way without any supplementary form of feeding.

Where additional feed to that obtained from the running system as outlined above is required, quite generous risers should be provided. Because of the long freezing range and pasty method of solidification, risers generally will have to be of greater section than the part of the casting they are feeding. Again, considerable economies can be obtained by utilising FEEDEX exothermic or KALMIN 44 insulating riser sleeves. These will effectively reduce heat loss through the walls of the riser, but having closed off that route it becomes increasingly important that loss to atmosphere is controlled as much as possible. All risers should therefore be covered with a layer (about 13 mm thick) of FEEDOL 16 or 17 immediately on completion of casting. It is better if the risers are short-poured slightly so that the optimum effect from the anti-piping compound is obtained.

CASTING OF CORED AND SOLID BRONZE STICKS

The first part of these notes is concerned with the casting of sticks up to 1 m in length and from 25 to 150 mm in diameter. This type of casting formerly formed a large part of the bronze foundry industry since it provided the starting point of the large extrusion and rolling sectors. While some cored stick is still made by these methods, the process has been very largely superseded by continuous and semi-continuous casting methods. These newer techniques demand much more capital outlay and technical process control, but the end product is far more consistent and reliable.

Moulds

Split moulds with spigot-joints, or one piece cannon moulds are both in general use, but the former type are preferred owing to the ease with which their surfaces can be cleaned and dressed. Moreover, unlike cannon moulds, which have to be tapered in order to make withdrawal of the casting easy, a split mould produces a parallel casting. A suitable dressing for applying to a warm mould is INGOTOL 16, but for the best results, particularly in the case of high phosphorus alloy, MOLDCOTE 13 applied in a similar way is recommended.

For moulds which will have a long life, high resistance to repeated casting and, hence, freedom from premature crazing, Hudson[*] suggests the following grey iron composition:

Total carbon	3.0–3.3%
Silicon	1.2–1.6%
Manganese	0.8–1.2%
Sulphur	0.12% max.
Phosphorus	0.3% max.

The soundness of a solid or cored chill casting depends largely on obtaining the ideal balance between pouring temperature and rate of pouring. Ideally, these factors should be so adjusted that as the mould is filling, solidification takes place in one direction; from the base of the mould upwards and not by process of simultaneous solidification in all parts of the casting. This demands relatively slow pouring rates which are achieved by allowing the melt to fall into the mould by way of a dozzle box in which a hole is drilled in the base.

Generally pouring rates are best found by practical experiment, but in the following table are given suggested pouring hole diameters suitable for various casting diameters.

[*]Hudson R. F. Non-ferrous Castings.—Chapman & Hall.

Description	Aperture diameter
1 in (2.54 cm) sticks	3/32 in (2.4 mm)
2 in (5.00 cm) sticks	5/32 in (4.0 mm)
3 in (7.62 cm) billets	3/16 in (4.8 mm)
4 in (10.2 cm) billets	7/32 in (5.6 mm)
6 in (15.2 cm) billets	1/4 in (6.4 mm)

Cores for Hollow Stick Castings.

Cores may be machined from mild steel but they must have a taper of at least 15 mm per metre length of the core. Various methods of coating the core surface are in use, to prevent sticking of the core to the casting and to make withdrawal of the core easy. One method is to strickle a coating of dry sand mixture around the core or wrap two layers of thin asbestos paper around it. In both cases, however, the core must be returned to a drying oven for either hardening of the sand coating or to drive off uncombined moisture from the asbestos paper. Following this a dressing is advised and MOLDCOTE 13 is particularly recommended.

Continuous and Semi-continuous Casting Methods

As indicated, these high production methods are rapidly superseding the older labour intensive craft methods where the demand can justify the considerable capital outlay involved.

In theory the processes are similar and quite simple. A bulk melting furnace(s) feeds molten metal into a separately heated holding furnace. From the base (semi-continuous casting) or wall (continuous casting) of the holding furnace, metal is fed out via a water-cooled copper and graphite die and withdrawn slowly and continuously as solid rod, strip, etc. In semi-continuous casting one or more billets, etc. (depending on the number of dies), are cast vertically, the limiting factors being working room under the die and furnace capacity; the process is usually operated on a batch basis. The continuous system, once in operation, is limited only by the supply of molten metal and the ability to handle the sections produced. A simplified schematic layout of a typical system is shown below.

In practice a large number of factors must be carefully balanced and stringently controlled in order that the processes can function at all.

240 THE FOUNDRYMAN'S HANDBOOK

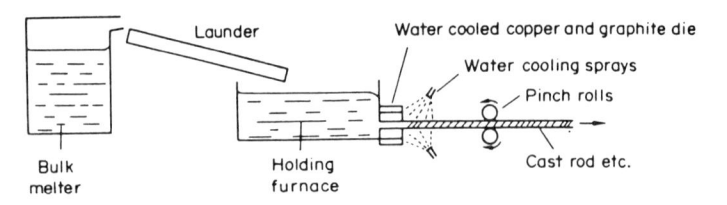

Fig. 5.6. Schematic diagram of a continuous casting unit.

A delicate temperature balance between the various parts is clearly essential. The only point at which metal treatment as such is possible or desirable is in the bulk melting unit, so it is important that correct melting, fluxing, and deoxidation techniques are used as a matter of course.

Suitable Test Bars for Use with Bronzes and Gunmetals

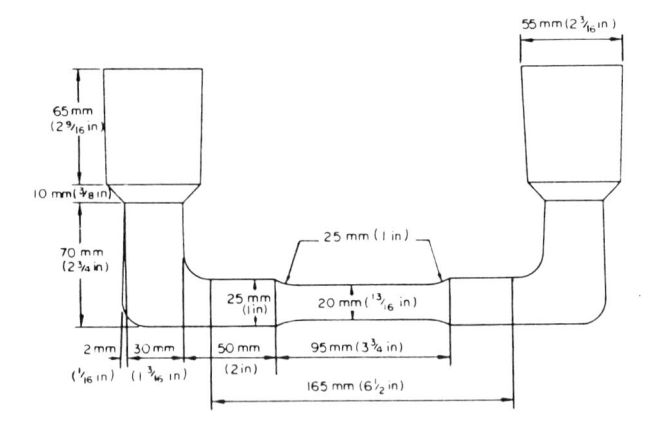

Fig. 5.7. BNFMRA "Cast to shape" test bar pattern dimensions (not to scale).

This test bar shall be used for the following alloys: LG1, LG2, LG4, G1.

This test bar may also be chosen when required for the following alloys: PB4, LPB1, LB2, LB4, PB1, PB2, CT1, LB5, LB1, G3.

TEST BARS SUITABLE FOR USE WITH BRONZES AND GUNMETALS

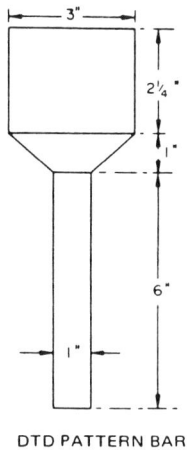

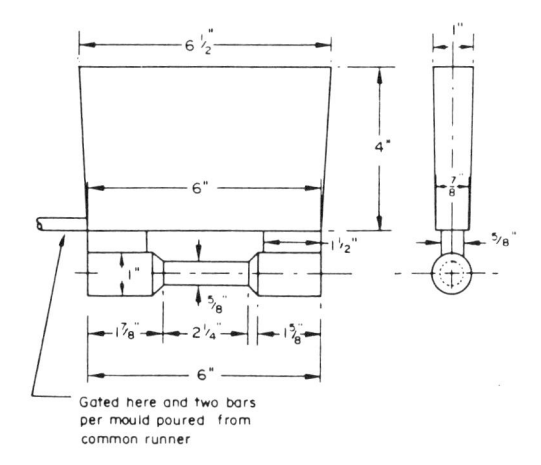

DTD PATTERN BAR

Gated here and two bars per mould poured from common runner

Fig. 5.8

THE RECOVERY OF NON-FERROUS SWARFS, SCRAPS, ETC.

There is a very considerable arising of finely divided swarf, turnings, etc., from all sectors of the non-ferrous metal industry, and it is important that as much metal as possible is recovered for re-use. A similar reclamation technique will be used whatever metal or alloy is being melted, but different flux compositions will be required as indicated by the various metal characteristics.

Furnace Types

The furnaces used are generally of the slowly rotating cylindrical reverberatory type holding 5 tonnes or more of metal when full. Induction furnaces are, however, being used more and more, and are particularly suitable for very finely divided scraps since there is a relatively static atmosphere immediately in contact with the metal. At the same time a good "boil" can ensure scrap is melted and assimilated quickly, thus keeping metal losses at a minimum.

Procedure

Whichever alloy or mixture of scraps is being recovered, the operating sequence is similar for both of the furnace types mentioned. Some solid or heavy scrap is first melted to produce a "heel" of metal in the furnace. This is then provided with a fluid cover of RECUPEX, and further quantities of the flux are scattered over the swarf, etc. waiting to be charged. Depending on the fineness of the scraps and their cleanliness, the amount of flux required can vary from around 2% by weight of the charge up to 5%. Some fluxes can attack refractories, etc., if used in excessive amounts, so it is always best to obtain advice beforehand. Charging can then proceed, the scrap being added to the furnace at approximately the same rate at which it is melting so that wide temperature fluctuations in the pool of molten metal do not occur. The amount of flux used should be just sufficient to maintain a slag cover of the requisite fluidity in order that the melt is protected at all times.

Contamination by Aluminium

After recovering incompletely segregated or mixed lots of scrap it is quite common that the resulting metal will be contaminated with aluminium. Unless the presence of this element is required in the alloy being produced it is usually regarded as a harmful contaminant, particularly if the metal is to be used for sand castings. Before casting, therefore, and provided the amount of aluminium present does not exceed approximately 0.5% the melt should be treated with ELIMINAL. Above 0.5% contamination it is preferable to dilute with aluminium-free metal until that level is reached or the amount of dross produced can become unwieldy. The temperature of the metal is raised to around 1150°C and the RECUPEX slag is drained off or removed into a slag pot. ELIMINAL is then applied to the bath of metal and brought into as intimate contact as possible by rabbling and stirring. The amount of aluminium remover required will depend on the degree of contamination but will generally be about four times the weight of aluminium to be removed. The treatment is best applied in stages using part of the ELIMINAL each time. Allow the flux to fuse on the surface before working it into the melt.

Choice of Materials

Alloy	Flux cover	Deoxidant
Brass 60/40, 65/35, 70/30	RECUPEX 108, 119	—
Bronzes and gunmetals	RECUPEX 118, 119 or 55	DS Tubes
HT Brass and Al bronze	ALBRAL 16	E Tubes
Nickel and cupro nickel	RECUPEX 118, 119 or 55	MG Tubes

PICKLING OF BRASS AND GUNMETAL CASTINGS TO IMPROVE APPEARANCE

There are various opinions among buyers as to the colour of brass and gunmetal castings which they purchase. It is important that the castings should have a good appearance and this may be achieved as follows:
 Make up a solution containing:
  66% concentrated nitric acid,
  17% water,
  17% concentrated sulphuric acid.
 The water should first be placed into a suitable vessel (stone jar). Then gradually add the sulphuric acid with constant stirring. Cool the solution, and slowly introduce the nitric acid, stirring the while. Cover the mixture with a suitable lid in order to prevent evaporation (metal vessels or covers cannot be used for this solution).

A wide-necked vessel is desirable, and side by side with it, should be a similar vessel containing clean water. Castings are fastened on to a wire and quickly dipped, first in the acid and then in the water, being well swilled till free from acid.

The castings, after drying in sawdust and polishing, have a pleasing appearance. It is important that the water is clean and changed often, because the casting must be free from acid after washing.

RECOMMENDED STANDARD MELTING AND FLUXING PROCEDURE FOR NON-FERROUS METALS AND ALLOYS

COPPER—HIGH CONDUCTIVITY

Copper of over 99% purity. Can be produced with electrical conductivity in excess of 85% I.A.C.S.

Treatment

Melt down under a reducing cover of CUPRIT 8. Deoxidise by plunging, at least twice, DEOXIDISING TUBES CB (containing calcium boride) or DEOXIDISING TUBES L (containing lithium) immediately before casting.

Quantity Required per 50 kg Melt

250–500 g CUPRIT 8.
1 DEOXIDISING TUBE CB5 (to be plunged), or 1 DEOXIDISING TUBE L1 (to be plunged).

Effect

Reducing atmosphere combined with a cleansing slag. Strong deoxidation without deterioration of conductivity properties.

Recommended Casting Temperatures

Light castings, under 13 mm section	1250°C
Medium castings, 13–38 mm section	1200°C
Heavy castings, over 38 mm section	1150°C

For the most effective removal of gas plunge LOGAS 50 briquettes.

BRASSES—COMMERCIAL

Non-specification. These alloys usually contain a small percentage of aluminium as an impurity.

Treatment

Melt down under a fluid cover of ALBRAL 2 and plunge a further quantity a few minutes before pouring. Alternatively, if aluminium is to be removed, use ELIMINAL 2. In either case finally plunge DEOXIDISING TUBES DS.

Quantity Required per 50 kg

0.5 kg ALBRAL 2 ($\frac{3}{4}$ as surface cover, $\frac{1}{4}$ to be plunged), or 0.5 kg ELIMINAL 2 (used as above).
1 DEOXIDISING TUBE DS2.

Effect

ALBRAL reduces oxide (alumina) formation while ELIMINAL removes aluminium, both giving a general cleansing action. Increased fluidity obtained by "deoxidation" treatment.

Recommended Casting Temperatures

Light castings, under 13 mm section	1150°C
Medium castings, 13–38 mm section	1100°C
Heavy castings, over 38 mm section	1070°C

BRASS ALLOYS (SAND CAST)

For example, 60/40 and 65/35 alloys containing no aluminium. For such applications as pipe-line and boiler fittings, etc.

Treatment

Melt down under a fluid cover of CUPRIT 49 or use CUPRIT 1 blocks. Finally plunge DEOXIDISING TUBES DS immediately before casting.

Quantity Required per 50 kg Melt

0.5 kg CUPRIT 49 (as surface cover), or 1×0.5 kg CUPRIT 1 block (at bottom of crucible).
1 DEOXIDISING TUBE DS2.

Effect

Fluid slag curtails zinc loss and has a general cleansing action. Increased fluidity obtained by "deoxidation" treatment.

Recommended Casting Temperatures

Light castings, under 13 mm section	1100°C
Medium castings, 13–38 mm section	1050°C
Heavy castings, over 38 mm section	1020°C

BRASS ALLOYS (DIE CAST)

For example, 60/40 alloy and silicon brass. Small percentages of aluminium (up to 0.5%) and silicon are usually present.

Treatment

Melt down under a non-fluid cover of ALBRAL 3 and plunge further small quantities as furnace is recharged from time to time. Dry dross can either be pushed aside during ladling out or skimmed off. Occasionally plunge DEOXIDISING TUBES DS.

Quantity Required per 50 kg Melt

0.5 kg ALBRAL 3 (as initial cover).
250 g ALBRAL 3 (to be added at pouring temperature).
1 DEOXIDISING TUBE DS2.

Effect

Oxide formation and zinc fuming curtailed together with an overall cleansing action. Increased fluidity obtained by "deoxidation" treatment.

NON-FERROUS CASTING ALLOYS 247

Recommended Casting Temperatures

Light castings, under 13 mm section	1100°C
Medium castings, 13–38 mm section	1100°C
Heavy castings, over 38 mm section	1100°C

BRAZING METAL

For example, 80/20 copper–zinc alloy.

Treatment

Melt down under a fluid cover of CUPRIT 49 or use CUPRIT 1 Blocks. Finally plunge DEOXIDISING TUBES DS immediately before casting.

Quantity Required per 50 kg Melt

0.5 kg CUPRIT 49 (as surface cover)
or 1×0.5 kg CUPRIT 1 block (in bottom of crucible).
1 DEOXIDISING TUBE DS2.

Effect

Maximum cleansing action with curtailment of zinc loss obtained by fluid slag. Increased fluidity obtained by "deoxidation" treatment.

Recommended Casting Temperatures

Light castings, under 13 mm section	1150°C
Medium castings, 13–38 mm section	1100°C
Heavy castings, over 38 mm section	1070°C

GUNMETAL (TIN BRONZE) ALLOYS

For example BS 1400-G1, i.e. 88/10/2 (Admiralty).
BS 1400-LG2, i.e. 85/5/5/5.
BS 1400-LG3, i.e. 86/7/5/2.

Treatment

Melt charge on top of CUPREX 1 block in bottom of crucible or melt down under a cover of CUPREX 100. Deoxidise by plunging DEOXIDISING TUBES DS immediately before casting.

Quantity Required per 50 kg Melt

1 × 0.5 kg CUPREX block 1 (in bottom of crucible), or 0.5 kg CUPREX 100 (as surface cover).
1 DEOXIDISING TUBE DS3.

Effect

The classical "oxidation-deoxidation" technique, i.e. oxidising and scavenging action, prevents hydrogen pick-up and cleans the melt, while deoxidation removes excess oxygen, reduces oxides and improves fluidity. (Residual phosphorus not less than 0.02%.)

Recommended Casting Temperature

Light castings, under 13 mm section   1200°C
Medium castings, 13–38 mm section   1160°C
Heavy castings, over 38 mm section   1120°C
For the most effective gas removal plunge LOGAS 50 briquettes.

PHOSPHOR BRONZE

For example, BS 1400-PB3. For general sand castings.
BS 1400-PB1. For bearings.
BS 1400-PB2. For gear blanks and bearings (good rigidity).

Treatment

Melt charge on top of CUPREX 14 blocks in bottom of crucible or melt down under a cover of CUPREX 140. Plunging of DEOXIDISING TUBES not necessary unless it is intended to increase phosphorus content.

Quantity Required per 50 kg Melt

1×0.5 kg CUPREX 14 block (in bottom of crucible), or 750 g CUPREX 140 (as surface cover).
Add one DEOXIDISING TUBE DS4 to compensate for loss of phosphorus during melting.

Effect

Scavenging action removes gas and has general cleansing effect. Tin "sweat" curtailed. DEOXIDISING TUBES DS increase phosphorus content where necessary.

Recommended Casting Temperatures

Light castings, under 13 mm section	1100°C
Medium castings, 13-38 mm section	1070°C
Heavy castings, over 38 mm section	1040°C

For the most effective gas removal plunge LOGAS 50 briquettes.

SILICON BRONZE

For example, DTD 355, Everdur and PMG metal.

Silicon	1.5-5%
Manganese	1.5% max.
Iron	2.5% max.
Zinc	5.0% max.
Copper	balance.

Treatment

Melt down under a fluid cover of ALBRAL 2 and when pouring temperature is reached plunge in further small quantities. Finally plunge DEOXIDISING TUBES E immediately before casting.

Quantity Required per 50 kg Melt

0.5 kg ALBRAL 2 ($\frac{3}{4}$ as a cover, $\frac{1}{4}$ to be plunged).
1 DEOXIDISING TUBE E3.

Effect

Curtails oxidation and reduces metallic oxides combined with general cleansing action. "Deoxidation" treatment coalesces suspended non-metallics and improves fluidity.

Recommended Casting Temperatures

Light castings, under 13 mm section	1200°C
Medium castings, 13–38 mm section	1160°C
Heavy castings, over 38 mm section	1120°C

For the most effective gas removal plunge LOGAS 50 briquettes.

MANGANESE BRONZE (High Tensile Brass)

For example BS 1400-HTB1.

Manganese	3% max.
Aluminium	2.5% max.
Nickel	1% max.
Iron	0.5–2%
Tin	1.5% max.
Lead	0.5% max.
Zinc	balance.
Copper	55–63%

Treatment

As for silicon bronze (above) but finally plunge two DEOXIDISING TUBES E1 per 50 kg melt.

Recommended Casting Temperatures

Light castings, under 13 mm section	1080°C
Medium castings, 13–38 mm section	1040°C
Heavy castings, over 38 mm section	1000°C

For the most effective gas removal plunge LOGAS 50 briquettes.

ALUMINIUM BRONZE

For example BS 1400-AB1 DTD 174A. Suitable for both sand and die casting:

Aluminium	9.5%
Nickel	1.0% max.
Manganese	1.0% max.
Iron	2.5%
Copper	balance.

Treatment

As for silicon bronze (above) but finally plunge one DEOXIDISING TUBE E3 per 50 kg melt.
If die casting use ALBRAL 3.

Recommended Casting Temperatures

Light castings, under 13 mm section	1250°C
Medium castings, 13–38 mm section	1200°C
Heavy castings, over 38 mm section	1150°C

For the most effective gas removal plunge LOGAS 50 briquettes.

COPPER NICKEL ALLOYS

For example, Nickel bronze—20% Ni, balance copper.
Cupro nickel—66% Ni, balance copper.

Treatment

Melt down under an oxidising cover of CUPREX 1 blocks. For pressure tight castings, plunge REGENEX or LOGAS 50 briquettes when approaching the pouring stage. Before pouring plunge DEOXIDISING TUBES NS together with DEOXIDISING TUBES MG. Finally, if degassing by means of REGENEX or LOGAS 50 has not been carried out, plunge DEOXIDISING TUBES L, containing lithium.

Quantity Required per 50 kg Melt

1 kg CUPREX 1 blocks.
250 g REGENEX.
One DEOXIDISING TUBE NS4 + 1 DEOXIDISING TUBE MG 4.
Two DEOXIDISING TUBES LO (if degassing not carried out previously.)

Recommended Pouring Temperatures

	Nickel bronze	Cupro nickel
Light castings under 13 mm section	1400°C 2550°F	1560°C 2840°F
Medium castings, 13–38 mm section	1350°C 2460°F	1530°C 2790°F
Heavy castings over 38 mm section	1280°C 2335°F	1500°C 2730°F

NICKEL-SILVER ALLOYS

(Nickel Brass)
For example, 18% Ni, 10% Zn, $2\frac{1}{2}$% Sn, up to 7% Pb, balance copper.

Treatment

Melt down under a flux cover of CUPREX 1 blocks. Before pouring plunge DEOXIDISING TUBES E followed by DEOXIDISING TUBES DS. The former provides the manganese deoxidant whilst the latter introduces phosphorus.

Quantity Required per 50 kg Melt

0.5 kg CUPREX 1 blocks.
Two DEOXIDIISING TUBES E⎫ or one DEOXIDISING TUBE NS4.
One DEOXIDISING TUBE DS4⎭ See notes at end of this section.

Recommended Pouring Temperatures

Light Castings, under 13 mm section	1300°C
Medium Castings, 13–38 mm section	1280°C
Heavy Castings, over 38 mm section	1250°C

Nickel Additions

It is often necessary to introduce further quantities of nickel into melts, either nickel bronze or nickel–silver alloy. Nickel is best added in the form of 50/50 copper–nickel shot or as a hardener alloy.

ALTERNATIVE DEOXIDANTS

Considering the chemical reactions involved in the deoxidation of cupro–nickel and nickel–silver alloys, the ideal deoxidant should function as follows:
1. It should combine with all oxygen present and reduce or combine with any oxides to form a fluid slag.
2. Deoxidation products should not be entrained in the solidified casting.
3. A residual deoxidant should not adversely affect physical or working properties of the metal and should prevent further oxidation during pouring.

Phosphorus seems to satisfy all of the above requirements but a residual content of 0.025% is regarded necessary to ensure adequate deoxidation. With phosphorus at or above this level, difficulties may arise during the working of the nickel silver or high nickel bearing alloys, due to embrittlement.

Alternative and readily available deoxidising agents are as follows:

MAGNESIUM:—This is very effective, particularly as it eliminates the harmful effects of sulphur, but the oxide formed tends to remain entrapped in the metal at grain boundaries, thereby causing embrittlement.

MANGANESE:—This is an excellent deoxidant and it also reacts with sulphur to some extent. It is present in DEOXIDISING TUBES E. Manganese is known to impart some grain refinement and it is also believed to effect some improvement in metal behaviour during rolling.

CALCIUM:—This is considered a very good deoxidant, although metal fluidity is slightly reduced.

SILICON:—This element could be classed as a satisfactory deoxidant, although it forms a solid oxide. Separation of the oxide from the metal is, however, rapid and may affect the surface appearance and pressure tightness of the casting.

BORON:—This should also be considered as a satisfactory deoxidant, particularly in view of its possibilities as a grain refining agent. Excess can however exert an embrittling effect due to segregation of borides at grain boundaries.

DEOXIDISING TUBES NS contain balanced proportions of many of the deoxidants mentioned above. They are a recent FOSECO development and are particularly recommended for producers of nickel–silver castings which are to be subsequently worked by rolling, etc. Tests with DEOXIDISING TUBES NS have brought about a significant reduction in rejects in the wrought nickel–silver industry. In the melting recommendations previously given for nickel–silver alloys, DEOXIDISING TUBES NS should therefore be used instead of the standard recommendations given, and are also applicable to sand castings production.

For the deoxidation of cupro–nickel alloys, magnesium stick is widely used in view of its deoxidation efficiency. It is suggested, nevertheless, that DEOXIDISING TUBES NS are used together with magnesium metal on an approximately 50/50 basis as given in previous recommendations.

SECTION VI
IRON CASTINGS

COMPOSITION OF TYPICAL CUPOLA CHARGE MATERIALS

Pig Iron. Grading of pig iron by fracture has now largely given place to grading by chemical analysis. The appearance of the fracture of sand cast pig iron is governed chiefly by the silicon content. Although the modern foundryman buys his pig irons to chemical specification, the old grading numbers still persist.

Silicon contents are fairly consistent throughout the grade numbers of different makers but the elements manganese and phosphorus vary to a considerable degree. The analyses given below are representative of the usual grade numbers with respect to silicon, which decreases as the grade number rises. Throughout any one make of pig iron the manganese and phosphorus do not change much.

Grade no.	Total carbon percentage	Graphite percentage	Combined carbon percentage	Silicon percentage	Manganese percentage	Sulphur percentage	Phosphorus percentage
Silky	3.0 to 3.3	3.0 to 3.26	0.04	4.0 to 5.0	1.0 to 1.4	0.02	1.0 to 1.5
No. 1	3.8	3.7	0.1	3.0 to 4.0	1.0 to 1.4	0.03	1.0 to 1.5
No. 2	3.7	3.5	0.2	2.5 to 3.5	1.0 to 1.4	0.035	1.0 to 1.5
No. 3	3.65	3.3	0.35	2.75	1.0 to 1.4	0.04	1.0 to 1.5
No. 4	3.5	3.1	0.4	2.0 to 2.5	0.9 to 1.3	0.05	1.0 to 1.5
No. 4 forge	3.3	2.8	0.5	2.25	0.8 to 1.2	0.07	1.0 to 1.5
No. 5	3.25	2.6	0.65	1.75	0.7 to 1.0	0.1	1.0 to 1.5
Mottled	3.1	1.6	1.5	1.0	0.5	0.12	1.0
White	3.25	0.05	3.2	0.6	0.3	0.15	0.8

Hematite pig irons are graded similarly, but their special feature is very low sulphur and phosphorus, usually under 0.05% for each element.

Cylinder and refined pig irons are graded according to their total carbon and silicon contents which are adjusted to give maximum strength and density of structure with good machining properties.

IRON CASTINGS

Scrap. Most cupola charges contain pig iron and scrap in various proportions. Scrap cast iron is graded under a number of headings. Although its composition may vary over fairly wide limits some average compositions are as follows:

Grade	Total carbon percentage	Silicon percentage	Manganese percentage	Sulphur percentage	Phosphorus percentage
Machinery	3.5	2.0	0.55	0.1	0.8
Railroad	3.3	2.0	0.60	0.1	0.7
Agricultural	3.5	2.25	0.60	0.09	1.0
Light	3.6	2.3	0.50	0.09	1.2
Malleable	2.0	0.75	0.30	0.15	0.15
Mild Steel	0.25	0.20	0.60	0.05	0.05

Sizes and Weights of Scrap for Cupolas. Charge materials should be within definite size limits which vary according to the internal diameter of the cupola. Uniform melting and working of the furnace depends greatly on correct charge weights and suitably graded components. Speaking generally, it is safest, if "hanging" up of charges is to be avoided, to restrict the overall length of the pig iron and scrap to a dimension equal to one-third of the internal diameter of the cupola. The appended tables give maximum dimensions which should not be exceeded.

CAST IRON

Diameter cupola inside lining	Maximum thickness (mm)	Minimum thickness (mm)	Maximum length (mm)	Maximum weight (kg)
30"	100	6	450	50
48"	100	6	600	50
60"	150	6	900	75
72"	150	6	1200	75
90"	200	6	1500	100

STEEL SCRAP

Diameter cupola inside lining	Maximum thickness (mm)	Minimum thickness (mm)	Maximum length (mm)	Maximum weight (kg)
30"	25	6	450	13
48"	25	6	600	22
60"	50	6	900	34
72"	75	6	1200	45
90"	100	6	1500	67

Coke. Cupola coke should be dense, hard, strong and of a silvery grey colour. It should be within the limits of composition shown below.

Fixed carbon not less than   90%
Ash not more than   10%
Sulphur not more than   1.0%

Some compositions typical of different kinds of coke are as follows:

	By-Product	Beehive	Patch Petroleum
Fixed carbon	90.0 to 92.0	93.0 to 95.0	98.0 plus
Ash	7.0 to 9.0	5.0	0.50
Sulphur	0.6	0.5	0.30
Volatile matter	0.6 to 0.9	0.9	1.0

Limestone. A good grade of Limestone suitable for fluxing the cupola should contain:
96 to 98% calcium and magnesium carbonates.
Less than 2% silica.

Cupola Melting Gains and Losses. In calculating metal charges the following average allowances should be made:
Silicon 10 to 15% loss.
Manganese 20 to 30% loss.
Sulphur 20 to 50% gain.
Phosphorus 1 to 2% gain.

These changes depend upon the conditions of operating the cupola and therefore may vary widely. Under properly controlled working the gains and losses occur within fairly well defined limits as shown.

The silicon loss applies to silicon contents between 1.5% to 3.5%. When silicon falls below 1% the melting loss is greater and may be as much as 50% of the silicon charged.

Influence of the Elements Normally Present in Cast Iron on the Physical Properties

Element	Fluidity	Softness	Shrinkage	Strength	Density	Chill	Sulphur	Combined Carbon	Graphite
Combined Carbon	Decreases	Decreases	Increases	Increases	Increases	Increases	Neutral	—	Decreases
Graphite	Increases	Increases	Decreases	Decreases	Decreases	Decreases	Neutral	Decreases	—
Silicon	Increases	Increases up to 1%	Decreases	Decreases	Decreases	Decreases above 1% Increases	Decreases	Decreases	Increases
Manganese	Increases	Increases	Little effect	Increases	Increases	Promotes	Decreases	Increases	Decreases
Sulphur	Decreases	Decreases	Increases	Decreases	Increases	Little effect	—	Increases	Decreases
Phosphorus	Increases	Decreases	Aggravates	Decreases	Neutral		Neutral	Tends to Increase	Neutral

INFLUENCE OF NORMAL CONSTITUENTS IN CAST IRON

When deciding upon a chemical composition for any given class of casting the effects of the main constituents have to be considered. These, summarised briefly, are as follows:

Carbon. In all cast iron compositions carbon is the most important element. Its total amount and state of existence in the structure, that is combined as carbide or free as graphite, exert a major influence on the physical properties.

For alloys of iron and carbon only, the eutectic, or lowest freezing point composition is that having 4.3% of carbon. With carbon contents above this figure the alloys are termed hyper-eutectic, and those below it hypo-eutectic alloys. The hyper-eutectic compositions are liable to have coarse open-grained structures and comparatively large flakes of "kish" graphite when cooled slowly as in ordinary sand moulds. On the other hand the hypo-eutectic alloys have relatively fine graphite and denser, tighter structures. The tendencies mentioned become more extreme as the carbon ratio rises in the hyper-eutectic, or falls in the hypo-eutectic compositions.

Total carbon content also has an important influence on liquid shrinkage and solid contraction. Irons containing less than 3% become more difficult to cast sound as the carbon falls below this value and methods of running and feeding approaching those used in steel casting practice become more necessary.

Commercial cast irons are not simply alloys of iron and carbon, but invariably contain proportions of silicon, manganese, phosphorus and sulphur as well. The presence of these elements affects the carbon solubility. It is known that silicon and phosphorus each reduce the amount of carbon required to form the eutectic alloy by 0.3% for each 1% of silicon and 1% of phosphorus present.

Therefore, to find the carbon equivalent value for any composition add together the silicon and phosphorus percentages and divide by 3. The result so found is added to the carbon percentage. If this then shows greater than 4.3, the composition will tend to give open-grained castings with large graphite flakes, but if less than 4.3 the structure will be denser, sounder and contain less graphite as the percentage falls from the eutectic value.

For example, assuming a composition having total carbon 3.1%, silicon 2.0% and phosphorus 1.0%. The carbon equivalent found as explained is 2.0 + 1.0 = 3.0 ÷ 3.0 = 1.0. Then 3.1 + 1.0 = 4.1, which is below the 4.3 value. Irons in the high duty class have carbon equivalent values of about 3.5 to 3.8.

Silicon. Next in importance to carbon, with respect to influence on the properties of cast iron, is silicon. The presence of silicon tends to throw the carbon out of combination or solution in the iron and make it form graphite. In so doing, it reduces both the hardness and the shrinkage and decreases strength and density. For the best mechanical properties with good resistance to pressure and wear carbon and silicon ratios must be decided in relation to each other. That is the silicon must be decreased as the total carbon rises and vice versa, so as to maintain carbon equivalent values within the desired limits.

Manganese. First neutralises the ill effect of sulphur by combining with it to form manganese sulphide in place of iron sulphide. Manganese sulphide is lighter than iron sulphide and so tends to float out of the molten iron as slag. Iron sulphide is retained in the melt and deposited around the grain boundaries rendering the metal hot short and likely to produce cracked castings.

Manganese in excess of that required to combine with the sulphur forms manganese carbide and increases the hardness and chilling tendency. As sulphur is a more powerful hardener than manganese, the first effect of adding manganese often causes softening due to removal of sulphur hardness. Also in neutralising sulphur, manganese raises the fluidity of molten cast iron.

In order to ensure a correct manganese-sulphur balance in grey irons, the percentage of manganese should be at least equal to the sulphur content $\times 1.7$ plus 0.3%. Thus, with 0.1% sulphur, the minimum manganese content is $0.1 \times 1.7 = 0.17$ plus $0.3 = 0.47\%$. Manganese up to 1.0% improves the density and strength of grey cast irons.

Sulphur. This element exerts the opposite effect to silicon, tending to prevent carbon deposition as graphite. Therefore, it has a marked hardening influence.

Apart from this, as explained under "Manganese," unbalanced sulphur makes molten iron sluggish, encouraging short running and pinhole defects due to entrapment of mould and other gases.

Phosphorus. The information given here is drawn largely from the summarised work of the British Cast Iron Research Association and refers to plain cast irons not specially alloyed, heat treated or ladle treated.

Mechanical strength is at a maximum at about 0.35% phosphorus. At about 0.65% the strength properties fall and continue to do so as the phosphorus further increases.

The Brinell hardness increases about four points for each 0.1% phosphorus added to a non-phosphoric iron.

Impact or shock strength diminishes with increase in phosphorus content.

Speaking generally, phosphorus increases wear resistance, other things being equal.

It is well known that phosphorus increases the fluidity and running power of all cast iron compositions. At the same time it renders more difficult the production of sound castings where ribs, internal bosses and abrupt variations in sectional thickness are encountered.

Compositional limits for adequate strength and soundness in phosphoric cast irons taking into account both silicon and total carbon, have been worked out by the British Cast Iron Research Association.

IRON CASTINGS

Cupola Operation Data (Metric)

Inside diameter at melting zone (cm)	Area at melting zone (cm²)	Total tuyere area (cm²)	Melting rate per hour tonnes	Coke charge (kg)	Iron charge (kg)	Limestone charge (kg)	Blast volume (m³ per min)	Blast pressure (cm W.G.)	Bed height above tuyeres (cm)	Fan force de cheval	Height from tuyeres to charge door sill (metres)
46	1640	320	1.0	11	100	3.6	17.5	30	100	2.0	2.5
61	2900	645	2.0	20	200	6.3	31	30	100	5.0	3.0
76	4500	1130	3.0	32	320	10.4	48	35	105	7.0	3.0
91	6550	1290	4.5	45	450	15	69	38	105	12.0	3.5
107	8900	1800	6.0	60	600	20	93	40	110	17.0	3.5
122	11700	1930	8.0	82	820	27	123	43	110	23.0	4.0
137	14700	2450	10.0	101	1010	32	156	45	110	30.0	4.0
152	18300	2570	12.5	125	1250	41	191	48	110	40.0	4.5
168	22000	3150	15.0	158	1580	50	232	51	110	50.0	5.0
183	24200	3225	18.0	180	1800	60	278	51	110	60.0	5.0
198	31000	3850	21.0	210	2100	68	326	51	110	70.0	5.5
213	35800	4450	25.0	250	2500	82	382	53	115	85.0	6.0
229	41000	5100	28.5	280	2800	90	439	56	115	110.0	6.0

The above figures represent good average practice. They are by no means critical values and can vary considerably according to individual conditions and circumstances. Their main purpose here is to act as a guide in cases where such help is required.

CUPOLA OPERATION DATA (AVOIRDUPOIS)

Inside Diameter at melting zone inches	Area at melting zone square inches	Total tuyere area square inches	Melting rate per hour tons	Coke charge lb	Iron charge lb	Limestone charge lb	Blast volume cubic feet per min.	Blast pressure inches W.G.	Bed height above tuyeres inches	Fan H.P.	Height from tuyeres to charge door sill
18	254	50	1.0	25	250	8	620	12	38	2.0	8' 0"
24	452	100	2.0	45	450	14	1100	12	39	5.0	9' 0"
30	707	175	3.0	70	700	23	1700	14	41	7.0	10' 0"
36	1018	200	4.5	100	1000	33	2450	15	41	12.0	11' 0"
					cwts.						
42	1385	280	6.0	130	12	44	3300	16	42	17.0	12' 0"
48	1810	300	8.0	180	16	60	4350	17	42	23.0	13' 0"
54	2290	380	10.0	225	20	70	5500	18	43	30.0	14' 0"
60	2827	400	12.5	275	24	90	6750	19	43	40.0	15' 0"
66	3421	490	15.0	350	30	110	8200	20	44	50.0	16' 0"
72	4072	500	18.0	400	35	130	9800	20	44	60.0	17' 0"
78	4778	600	21.0	465	40	150	11500	20	44	70.0	18' 0"
84	5542	690	25.0	540	48	180	13500	21	45	85.0	19' 0"
90	6362	790	28.5	620	55	200	15500	22	45	110.0	20' 0"

The above figures represent good average practice. They are by no means critical values and can vary considerably according to individual conditions and circumstances. Their main purpose here is to act as a guide in cases where such help is required.

Influence of Some Alloying Elements Used in the Production of Cast Iron Upon its Structure

Element	Percentage most frequently used	Effect on chill	Effect on structure	Comments
Aluminium	Up to 2.0	Reduces. 1.0% is approximately equivalent to 0.5% Silicon.	Stabilises Ferrite. Increases and coarsens the Graphite. Decreases the hardness.	Generally used in small percentages as a de-oxidant and scavenger only.
Chromium	0.15 to 1.00	Increases. 1.0% approximately neutralises the graphitising effect of 1.0% Silicon.	Stabilises Cementite. Reduces and refines the Graphite. Increases the hardness.	Used for hardness, chilling power and wear resistance.
Copper	0.5 to 2.0	Decreases. 1.0% is approximately equivalent to 0.35% Silicon. Assists in control of chill depth.	Tends to increase and refine the Graphite.	Toughens the matrix and increases the fluidity.
Manganese	0.3 to 1.25	By first combining with Sulphur it tends to reduce the chill. In excess of this amount it increases the chill. 1.0% of Manganese neutralises about 0.25% Silicon.	Stabilises Austenite. Refines the Graphite and Pearlite.	Also acts as a deoxidiser. Gives grain refinement, density and increased fluidity.
Molybdenum	0.30 to 1.00	Increases. 1.0% is as effective as about 0.33% Chromium and neutralises the effect of 0.35% Silicon.	Refines the Graphite and Pearlite.	Used chiefly in combination with Nickel, Copper and Chromium in the production of high strength irons.

Influence of Some Alloying Elements Used in the Production of Cast Iron Upon its Structure—(continued)

Element	Percentage most frequently used	Effect on chill	Effect on structure	Comments
Nickel	0.10 to 3.00	Decreases. 1.0% is about equal to 0.33% Silicon and offsets the chilling effect of about 0.33% Chromium.	Stabilises Austenite. Refines the Pearlite and Graphite.	Improves the density and toughness. Evens out the hardness between light and heavy sections.
Silicon	0.5 to 3.50	Reduces.	Stabilises Ferrite. Increases the quantity and coarseness of the Graphite.	Softens, weakens and imparts an open grained structure.
Titanium	0.05 to 0.10	Decreases powerfully.	Increases but refines the Graphite.	Used chiefly as a de-oxidiser and degasser. Improves fluidity.
Vanadium	0.15 to 0.50	Increases strongly. 1.0% Vanadium offsets the chill reducing influence of about 1.75% Silicon.	Stabilises Cementite and improves the structure of the chill.	Increases hardness and resistance to wear and heat.
Zirconium	0.10 to 0.30	Mildly reduces.	Assists formation of Graphite	Reduces hardness. De-oxidises and improves the fluidity and density.

CUPOLA CHARGE CALCULATIONS

A wide variety of materials can be used in order to obtain almost any required analysis. In consequence it is largely a matter of experience to determine whether the total carbon content will show a gain or a loss on the calculated content and to what degree.

Under the normal conditions of melting the tendency is for the total carbon to attain the eutectic equivalent. If the quantity charged is above this value, a loss may be expected. On the other hand, where the charge contains less than the eutectic value, the trend is towards a carbon pick-up. For total carbon contents between 3.2% an 3.5% as charged, little change is likely to occur during melting. In view of the element of judgement required in assessing total carbon it will be omitted from the calculations given as examples.

Suppose the materials available are a No. 1 grade foundry pig iron and some cast iron scrap of the agricultural type. Chemical analyses of these two materials are as follows:

	Silicon	Manganese	Sulphur	Phosphorus
Pig iron	3.5	1.2	0.03	1.0
Scrap	2.25	0.6	0.09	1.0

It is required to make up a charge from these two materials which on melting will have a composition between the following specified limits:

Silicon	2.20 to 2.4%
Manganese	0.60 to 0.80%
Sulphur	0.08 to 0.10%
Phosphorus	0.75 to 1.00%

Taking silicon as the controlling element, it is seen that the two items charged in equal proportions yield a silicon content of 2.87% or allowing a 10% melting loss, say 2.6%, which is above the top limit specified.

Rough preliminary trials along these lines will soon suggest likely proportions before going to the trouble of making the calculation in greater detail. Thus as a charge composed of equal proportions of pig iron and scrap iron gives too high a silicon content it is obvious that less than 50% of the pig iron and more than 50% of the scrap iron must be taken. As the No. 1 pig iron has a good scrap carrying capacity by reason of its high silicon percentage, advantage can be taken of this to try out a mixture of 30% pig iron and 70% scrap. The full calculation is then as follows:

THE FOUNDRYMAN'S HANDBOOK

Ratio	Material	Silicon	Silicon × per cent	Manganese	Manganese × per cent	Sulphur	Sulphur × per cent	Phosphorus	Phosphorus × per cent
30%	No. 1 Pig Iron	3.5	105.0	1.2	36.0	0.03	0.9	1.0	30.0
70%	Cast Iron Scrap	2.25	157.5	0.6	42.0	0.09	6.3	1.0	70.0
100%	Totals		262.5		78.0		7.2		100.0
	Divide by 100		2.625		0.78		0.072		1.0
	Adjust for gains and losses		−0.26		−0.2		+0.022		Little change
	Theoretical Analysis		2.36		0.58		0.094		1.0

The theoretical chemical analysis thus comes to be:

Silicon   2.36% allowing for a 10% loss.
Manganese 0.58%   ,,   ,, 25% ,,
Sulphur   0.094%   ,,   ,, 30% gain.
Phosphorus 1.00%   ,,   ,, no change.

These results are within the specified requirements, except perhaps for the manganese, which is a little low. This could be brought up by adding manganese to the charge in the form of ferro-alloy briquettes, or by means of manganese ladle additions. Silicon and phosphorus can, where necessary, be increased in a similar manner.

CURTAILMENT OF SULPHUR PICK-UP DURING CUPOLA MELTING OF IRON CHARGES

High Scrap Ratios

When conditions make it necessary to use high proportions of returned or purchased scrap in the cupola charges, the sulphur content of the resulting cast iron tends to become high. It is therefore desirable that all steps should be taken to keep the sulphur pick up within safe and reasonable limits. In grey cast irons sulphur fulfils no useful purpose but its presence is unavoidable and it exerts little ill effect provided it is not allowed to increase unduly and is offset by an adequate percentage of manganese or other element capable of neutralising its adverse effect.

Detrimental Effect of High Sulphur

Sulphur in cast iron exerts a carbide stabilising or chill inducing tendency. Therefore, if it is allowed to increase without check it will produce irons which are sluggish when molten and give rise to trapped gases and unmachinability in the castings. Irons with a high unbalanced sulphur are also liable to crack in thin sections, and may show inverse chill. They are most unsuitable for castings which are required to be subsequently vitreous enamelled because they tend to produce blisters and pin-holing in the enamel coating.

Conditions Favouring High Sulphur Pick-Up

Every time pig iron or scrap iron is melted in the cupola the sulphur is picked up from the coke and this is especially the case if the coke is of poor quality and high in sulphur.

Oxidising conditions of melting are favourable to a high sulphur pick-up. This is because ferrous oxide is formed which passes into the slag, and a slag of high iron content will not take up sulphur.

Oxidising Conditions of Melting

These result when the iron charge melts too near the mouth of the tuyeres. A number of causes are responsible for this, namely:

1. Bed coke too low, either as charged originally, or allowed to become so during the course of the blow.

2. Use of too heavy a metal charge for the size of the cupola.
3. Soft coke of a highly reactive nature and coke of small size which allows the bed to burn away too rapidly.
4. Too little coke used in between each metal charge.

It is also possible to obtain oxidised metal by using in the charges a heavy proportion of badly rusted or burnt scrap iron.

Curtailment of Sulphur Pick-Up by Correct Fluxing Technique

The chemistry of fluxing reactions is highly complex, and much has yet to be learned of the way in which slag forms. Neither lime nor magnesia, by themselves, melt at temperatures generated within the cupola. It is chiefly as a result of their combination chemically with silica that they form liquid slags. Such reactions do not take place until the materials reach the melting zone.

Therefore, it is of importance to make additions to the flux such as will allow fluxing reaction to start somewhat higher up in the cupola stack. By so doing, liquid flux will trickle down over the coke and limestone, thereby speeding up combustion and refining throughout the melting zone. At the same time, a hotter and more active slag will be produced. In this way, the coke is constantly flushed in the melting zone, allowing more intimate contact of the blast and of the droplets of iron with the surface of the incandescent fuel. Thus, better melting is promoted and higher temperatures generated which result in an increased carbon pick-up and molten iron of high fluidity and freedom from contained gas.

The Function of BRIX 555 blocks

Although individual cases may need special consideration, normally, a limestone addition of about 3% is common practice but this can be reduced by one-third approximately, i.e., to 2% when introducing BRIX to the cupola. For the first half dozen charges BRIX should be added at the rate of 1.5 kg per 1000 kg. Thereafter a lower regular input will suffice and about 1 kg per 1000 kg is suggested as a general guide. This will ensure a clean drop and produce better quality iron during the blow down.

BRIX blocks activate the fluxing reactions and speed up the rate of combustion by fluxing away coke ash and leaving a cleaner coke surface at the face of which combustion takes place. So the melting and superheating zone is restricted to a shorter, hotter area giving a higher temperature and reducing the time during which the descending droplets of iron are exposed to the blast in their passage to the well of the cupola. Melting losses are also reduced due to the tendency for a protective coating of reactive slag to surround the molten drops as they trickle through the coke bed.

BRIX blocks are so constituted that they react with silica to release calcium oxide. The calcium oxide is formed in the nascent condition, and is therefore extremely active as a flux. The beneficial effect of BRIX in this respect is therefore two-fold, it eliminates silica directly and also provides a particularly active fluxing agent. It is easy to realise its advantage when it is necessary to use high ash coke and poor grade limestone.

Additional benefits resulting from carrying out the fluxing technique described here, are the increased fluidity of the slag, which contributes to greater ease and smoother working of the furnace. For example, it is easier to keep the tuyeres clean and free from slag curtains. At the end of the blow the drop will be easier and cleaner, and a lower content of entrapped fine particles of metallic iron will be lost in the slag.

Maintenance of a Slag Blanket

Higher sulphur contents are often obtained in the iron resulting from the first one or two charges melted. This is because the molten iron first produced absorbs sulphur from the bed coke. It is possible to minimise this sulphur pick-up by charging limestone and BRIX blocks along with the topmost layers of the bed coke. For this purpose it is recommended that three times the amount of limestone and BRIX blocks normally employed between each charge should be added to the bed coke.

Without a flux in the bed coke no appreciable quantity of slag is formed until after the first few charges have melted. Therefore, the first molten iron to collect in the well of the furnace is not only higher in sulphur but it is exposed to the impingement of the blast from the tuyeres and is often delivered lower in temperature than charges melted subsequently and protected by a slag layer. By ensuring preliminary slag formation by fluxing the bed coke, not only is the first iron melted lower in sulphur but it is also higher in temperature.

If the maximum desulphurising effect of the slag resulting from the flux is desired it should be the aim to maintain a slag layer or blanket constantly in the well of the cupola. As pointed out, this protects the molten iron from the action of the blast and promotes a further refining action in that all the droplets of molten iron have to pass through it. On long blows it will be sufficient to run off the surplus slag periodically and thereby obtain the additional benefit which is lost where the slag is allowed to run away continuously.

CHEMICAL COMPOSITION OF UNALLOYED IRONS SUITABLE FOR DIFFERENT CLASSES OF CASTINGS

Where it is intended to make alloy additions to obtain improved properties, the analysis given may be taken as that of the base iron. To obtain the maximum benefit from the alloys it is intended to add, it will be necessary to adjust the base composition, before making the additions, so as to balance it with respect to the influence of the elements to be added. The information necessary for this purpose is given in the table on pages 265 and 266.

Class of Casting	Size	Total Carbon	Silicon	Manganese	Sulphur	Phosphorus
Acid Resisting	Thin	3.25	2.0	0.75	0.05	0.4
	Medium	3.00	1.5	1.0	0.05	0.3
	Thick	3.00	1.0	1.25	0.05	0.2
Agricultural	Thin	3.75	3.0	0.60	0.06	1.3
	Medium	3.50	2.5	0.70	0.08	1.0
	Thick	3.25	2.0	0.80	0.10	0.75
Air Cylinders	Thin	3.40	1.8	0.75	0.08	0.6
	Medium	3.20	1.50	1.0	0.1	0.4
	Thick	3.0	1.0	1.0	0.1	0.3
Annealing Boxes	1″ sect.	2.9	0.75	0.5	0.08	0.2
Balls for Mills	Medium	3.0	0.90	1.0	0.15	0.6
Bed Plates	Thin	3.5	2.0	0.70	0.08	1.0
	Medium	3.3	1.75	0.75	0.10	0.75
	Thick	3.2	1.5	0.8	0.10	0.50
Brake Shoes	Medium	3.25	1.5	1.50	0.10	0.60
Car Wheels	Thick	3.0	1.0	1.0	0.12	0.3
Do.—Chilled	Thick	3.3	0.7	0.8	0.13	0.3
Caustic Pots	Medium	3.25	1.25	0.75	0.08	0.3
Chills	Medium	3.5	2.5	0.75	0.05	0.2
Chilled Rolls	Small	3.3	0.8	0.3	0.05	0.5
	Medium	3.2	0.7	0.3	0.06	0.4
	Large	3.0	0.6	0.3	0.07	0.3
Couplings	Medium	3.3	1.75	0.7	0.08	0.5
Crusher Jaws	Medium	3.4	1.0	1.0	0.15	0.5
	Large	3.2	0.8	1.2	0.15	0.4
Cylinders—Ammonia	Medium	3.0	1.5	1.0	0.1	0.8
Do.—Automobile	Small	3.25	1.9	0.8	0.1	0.4
	Medium	3.0	1.75	1.0	0.1	0.3
Do.—Gas Engine and Diesel	Small	3.5	2.0	0.6	0.08	0.5
	Medium	3.25	1.50	0.8	0.09	0.4
	Large	3.0	1.25	1.0	0.10	0.3
Cylinders—Hydraulic	Medium	3.25	1.5	0.8	0.1	0.40
	Thick	3.0	1.0	1.0	0.12	0.30
Do.—Locomotive	Medium	3.5	1.75	0.8	0.09	0.75
Do.—Steam	Thick	3.25	1.5	1.0	0.1	0.50
Do.—Motor Cycle	Thin	3.3	1.8	1.0	0.1	0.80
Dynamo Frames	Small	3.25	3.0	0.5	0.05	0.60
	Large	3.00	2.5	0.5	0.06	0.50
Electric Work	Thin	3.40	3.25	0.5	0.06	1.25
	Medium	3.25	2.50	0.6	0.08	1.00
Engine Frames	Medium	3.50	2.0	1.0	0.08	1.00
	Large	3.25	1.5	1.0	0.1	0.75
Fan Cases	Medium	3.3	2.25	1.0	0.08	1.0
Fly Wheels	Thin	3.4	2.0	0.6	0.07	0.7
	Medium	3.2	1.75	0.8	0.08	0.5
	Thick	3.0	1.25	1.0	0.10	0.3

CHEMICAL COMPOSITION OF UNALLOYED IRONS SUITABLE FOR DIFFERENT CLASSES OF CASTINGS—*(continued)*

Class of Casting	Size	Total Carbon	Silicon	Manganese	Sulphur	Phosphorus
Friction Clutches	Thin	3.6	2.75	0.6	0.06	0.7
	Medium	3.5	2.50	0.7	0.08	0.5
Gears	Small	3.5	2.25	1.0	0.08	1.0
	Medium	3.25	1.75	1.0	0.09	0.7
	Large	3.00	1.50	1.0	0.10	0.5
Glass Moulds	Small	3.50	2.50	0.5	0.07	0.5
	Medium	3.30	2.0	0.75	0.08	0.3
	Large	3.20	1.75	1.0	0.10	0.2
Grate Bars	1" Thick	3.20	1.25	1.0	0.12	0.3
Grinding Balls	Small	3.25	1.5	0.3	0.10	0.3
Grinding Plates	Medium	3.5	0.6	0.6	0.12	0.3
Hardening Pots	Thin	3.2	1.0	0.6	0.06	0.2
Hardware	Thin	3.5	2.75	0.6	0.08	1.5
Heat Resisting	Medium	3.3	1.6	1.0	0.08	0.25
	Thick	3.0	1.2	1.0	0.10	0.10
Hollow Ware	Thin	3.5	2.7	1.0	0.07	1.5
Ingot Moulds	Medium	3.6	1.5	1.0	0.06	0.1
	Large	3.3	1.2	1.0	0.05	0.1
Machine Tools	Thin	3.50	2.3	0.75	0.08	1.0
	Medium	3.25	2.0	1.0	0.1	0.75
	Thick	3.00	1.2	1.0	0.1	0.50
Mine Car Wheels	Medium	3.1	1.0	0.75	0.1	0.40
Mowers	Thin	3.5	2.5	0.50	0.1	1.0
Ornamental Work	Thin	3.6	3.0	0.50	0.05	1.25
	Medium	3.5	2.75	0.60	0.05	1.00
	Thick	3.25	2.50	0.70	0.06	0.75
Permanent Moulds	Medium	3.5	2.3	0.90	0.07	0.4
Do.—M. Castings	Medium	3.2	3.0	1.0	0.06	0.8
Piano Frames	Medium	3.4	2.5	0.8	0.08	0.6
Pipes (Water)	Thin	3.5	2.5	0.6	0.1	1.0
	Medium	3.3	2.25	0.8	0.1	1.0
Do.—(Steam)	Medium	3.25	1.50	0.9	0.07	0.3
Pistons (Automobile)	Thin	3.3	1.8	0.7	0.1	0.7
Do.—Rings	Thin	3.5	2.0	0.7	0.1	0.7
	Medium	3.25	1.75	0.8	0.1	0.5
Plough Points	Thin	3.25	1.0	0.9	0.08	0.3
Pulleys	Thin	3.6	2.75	1.0	0.07	1.0
	Medium	3.3	2.25	1.0	0.08	0.75
	Thick	3.0	1.75	1.0	0.10	0.6
Radiators	Thin	3.5	2.5	0.7	0.06	0.8
Rolls	Medium	3.2	0.7	0.3	0.1	0.4
	Large	3.0	0.6	0.5	0.1	0.2
Soft Castings	Thin	3.7	2.75	0.5	0.05	0.6
	Medium	3.5	2.50	0.6	0.06	0.5
Slag Pots	Medium	3.3	1.70	0.8	0.07	0.2
Stove Plates	Thin	3.5	2.7	0.5	0.06	1.0
	Medium	3.25	2.25	0.6	0.08	1.0
Typewriter Frames	Thin	3.5	2.30	0.5	0.07	1.0
Valves	Thin	3.3	2.2	0.6	0.07	0.6
	Medium	3.0	1.8	0.8	0.08	0.5
	Thick	2.8	1.2	1.0	0.1	0.3
White Iron Castings	Medium	3.0	0.8	0.4	0.15	0.5

BRITISH STANDARD SPECIFICATIONS FOR CAST IRON

CHEMICAL COMPOSITION AND MECHANICAL TESTS

None of the British Standard Specifications No. 1452/1977 covering general grey iron and high duty iron castings respectively, ask for definite chemical composition. Indeed they clearly state that "The composition of the irons as cast shall be left to the discretion of the manufacturers but the maximum percentage of phosphorus may be specified by the engineer or purchaser if he so desires." This is an advantage in that it leaves the foundryman free to use whatever materials suit him best, provided he can obtain the required mechanical test results from the cast bars. On the other hand, D.T.D. and some other specifications covering cast iron for Air Ministry requirements do stipulate limits for chemical composition as well as the mechanical tests.

British Standards 1452: 1961 and 1452: 1977

Grade numbers are arranged to indicate the tensile strength required on the 1.2 inch (30–32 mm) diameter as cast bar.

Although the 1961 specification has been long superseded by 1452: 1977 it is still commonly used and reference to its requirements is frequent in casting specifications. For this reason both tables are given in the following pages. The requirements of the International Organisation for Standardisation ISO R185. 1961 are also given for reference. In all the tables the original units in which the specification was issued are printed in bolder type.

IRON CASTINGS

BS 1452:1961

Grade	Nominal section thickness	Dia. of as-cast test bar	Tensile strength, R_m, min		
	mm	mm	N/mm²	kgf/mm²	tonf/in²
10	< 9.5	15.2	170	17.3	11.0
	9.5–19.0	22.2	162	16.5	10.5
	19.0–29.6	30.5	154	15.7	10.0
	29.6–41.3	40.6	147	15.0	9.5
	>41.3	53.3	139	14.2	9.0
12	< 9.5	15.2	201	20.5	13.0
	9.5–19.0	22.2	193	19.7	12.5
	19.0–29.6	30.5	185	18.9	12.0
	29.6–41.3	40.6	178	18.1	11.5
	>41.3	53.3	170	17.3	11.0
14	< 9.5	15.2	247	25.2	16.0
	9.5–19.0	22.2	232	23.6	15.0
	19.0–29.6	30.5	216	22.0	14.0
	29.6–41.3	40.6	209	21.3	13.5
	>41.3	53.3	201	20.5	13.0
17	< 9.5	15.2	293	30.0	19.0
	9.5–19.0	22.2	278	28.3	18.0
	19.0–29.6	30.5	263	26.8	17.0
	29.6–41.3	40.6	247	25.2	16.0
	>41.3	53.3	232	23.6	15.0
20	< 9.5	15.2	340	34.6	22.0
	9.5–19.0	22.2	324	33.1	21.0
	19.0–29.6	30.5	309	31.5	20.0
	29.6–41.3	40.6	293	30.0	19.0
	>41.3	53.3	278	28.3	18.0
23	< 9.5	15.2	386	39.4	25.0
	9.5–19.0	22.2	371	37.8	24.0
	19.0–29.6	30.5	355	36.2	23.0
	29.6–41.3	40.6	340	34.6	22.0
	>41.3	53.3	324	33.1	21.0
26	< 9.5	15.2	432	44.1	28.0
	9.5–19.0	22.2	417	42.5	27.0
	19.0–29.6	30.5	402	41.0	26.0
	29.6–41.3	40.6	386	39.4	25.0
	>41.3	53.3	371	37.8	24.0

BS 1452:1977

Grade	Dia. of as-cast test bar	Tensile strength, R_m, min		
	mm	N/mm^2	kgf/mm^2	tonf/in^2
150	30–32	150	15.3	9.7
180	30–32	180	18.4	11.7
220	30–32	220	22.4	14.2
260	30–32	260	26.5	16.8
300	30–32	300	30.6	19.4
350	30–32	350	35.7	22.7
400	30–32	400	40.8	25.9

International Organisation for Standardisation ISO R185 1961

Grade	Dia. of as-cast test bar	Tensile strength, R_m, min			
	mm	N/mm^2	kgf/mm^2	tonf/in^2	lbf/in^2
10	30–32	98	10	6.3	14200
15	30–32	147	15	9.3	21300
20	30–32	196	20	12.7	28400
25	30–32	245	25	15.9	35600
30	30–32	294	30	19.0	42700
35	30–32	343	35	22.2	49800
40	30–32	392	40	25.5	56900

Tensile Test Bars

Cast-on test bars may be specified where the design of the casting, the grade of iron and the method of running permit.

Separately cast bars shall be poured at the same time and from the same ladle of metal as the casting or castings they represent.

Standard Transverse Test Bars BS1452: 1961

Transverse test bars have standard diameters adjusted to relate to the average sectional thickness of the casting represented, as indicated in table below.

Dimensions of Transverse Test Bars related to Main Cross-sectional Thickness of Casting

Diameter of test bar (in)	Overall length (in)	Main cross-sectional thickness of casting represented (in)
0.6	10	Not exceeding $\frac{3}{8}$
0.875	15	Over $\frac{3}{8}$ and not exceeding $\frac{3}{4}$
1.2	21	Over $\frac{3}{4}$ and not exceeding $1\frac{1}{4}$
1.6	21	Over $1\frac{1}{4}$ and not exceeding $1\frac{5}{8}$
2.1	27	Over $1\frac{5}{8}$

For castings of more than 2 inches main sectional thickness a test bar of larger diameter than 2.1 inches may be used by agreement between the manufacturer and purchaser.

Formerly tensile and transverse tests were on an equal basis. This no longer applies, the transverse test being subsidiary.

Transverse Test

Transverse test requirements are specified as breaking load in pounds and as transverse rupture stress. The latter is used rather than the actual breaking load, because it takes into account the variations in span and diameter, thus making the test results more capable of direct comparison when obtained from bars of different dimensions. Variations due to rate of cooling, however, cannot be allowed for in the calculations.

Transverse rupture stress is obtained from the formula

$$\frac{\text{Breaking load in tons} \times \text{span in inches}}{4 \times \text{modulus of section}}$$

Since we are concerned with round bars only, the modulus of section of a round bar is obtained from the expression $0.0982 \times \text{diam.}^3$, where the diameter is measured in inches. As in all standard bars, span and section for each are constant; the formula shown above can be reduced to a single factor for each. These factors are given in the table below, and it is necessary only to multiply the transverse breaking load in pounds by the factor for the bar concerned to obtain the transverse rupture stress in tons per square inch.

Factors for Converting Transverse Breaking Load in Pounds to Transverse Rupture Stress in Tons/in² for Standard Round Bars

Bar diameter (in)	Breaking span (in)	Factor (in)
0.6	9	0.0475
0.875	12	0.0204
1.2	18	0.0118
1.6	18	0.005
2.1	24	0.00295

Deflection has an important bearing on the toughness and quality of cast iron and forms part of the transverse specification. A combined summary of the tensile and transverse requirements for all the grades is given in the table appended.

IRON CASTINGS

SUMMARY OF TRANSVERSE AND TENSILE TEST REQUIREMENTS B.S. 1452/1961

Grade	Bar diameter inches	Span inches	Load lb	Rupture stress tons/in^2	*Deflex inches	Gauge diameter	Area	Minimum tensile tons/in^2
Grade 10	0.6	9	430	20.5	Not specified	0.399	0.125	11.0
	0.875	12	975	19.9		0.564	0.25	10.5
	1.2	18	1640	19.3		0.798	0.50	10.0
	1.6	18	3750	18.7		1.128	1.00	9.5
	2.1	24	6100	18.0		1.493	1.75	9.0
Grade 12	0.6	9	485	23.1	Not specified	0.399	0.125	13.0
	0.875	12	1100	22.4		0.564	0.25	12.5
	1.2	18	1850	21.8		0.798	0.50	12.0
	1.6	18	4250	21.2		1.128	1.00	11.5
	2.1	24	6950	20.5		1.493	1.75	11.0
Grade 14	0.6	9	565	26.9	Not specified	0.399	0.125	16.0
	0.875	12	1255	25.6		0.564	0.25	15.0
	1.2	18	2060	24.3		0.798	0.50	14.0
	1.6	18	4750	23.7		1.128	1.00	13.5
	2.1	24	7830	23.1		1.493	1.75	13.0
Grade 17	0.6	9	645	30.7	0.09	0.399	0.125	19.0
	0.875	12	1440	29.4	0.13	0.564	0.25	18.0
	1.2	18	2380	28.1	0.20	0.798	0.50	17.0
	1.6	18	5390	26.9	0.16	1.128	1.00	16.0
	2.1	24	8680	25.6	0.22	1.493	1.75	15.0
Grade 20	0.6	9	725	34.4	0.10	0.399	0.125	22.0
	0.875	12	1625	33.2	0.15	0.564	0.25	21.0
	1.2	18	2700	31.9	0.23	0.798	0.50	20.0
	1.6	18	6150	30.7	0.19	1.128	1.00	19.0
	2.1	24	9970	29.4	0.25	1.493	1.75	18.0
Grade 23	0.6	9	805	38.2	0.10	0.399	0.125	25.0
	0.875	12	1815	37.0	0.15	0.564	0.25	24.0
	1.2	18	3030	35.7	0.23	0.798	0.50	23.0
	1.6	18	6900	34.4	0.19	1.128	1.00	22.0
	2.1	24	11250	33.2	0.25	1.493	1.75	21.0
Grade 26	0.6	9	885	42.0	0.10	0.399	0.125	28.0
	0.875	12	1995	40.7	0.15	0.564	0.25	27.0
	1.2	18	3350	39.5	0.23	0.798	0.50	26.0
	1.6	18	7660	38.2	0.19	1.128	1.00	25.0
	2.1	24	12540	37.0	0.25	1.493	1.75	24.0

*These values represent minimum deflections.

RELATIONSHIP BETWEEN CARBON EQUIVALENT AND TENSILE STRENGTH OF GREY CAST IRON

BS 1452: 1956 quotes seven grades of cast iron, the basis being the minimum tensile strength of a test bar machined to standard dimensions from a 1.2 inch diameter as cast bar. There is no requirement in terms of composition of individual irons, and the foundryman is free to make his own choice based on requirements for the particular casting, section size, etc. To help in this a relationship between composition of the iron and expected tensile strength has been worked out. The elements that have the greatest effect on strength of flake irons are carbon, silicon, and phosphorus, and the influence of each is dependent on the presence of the others. The matter would be much simplified if only one factor had to be taken into consideration, and fortunately a simple equation allows this to happen. If, to the total carbon content of the iron, is added one-third of the silicon content plus one-third of the phosphorus content, a figure known as the carbon equivalent (C.E.) is obtained. This is the figure equivalent to the carbon content of a simple iron-carbon alloy if no silicon or phosphorus was present to complicate matters. The equation can be expressed as:

$$\text{Carbon Equivalent (C.E.)} = \text{T.C.}\% + \frac{\text{Si}\% + \text{P}\%}{3}$$

Since the structure (and hence the strength) of flake irons is a function of composition a knowledge of the C.E. of an iron can give an approximate indication of the strength to be expected in any sound section. This relationship is conveniently expressed in graphical form, Fig. 6.1.

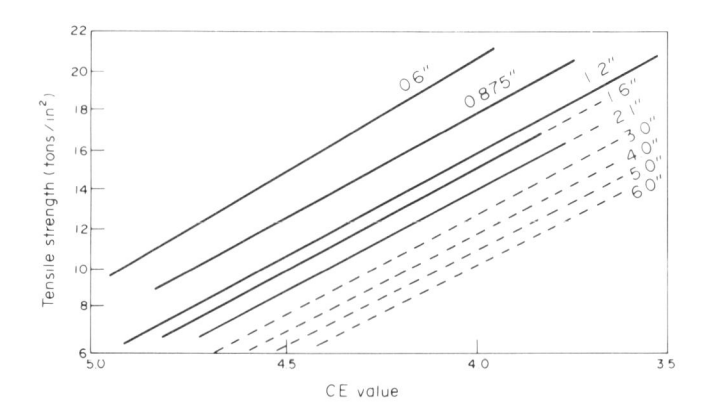

Fig. 6.1. Relationship between tensile strength and carbon equivalent value for various bar diameters.

Thus an iron with a C.E. value of 4.0 should give a tensile strength of 18 tons/in^2 in a 0.875 inch diameter sand cast bar or 16 tons/in^2 in a 1.2 inch diameter bar. It is important to remember that each line of the graph represents an average figure and a band of uncertainty exists on either side.

From Fig. 6.1 it is possible to construct a series of curves relating strength and section and these are shown in Fig. 6.2 for the five basic gradings of BS 1452: 1956.

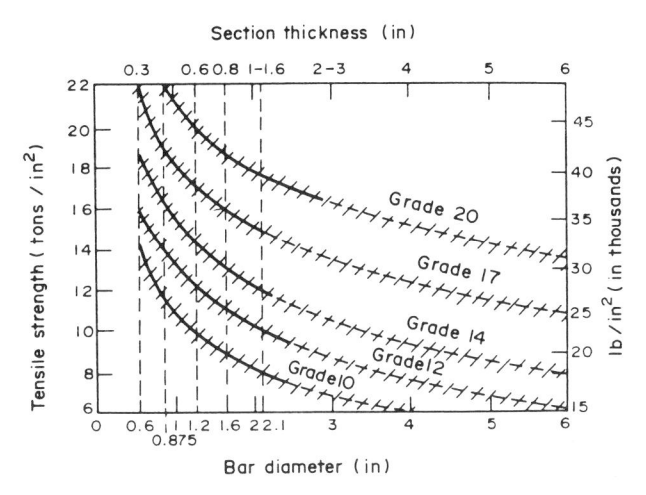

Fig. 6.2. Variation of tensile strength with bar diameter or section thickness for several grades of iron.

GREY CAST IRON—CLASSIFICATION OF GRAPHITE FLAKE SIZE AND SHAPE

When a polished microspecimen of a flake iron is examined under the microscope, the form, distribution, and size of the graphite can be classified. This is done by comparing the specimen at a standard magnification of 100 diameters with a stylised series of standard diagrams and allocating letters and numerals to indicate form, shape, and size of the graphite. The standard reference data sheet for this purpose has been agreed by an international committee and is that proposed by the American Society for the Testing of Metals A247–47.

Certain requirements must obviously be met before a sample is evaluated. Attention must clearly be paid to the location of the microspecimen in relation to the rest of the casting, to the wall thickness, to the proximity of chills, denseners, etc., and to the distance from the as-cast surface. Some care is also necessary in grinding and polishing so that as much graphite as possible is retained in a truly representative cross-section. The specimen is normally examined in the as polished, unetched condition at a magnification of 100 diameters and with a field of view of approximately 80 mm diameter.

Graphite form is indicated by a Roman numeral, e.g. Type I.

Graphite distribution is indicated by reference to letters A–E, e.g. Type 1C.

Graphite size is indicated by numbers from 1–8, e.g. Type 1C4.

An indication of the microstructure of a flake iron can therefore be given by, for example, a description such as graphite Type IIIB4, suggesting this is a rosetted graphite structure with the longest flakes between $\frac{1}{2}$–1 inch in length at × 100 magnification. The Roman numeral (see Fig. 6.3) indicates the graphite form and may not always be used if the type of iron is otherwise clearly defined.

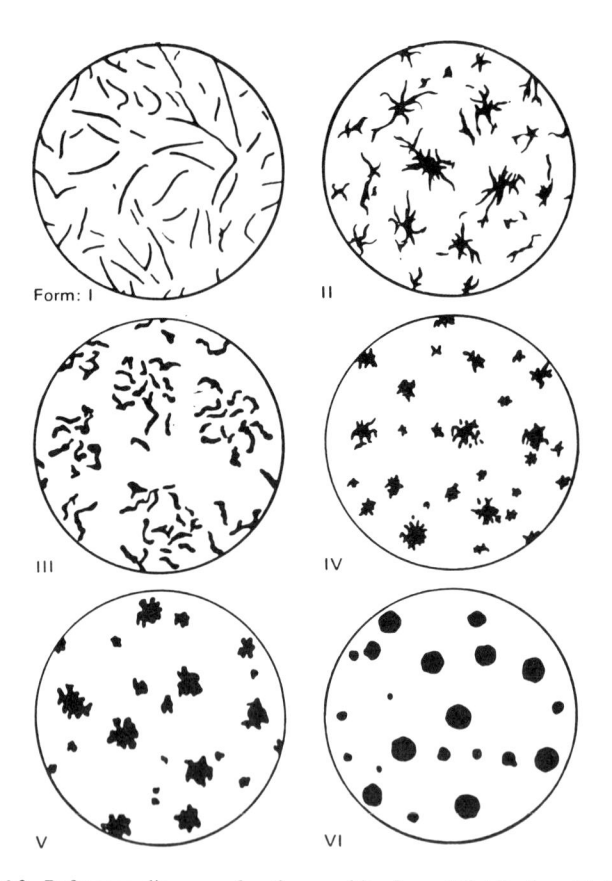

Fig. 6.3. Reference diagrams for the graphite form (Distribution A). The diagrams show only the outlines and not the structure of the graphite.

IRON CASTINGS 283

DIMENSIONS OF THE GRAPHITE PARTICLES FORMS I TO VI.

Reference number	Dimensions of the particles observed at ×100 (mm)	True dimensions (mm)
1	>100	>1
2	50 −100	0.5 −1
3	25 − 50	0.25 −0.5
4	12 − 25	0.12 −0.25
5	6 − 12	0.06 −0.12
6	3 − 6	0.03 −0.06
7	1.5 − 3	0.015 −0.03
8	< 1.5	<0.015

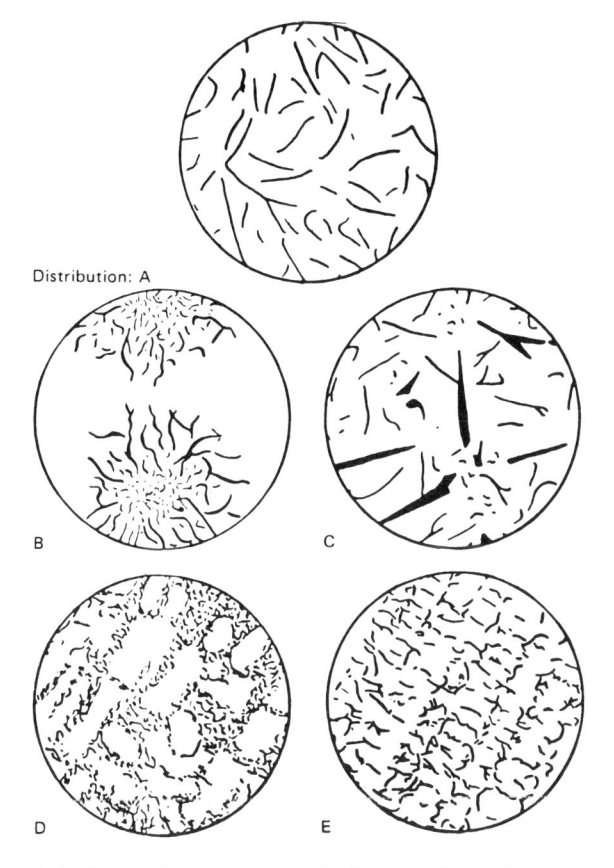

Fig. 6.4. Reference diagrams for the distribution of graphite (Form 1). The diagrams show only the outlines and not the structure of the graphite.

284   THE FOUNDRYMAN'S HANDBOOK

Size 1. Longest flakes 4 in or more in length.

Size 2. Longest flakes 2—4 in length.

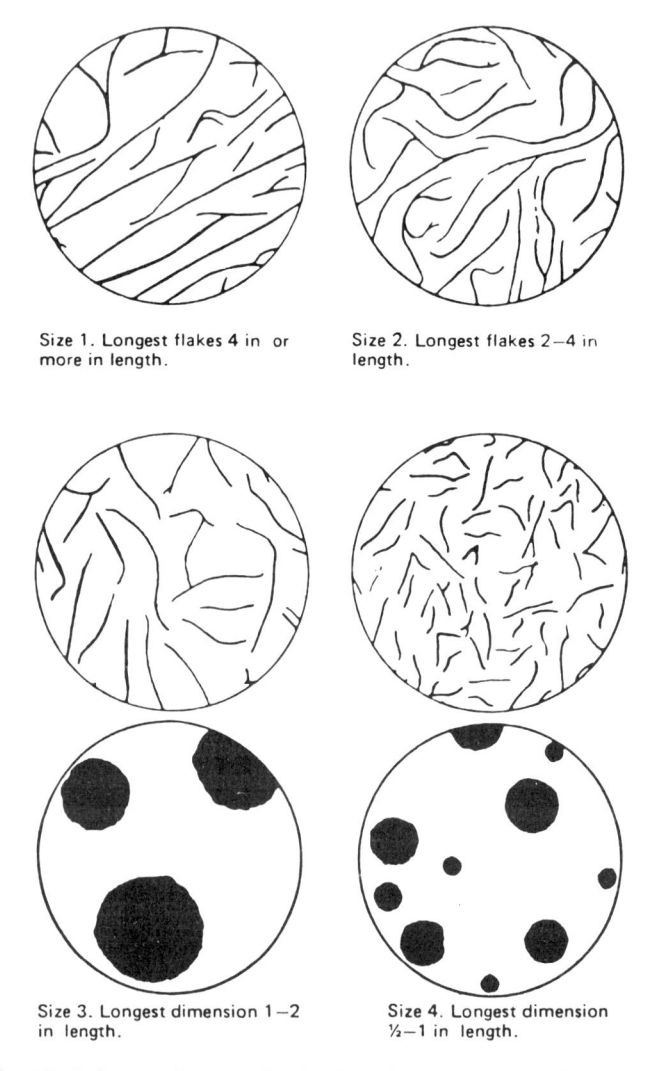

Size 3. Longest dimension 1—2 in length.

Size 4. Longest dimension ½—1 in length.

Fig. 6.5. Reference diagrams for the size of graphite flakes. The diagrams show only the outlines and not the structure of the graphite.

Size 5. Longest dimension
¼–½ inch length.

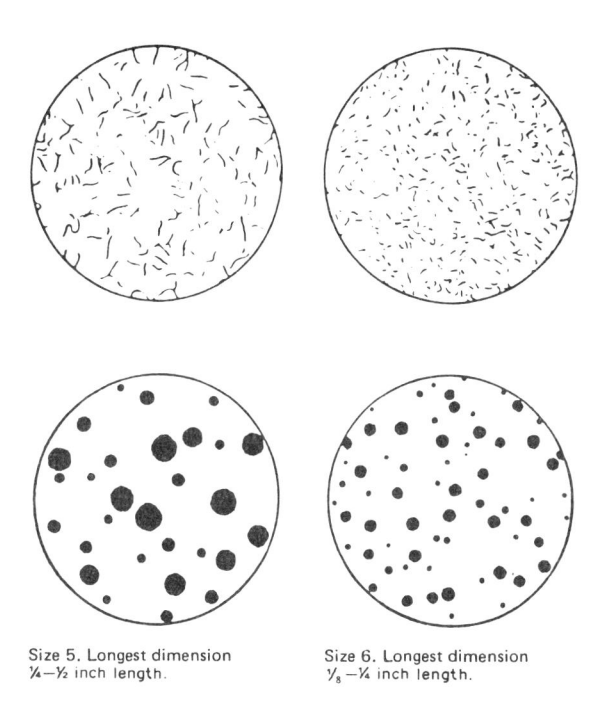

Size 6. Longest dimension
⅛–¼ inch length.

Fig. 6.5. (*cont.*)

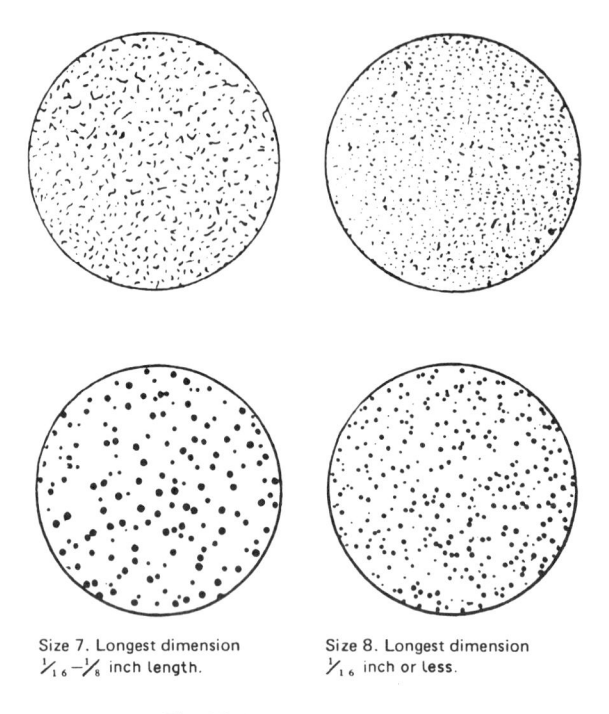

Size 7. Longest dimension
$\frac{1}{16}-\frac{1}{8}$ inch length.

Size 8. Longest dimension
$\frac{1}{16}$ inch or less.

Fig. 6.5. (*cont.*)

MALLEABLE CAST IRON

Although a number of types of malleable cast iron have been developed, there is one feature common to all. The castings must possess an all white, graphite free structure as cast. Then an annealing heat treatment is applied in order to decompose the hard, brittle carbide and to develop the properties of malleability and ductility from which the product derives its name.

TYPES OF MALLEABLE CAST IRON

There are two main kinds known as whiteheart and blackheart. The names have reference to the colour of the fracture of the annealed material. Special malleable cast irons are modifications of these two original types.

CHEMICAL ANALYSIS

Castings for whiteheart malleable are produced mainly from cupola melted charges. The chemical analysis varies with the size and kind of casting between the following limits:

Average Limits of Chemical Composition for Whiteheart Malleable

	Before Annealing	After Annealing
Total carbon	3.0–3.7%	0.5–2.0%
Silicon	0.4–0.8%	0.4–0.7%
Manganese	0.1–0.4%	0.1–0.4%
Sulphur	0.1–0.3%	0.1–0.3%
Phosphorus	0.1% max.	0.1% max.

Iron for blackheart castings is melted in air, rotary or electric furnaces, although cupola melting is also performed. Duplexing, by first melting in a cupola followed by refining and adjustment of composition in some kind of hearth furnace is also carried out. Normal limits of chemical composition:

Average Limits of Chemical Composition for Blackheart Malleable Before Annealing

	Air Furnace	Cupola
Total carbon	2.2–2.8%	3.0–3.3%
Silicon	0.8–0.95%	0.73–0.8%
Manganese	0.2–0.5%	2×sulphur %
Sulphur	0.2% max.	0.2% max.
Phosphorus	0.12% max.	0.12% max.

Average Limits of Chemical Composition for Blackheart Malleable After Annealing

	Air Furnace	Cupola
Total carbon	2.1–2.4%	2.5–3.0%
Combined carbon	0.1%	0.1%
Silicon	0.8–0.95%	0.7–0.8%
Manganese	0.2–0.5%	2×sulphur %
Sulphur	0.2% max.	0.2% max.
Phosphorus	0.12% max.	0.12% max.

The compositions given here are average values and intended only as a guide. In practice the limits may vary somewhat more widely.

Mechanical Properties of Malleable Cast Iron

Three British Standard Specifications cover the three main malleable iron groups as follows:
BS 310: 1972 Blackheart malleable iron castings
BS 309: 1972 Whiteheart malleable iron castings
BS 3333: 1972 Pearlitic malleable iron castings

Blackheart Malleable Iron

Because of the method of manufacture these irons have quite uniform and consistent properties underneath the immediate decarburised casting surface. The matrix is fully ferritic and they are the most ductile of all the malleable irons. Elongation values are high as are the notched impact properties although the latter form no part of the specification.

Whiteheart Malleable Iron

The microstructure of whiteheart malleable iron is variable through the section with a progressive increase in carbon content (present either as graphite or combined as pearlite) towards the centre. Mechanical properties will therefore be somewhat variable but good average results can be obtained consistently with good practice.

Pearlitic Malleable Iron

Of relatively recent introduction these irons are obtained by interrupting a normal blackheart anneal and air or oil quenching, or by heat treating and quenching subsequent to a normal blackheart anneal. The retained pearlitic matrix can also be obtained by varying the composition so that the required structure persists after a normal annealing cycle. These pearlitic malleable irons will give the highest mechanical properties available from this type of cast iron.

Tensile Properties

The three BS specifications given at the head of this section are further sub-divided into individual grades of iron as follows:
BS 310: 1972 covering blackheart malleable irons includes grades B290/6, B310/10 and B340/12
BS 309: 1972 covering whiteheart malleable irons includes grades W340/3 and W410/4

BS 3333: 1972 covering pearlitic malleable irons includes grades P440/7, P510/4, P540/5, P570/3 and P690/2

It will be noted that the prefix letter (B, W or P) indicates the type of malleable iron while the subsequent numbers indicate the minimum tensile requirement in Newtons/mm^2 followed by the elongation. In the following tables the original specification requirements are given in bolder type.

MECHANICAL PROPERTIES OF MALLEABLE CAST IRONS

Specification No.	Grade	Diameter of test bar (mm)	Tensile strength, R_m, min N/mm²	kgf/mm²	tonf/in²	0.5% Proof stress, $R_{p0.5}$, min N/mm²	kgf/mm²	tonf/in²	0.2% Proof stress, $R_{p0.2}$, min N/mm²	kgf/mm²	tonf/in²	Elongation, A, min %	Hardness HB
BS 309: 1972 Whiteheart	W340/3	9	270	27.5	17.5	—	—	—	—	—	—	7	
		12	310	31.6	20.1	—	—	—	—	—	—	4	
		15	340	34.7	22.0	—	—	—	—	—	—	3	
	W410/4	9	350	35.7	22.7	190	19.4	12.3	170	17.3	11.0	10	
		12	390	39.8	25.3	230	23.5	14.9	210	21.4	13.6	6	
		15	410	41.8	26.5	250	25.5	16.2	220	22.4	14.2	4	
BS 310: 1972 Blackheart	B290/6	15	290	29.6	18.8	170	17.3	11.0	—	—	—	6	
	B310/10	15	310	31.6	20.1	190	19.4	12.3	180	18.4	11.7	10	
	B340/12	15	340	34.7	22.0	200	20.4	12.9	190	19.4	12.3	12	
BS 3333: 1972 Pearlitic	P440/7	15	440	44.9	28.5	270	27.5	17.5	250	25.5	16.2	7	149–197
	P510/4	15	510	52.0	33.0	310	31.6	20.1	290	29.6	18.8	4	170–229
	P540/5	15	540	55.1	35.0	340	34.7	22.0	320	32.6	20.7	5	179–229
	P570/3	15	570	58.1	36.9	420	42.8	27.2	400	40.8	25.9	3	197–241
	P690/2	15	690	70.4	44.7	540	55.1	35.0	520	53.0	33.7	2	241–285

INTERNATIONAL ORGANISATION FOR STANDARDISATION (ISO)

Specification No.	Grade	Diameter of test bar mm	Tensile strength, R_m, min N/mm²	Tensile strength, R_m, min kgf/mm²	Tensile strength, R_m, min tonf/in²	Proof stress, $R_{p0.2}$, min N/mm²	Proof stress, $R_{p0.2}$, min kgf/mm²	Proof stress, $R_{p0.2}$, min tonf/in²	Elongation, A, min %	Hardness HB
ISO 5922–1981 Whiteheart	W35–04	9	340	34.7	22.0	—	—	—	5	*230 max.*
		12	350	35.7	22.7	—	—	—	4	
		15	360	36.7	23.3	—	—	—	3	
	W38–12	9	320	32.6	20.7	*170*	*17.3*	*11.0*	15	*200 max.*
		12	380	38.7	24.6	*200*	*20.4*	*13.0*	12	
		15	400	40.8	26.0	*210*	*21.4*	*13.6*	8	
	W40–05	9	360	36.7	23.3	*200*	*20.4*	*13.0*	8	*220 max.*
		12	400	40.8	26.0	*220*	*22.4*	*14.2*	5	
		15	420	42.8	27.2	*230*	*23.4*	*14.9*	4	
	W45–07	9	400	40.8	26.0	*230*	*23.4*	*14.9*	10	*150 max.*
		12	450	45.9	29.1	*260*	*26.5*	*16.8*	7	
		15	480	48.9	31.1	*280*	*28.5*	*18.1*	4	
Blackheart	B30–06*	12 or 15	300	30.6	19.4	—	—	—	6	*150 max.*
	B32–12	12 or 15	320	32.6	20.7	*190*	*19.4*	*12.3*	12	
	B35–10	12 or 15	350	35.7	22.7	*200*	*20.4*	*13.0*	10	
Pearlitic	P45–06	12 or 15	*450*	45.9	29.1	*270*	*27.5*	*17.5*	6	*150–200*
	P50–05	12 or 15	*500*	51.0	32.4	*300*	*30.6*	*19.4*	5	*160–220*
	P55–04	12 or 15	*550*	56.1	35.6	*340*	*34.7*	*22.0*	4	*180–230*
	P60–03	12 or 15	*600*	61.2	38.9	*390*	*39.8*	*25.3*	3	*200–250*
	P65–02	12 or 15	*650*	66.3	42.1	*430*	*43.8*	*27.9*	2	*210–260*
	P70–02	12 or 15	*700*	71.4	45.3	*530†*	*54.0*	*34.3*	2	*240–290*
	P80–01	12 or 15	*800*	81.6	51.8	*600*	*61.2*	*38.9*	1	*270–310*

*This grade is intended particularly for applications in which pressure tightness is more important than a high degree of strength or ductility.
†If grade P70–02 is air-quenched and tempered, 0.2% proof stress shall be 430 N/mm².

NOTES ON BS SPECIFICATIONS FOR MALLEABLE CAST IRON

As in the previous standards, minimum mechanical properties specified for each grade are determined using test bars representing the material quality of the castings. In addition to the use of metric units the following principal differences exist between these and the previous standards:

(a) BS 309 and BS 310 include a proof stress determination replacing a yield point measurement. Bend testing is not required.
(b) In all standards 0.5% proof stress values are required: the 0.2% proof stress values are given for information only.
(c) Brinell hardness values are included for guidance and are not part of the specification.

NODULAR CAST IRON
BS 2789-1961

This British Standard contains six grades of material instead of the three types specified in BS 2789: 1956.

TYPE SNG 24/17.—Those with a ferritic matrix in which resistance to impact is of paramount importance.

TYPE SNG 27/12.—Those with a mainly ferritic matrix of moderately high tensile strength in which high ductility and toughness are of great importance.

TYPE SNG 32/7.—Those with a ferritic-pearlitic matrix combining strength with reasonable ductility and forming an intermediate grade.

TYPES SNG 37/2, SNG 42/2 and SNG 47/2.—Those with a mainly pearlitic matrix, characterised by high tensile strength, but in which ductility and resistance to impact are of less importance.

The Brinell Hardness No. of Grade SNG 27/12 should not exceed 187, and that of Grade SNG 24/17, not to exceed 171.

British Standard 2789: 1956 has additionally been revised and enlarged and re-issued as BS 2789: 1973. In its new form it too contains specifications for six grades of spheroidal graphite (S.G.) or nodular iron, each relating to the six grades in BS 2789: 1961 and showing little real variation in terms of strengths required, ductility, etc. as will be seen from the following tables.

Mechanical Properties of Nodular Cast Iron British Standard 2789: 1961

Nodular cast iron or Spheroidal Graphite (S.G.) iron may be defined as a cast iron having a microstructure containing at least 90% of the graphite in the truly spheroidal form. Several elements are known to be capable of developing this type of structure, magnesium and cerium being the elements most widely used for this purpose. Other elements however, such as calcium and lithium, are also known to develop nodular type structures.

THE MAGNESIUM PROCESS FOR MAKING S.G. IRON

Of the elements known to be capable of developing nodular graphite structures when added to molten cast iron, magnesium offers the best industrial interest and seems to be able to develop the best mechanical properties. A further advantage of the magnesium process over the cerium, is the somewhat wider range of base iron composition permissible with it.

BS 2789: 1961

Grade	Tensile strength, R_m, min			0.5% permanent set stress.**			Elongation A, min.	Impact value Mean value (3 tests)			Individual value		
	N/mm²	kgf/mm²	tonf/in²	N/mm²	kgf/mm²	tonf/in²	%	J	kgf m	ft lbf	J	kgf m	ft lbf
SNG24/17	371	37.8	24	232	23.6	15	17	13.6	1.4	10	12.2	1.26	9
SNG27/12*	417	42.5	27	278	28.3	18	12						
SNG32/7	494	50.4	32	340	34.6	22	7						
SNG37/2	571	58.3	37	386	39.4	25	2						
SNG42/2	649	66.1	42	432	44.1	28	2						
SNG47/2	726	74.0	47	463	47.2	30	2						

*Impact test for Grades SNG27/12. When so required by the purchaser, Grade SNG27/12 shall show an average impact value of not less than 6 ft lbf (8.4 J; 0.84 kgfm/cm²) for a set of three individual values, and no individual test shall show an impact value less than 4 ft lb (5.6 J; 0.56 kgf m/cm²).
**The following values for 0.2% proof stress may be expected, and are given for information only.

0.2% proof stress, min.

Grade	N/mm²	kgf/mm²	tonf/in²
SNG24/17	232	23.6	15
SNG27/12	278	28.3	18
SNG32/7	324	33.1	21
SNG37/2	355	36.2	23
SNG42/2	386	39.4	25
SNG47/2	417	42.5	27

Values in units originally specified are in **bold** type.

IRON CASTINGS

BS 2789: 1973

| Grade | Tensile strength, R_m, min. ||| Proof stress $R_{p0.2}$, min. ||| Elongation A, min. | Impact value |||||| Hardness | Structure |
| | N/mm² | kgf/mm² | tonf/in² | N/mm² | kgf/mm² | tonf/in² | % | Mean value (3 tests) ||| Individual value ||| HB | |
								J	kgf m	ft lbf	J	kgf m	ft lbf		
370/17	**370**	37.7	24.0	**230**	23.5	14.9	**17**	(V)**13.0** (U)**15.0**	1.3 1.5	9.6 11.1	**12.0** **13.0**	1.2 1.3	8.9 9.6	≤179	Ferrite
420/12	**420**	42.8	27.2	**250**	25.5	16.2	**12**	—	—	—	—	—	—	≤201	Ferrite
500/7	**500**	51.0	32.4	**310**	31.6	20.1	**7**	—	—	—	—	—	—	170–241	Pearlite & ferrite
600/3	**600**	61.2	38.8	**350**	35.7	22.7	**3**	—	—	—	—	—	—	192–269	Pearlite & ferrite
700/2	**700**	71.4	45.3	**400**	40.8	25.9	**2**	—	—	—	—	—	—	229–302	Pearlite
800/2	**800**	81.6	51.8	**460**	46.9	29.8	**2**	—	—	—	—	—	—	248–352	Pearlite

Composition of Base Iron

Sulphur must be of a low order in the iron to be treated. Magnesium is also an active desulphurising agent and its first effect is to remove sulphur down to a level of about 0.01%. Until the sulphur is reduced to something like this limit, the magnesium exerts little influence on the graphite formation. In the formation of magnesium sulphide, 0.1% sulphur requires 0.075% magnesium.

Apart from the sulphur aspect, the preferred composition of the base iron is within the following limits:

Total carbon   3.4–4.2%
Silicon   1.8–2.5%
Manganese   0.4% maximum
Phosphorus   0.2% maximum

Whilst manganese and phosphorus do not prevent the formation of graphite nodules, they do reduce ductility very considerably and since this property is an important feature of nodular cast iron, it is essential to maintain these two elements at as low a level as possible.

Introduction of the Magnesium

One of the greatest practical difficulties is getting the required amount of magnesium into the melt with the necessary degree of consistency and assurance, combined with freedom from risk of danger. Magnesium boils at about 1120°C, and when plunged into molten cast iron at 1400°C, magnesium metal melts and vaporises instantaneously, escaping with violence and carrying some of the cast iron with it.

Successful magnesium treatment requires an emulsion of magnesium vapour and liquid cast iron. To obtain an intimate emulsion, liberation of the magnesium should occur over a period of time and at many places in the molten iron. This can be accomplished by using alloys of magnesium with one or more elements which are soluble in iron or a specially made impregnate such as MAG-COKE.

Nodularising Agents

Since magnesium metal boils at around 1100°C plunging it into molten iron at about 1400°C can provide some quite remarkable practical difficulties. Notwithstanding this a number of processes have been suggested based on the use of the pure metal and which should produce very economical systems. Despite the use of tilting pocketted ladles, automatic plungers, pour-over systems, converters, magnesium ingot protected with sand except for one area, etc. the practical difficulties are quite enormous due to the speed of reaction and the accompanying often unpredictable violence. For these reasons the most popular methods of producing S.G. iron usually rely on some form of carrier or indirect method of introducing the magnesium. A number of the most successful techniques are illustrated on pp. 307 et. seq. (Line drawings at end of this section).

An alloy of nickel and magnesium (5 or 15% magnesium) is perhaps the most efficient as it sinks in molten iron and the reaction is relatively quiet, especially with the 5% alloy. However, it is expensive and the simultaneous addition of nickel is not always welcomed. A range of magnesium ferrosilicons is also available containing from 3 to about 15% of magnesium with approximately 45% of silicon. Other materials include a magnesium impregnated coke (MAG-COKE) or compacts formed from granular mixtures of iron and magnesium (NODULANT). These latter materials avoid the progressive pick-up of silicon unavoidable with magnesium ferro-silicon and which eventually leads to the accumulation of large amounts of unusable back-scrap. All these materials can however be used in any of the alloying systems referred to and also in the bubbling ladle technique which is perhaps the best method of producing S.G. straight from a cupola as a simultaneous de-sulphurising system based on calcium carbide can be incorporated. One other method that should be mentioned is an in-mould system in which a mixture of magnesium ferro-silicon and inoculant is contained in a special reaction chamber within the mould itself. Great care has to be taken however to ensure accurate amounts of materials and also to prevent ingress into the mould cavity proper of reaction product, oxides, etc.

Of all these systems, that based on NODULANT compacts has the greatest practical flexibility both from the aspects of easy control of active ingredients (including the simultaneous addition of other agents such as cerium or calcium) and the lack of "contamination" with carrier elements. In particular the problems of silicon build up from magnesium ferro-silicon and the following post-inoculation treatment (always required with S.G. iron) are avoided. The yield of magnesium obviously varies with the treatment technique but it is at least as good and often better than that obtained with the other nodularisers. Because of these considerations the user has greater choice and flexibility of charge materials, better economics and some alleviation of possible problems with undesirable aluminium levels.

The simplest and most convenient method of producing S.G. iron with NODULANT (which contains 10% of magnesium in its standard form) is via a conventional sandwich or pour-over system. The NODULANT, either on its own or as a replacement for a proportion of magnesium ferro-silicon (where a silicon increase is required), is placed in a ladle pocket and covered with steel punchings, etc. The ladle should be about one quarter full before the reaction starts and reaction time should be about 90–120 seconds. Ladle geometry should be at least H = 2D.

The efficiency of magnesium recovery depends on a number of factors such as temperature, ladle size, method of use, sulphur content of the iron, required residual magnesium, etc. The amount of NODULANT (10% Mg) required may be calculated as follows:

$$\% \text{ Mg to add} = \frac{\text{Mg required } \%}{\text{Mg recovery \%—base iron S content \%}}$$

For a treatment weight of 10 hundredweight and a base sulphur level of 0.01% or less, magnesium recovery will generally fall within the range 25–50%. The lower figure will obtain at around 1500°C metal temperature and the higher at 1400°C using good treatment practice. Increased melt weight gives higher recoveries. Temperature loss when treating 1 ton of iron with 0.5% MAG-COKE will be of the order of 30–35°C only, provided that the plunger, etc., is strongly preheated.

Inhibiting Elements

Certain constituents or elements that may be present in cast iron have an inhibiting effect on nodule formation. Experimental work on approximately 1 inch thick castings indicates that the following elements in the amounts shown have an inhibiting effect:

Aluminium	above 0.13%	
Arsenic	above 0.09%	
Bismuth	above 0.002%	
Lead	above 0.005%	completely suppresses formation of nodules
Tin	above 0.04%	
Titanium	above 0.04%	

To this list can also be added antimony, tellurium and selenium although the limiting figures are not known.

All these exert a detrimental effect on the development of the nodular graphite structure. Also the combined effect of these elements is at least additive and may be even more potent. So when two or more are present in amounts which would not individually affect the structure, they may together prevent the formation of the nodular graphite. Some of the elements which cause interference are also capable of producing limited nodularisation, namely, bismuth, selenium, tellurium and titanium. This anomaly suggests that nodularisation may take place by two or more different mechanisms. Cerium added to a magnesium treated iron will neutralise the effect of interfering elements.

Quantity of Magnesium Required

During the magnesium treatment, part of the element is lost by evaporation. Part oxidises and reacts with the molten iron, forming compounds, both soluble and insoluble, some of which enter the slag. The chief slag forming reaction is that producing magnesium sulphide. It is only the residual magnesium which remains in the cast iron which is effective in promoting nodularisation of the graphite.

The residual magnesium in nodular cast iron is only a few hundredths of 1%. Limiting values which have been noted are 0.025% to 0.12%. C. K. Donoho considers that the best mechanical properties are obtained with a residual magnesium content of about 0.06%.

Like cerium, magnesium exerts a carbide stabilising action and when present much beyond 0.1% it is likely to give white or mottled structures, particularly where thin sections and rapid cooling are concerned.

Efficiency of magnesium recovery varies considerably but it seems to be greatest in the case of NODULANT and the nickel and copper alloys carrying 20% or less of magnesium. Recoveries reported vary from 25% up to over 60%. A magnesium ferro-silicon containing approximately 7% magnesium, approximately 45% silicon, the balance mainly iron showed the following recoveries:

Pouring temperature	Magnesium recovery
1454°C.	35%
1399°C.	42.6%
1343°C.	65.0%

Inoculation Treatment

Immediately following the addition of the magnesium bearing alloy, the melt has to be inoculated with from 0.5% to 0.75% of a graphitising inoculant such as INOCULIN 25. This is to offset the carbide stabilising action of the magnesium which operates in a similar manner to that of cerium.

Some claims have been made that when using the high silicon magnesium alloys such as magnesium ferro-silicon, no further inoculation is required. Whilst some corroboration of this has been obtained, it is not possible to be quite positive one way or the other in the light of present knowledge.

Influence of Other Elements

As with the cerium process, nickel, copper and molybdenum may be present in any proportions, either separately or combined. In this way, special austenitic, martensitic or acicular matrix structure can be obtained.

On the other hand, there are a number of elements which are known to inhibit development of the graphite nodules. These have already been mentioned and how their deleterious effects can be overcome by the combined use of magnesium and cerium.

Heat Treatment of Nodular Cast Irons

Successful treatment with cerium results in graphite nodules in a mixed matrix of ferrite and pearlite. Tensile strength is about 35 ton/in^2 with 2-4% elongation.

The magnesium irons in the "as cast" state, have the graphite nodules in a matrix more predominately pearlitic. Consequently tensile strength is higher, up to 45 ton/in^2, with less elongation, at about 2%.

By comparatively short annealing treatment at 850–900°C for a time which varies with composition and section thickness, all nodular irons can be converted to the all ferritic matrix. In this condition they resemble blackheart malleable in structure and will record a tensile strength of about 30 ton/in^2 and elongation of 10%.

Thus there are three types of nodular cast iron available.

1. The cerium iron, mainly ferritic as cast.
2. The magnesium iron, mainly pearlitic as cast.
3. Either of these rendered fully ferritic by annealing. All have some degree of ductility, and impact values greatly exceeding those of any form of grey cast iron hitherto produced.

THE CERIUM PROCESS

The Cerium Addition

Cerium may be added to cast iron as a wide variety of alloys. Pure cerium, iron-cerium, nickel-cerium, copper-cerium, silicon-cerium, manganese-cerium, and aluminium-cerium all dissolve quite easily in molten cast iron. On account of its ready commercial availability and its relatively low cost, a cerium alloy known as mischmetall is most frequently employed. Mischmetall contains approximately 50% of cerium, the balance being largely other rare earth elements, the most important of which is lanthanum. A typical analysis is as follows:

Cerium	45–53%
Lanthanum	22–25%
Neodymium	15–17%
Other rare earth elements	8–10%
Iron	Up to 5%

It is a relatively low melting point alloy of a high density, fairly soft and somewhat brittle, but it cannot be crushed easily into small pieces. A hacksaw may be used to cut it, but sparking will occur due to the pyrophoric nature of the alloy. (Cerium is an important constituent in lighter flints.)

For adding to molten cast iron, the separate pieces of mischmetall should not weigh less than about 20 g or more than about 300 g. The important point to be observed is that mischmetall should not be finely divided in the form of powder, as loss by oxidation may occur. On the other hand, the pieces should not be so big that difficulty in obtaining uniform solution is experienced.

Influence of Cerium

Cerium is a powerful desulphuriser. When the sulphur content of a cast iron exceeds about 0.02%, the effect of adding cerium is first to reduce the sulphur content. Cerium combines preferentially with sulphur even in the presence of manganese and the cerium sulphide so formed rises to the surface of the molten metal.

When the sulphur content of the iron is 0.02% or less, the cerium will function as a carbide stabiliser and will give nodular graphite structures under the appropriate conditions. As a general rule cerium in cast iron will not function as a carbide stabiliser and will not give nodular graphite unless the final analysis of the solidified casting shows less than about 0.02% sulphur and more than about 0.02% cerium.

The higher the sulphur content of the molten iron into which the cerium is added, the greater will be the amount of cerium required. As cerium is a relatively expensive addition in any form, it is obviously of great importance to limit the amount of the element required by maintaining the sulphur content of the iron to be treated, at the lowest possible level.

Composition of the Iron to be Treated

Base iron composition must be such as will solidify grey before treatment. The following points should be observed:

1. TOTAL CARBON CONTENT. This should be such that when the percentages of silicon and phosphorus are added together and divided by 3 and the result added to the total carbon content, the value resulting should be more than 4.3%.
2. SILICON CONTENT. After the treatment with cerium and the inoculant, silicon should be in excess of 2.3%, but not greater than 3.3%. When nickel and copper are present, the minimum silicon may be slightly less, but not lower than 1.8%, provided sufficient nickel and/or copper is present to give an equivalent silicon content of 2.3%, assuming 1% silicon equals 3% of nickel or copper.
3. MANGANESE CONTENT. The manganese content can have any value. However, apart from austenitic compositions it should be kept below 0.5% mainly in the interest of obtaining maximum ductility. The low sulphur of the treated iron allows the manganese present to exert its full carbide stabilising influence.
4. PHOSPHORUS CONTENT. Phosphorus has an embrittling effect on nodular cast irons. On no account should it exceed 0.5% for in excess of this the solubility of the cerium is reduced to a figure below the minimum required for the formation of nodular graphite. For development of the best mechanical properties the phosphorus should be kept at 0.4% maximum, or less.
5. SULPHUR CONTENT. The importance of keeping this low has already been stressed. After treatment with cerium it should be below 0.02%.

Control of Sulphur Content

With crucible or electric furnaces, sulphur content can be kept low by the selection of charge materials containing little of this element. Hematite pig irons can be obtained with sulphur contents below 0.03%. When such melting stock is used, no preliminary desulphurising treatment need be applied.

Although a similar technique can be applied to cupola melting, there is always some sulphur pick-up from the coke. Thus, in most cases, a desulphurising treatment will be necessary before making the cerium addition.

Addition of Cerium and Inoculant

Cerium in the form of mischmetall dissolves very readily in molten cast iron. It may be added to the molten metal in the ladle and then plunged or stirred in with a refractory plunger. Alternatively, it may be placed on the bottom of the ladle and base iron of correct composition teemed on to it. In most cases the "double treatment" technique will be adopted. This involves the use of a graphitising inoculant added either simultaneously with, or preferably immediately following, the cerium addition. INOCULIN 1 is recommended in a ratio of from 0.5% to 0.75% of the weight of the iron being treated.

When low sulphur or desulphurised iron is tapped direct on to the mischmetall, the inoculant can be added to the surface of the molten metal and stirred in. In many cases, it may be more convenient to place both the inoculant and the mischmetall in the bottom of the ladle, so that both are added simultaneously and stirring of the latter is not required. This method gives uniform results and good recovery of cerium, but the inoculant must not be added before the cerium. As the mischmetall probably dissolves more rapidly than the inoculant, the correct order of solution may be obtained in this way.

Cerium, as pure cerium, mischmetall, or any of the cerium alloys mentioned, dissolves in molten cast iron without any violent reaction, and the process is not hazardous in any way.

Scum Formation

Following the addition of the mischmetall and the inoculant, a scum of cerium sulphide tends to form on the surface of the melt. The quantity is dependent on the initial sulphur content. Since this slag is usually formed, precautions should be taken to prevent it entering the castings. Resort can be made to careful skimming, teapot spout ladles, the use of slag trap runners, and slag coagulants specifically developed for S.G. iron.

The best and most positive method of control without a doubt is to use filters. These are ceramic foams or woven gauze cloths that are introduced at strategic locations in the running/ingating system and effectively filter and smooth out incoming metal. Control of non-metallics entering the mould cavity is certain and the improvements in casting appearance and mechanical properties are quite remarkable. Elsewhere in this Handbook there is a section on metal filtration and reference should be made to it.

Metal Temperature

When adding the cerium and inoculant, metal temperature should be between 1350°C and 1500°C. When cerium is added to hyper-eutectic irons at temperatures below 1300°C there is a danger that some flake graphite will already have formed in the melt before the addition is made. Any flakes so formed will be unaffected by the cerium addition and will subsequently cause inferior mechanical properties.

After making the addition any convenient pouring temperature may be used.

Residual Cerium

When the sulphur content of the metal is below 0.02% in the base iron, the amount of cerium found by analysis in the solidified casting should be between 0.02% and 0.1%. For most purposes, it is sufficient to have between 0.03% and 0.06% of cerium. If the initial sulphur content is higher than 0.02%, too much of the cerium may be expended in forming cerium sulphide and the nodular structure will not be obtained.

Cerium dissolved in cast iron has a powerful carbide stabilising effect, but when the cerium treatment is followed by the addition of graphitising inoculant this effect is largely neutralised. Provided the cerium content falls within the range 0.03% to 0.06% accurate control is not necessary. For thin sections of 5 mm or less, it is advisable to keep the cerium at 0.04% maximum.

Quantity of Cerium

In most cases the correct amount of cerium to be added must be found by experiment. It will vary according to the method of melting and the sulphur content of the iron to be treated. The following examples serve to indicate the range of additions necessary when using mischmetall containing about 50% of cerium.

1. A hematite pig iron having 0.015–0.025% sulphur when crucible melted will require a minimum addition of mischmetall equivalent to 0.15–0.25% of the weight of the metal being treated.
2. Cupola melted hematite iron not desulphurised, with a sulphur content of 0.06% will need a mischmetall addition equivalent to 0.45–0.75% of the weight of the metal being treated.
3. Cupola melted hematite iron desulphurised, and having, after desulphurising, a sulphur content of 0.03%–0.04% will require an addition of

mischmetall equivalent to 0.3–0.5% of the weight of the metal to be treated.

Material which exhibits a predominantly flake graphite structure after treatment indicates a deficiency of cerium, and more mischmetall should be added. On the other hand a white or mottled fracture—except in the case of chilled castings—shows that the amount of mischmetall added needs to be reduced.

COMPACTED GRAPHITE (C.G.) CAST IRON

Although a graphite form intermediate between flake and spheroidal has been known for some time, mainly as a result of inadequately treated S.G. iron, it has only recently emerged as an engineering material with a more certain and reproducible production route. Because the graphite shape is neither flake nor spheroid, cast irons in this class have been called compacted graphite (C.G.) irons which more or less describes the graphite form.

As might be expected the physical properties lie somewhere between those of S.G. and grey iron. Strength at about 350 n/mm^2 is close to that obtainable from S.G. with a similar significant ductility of around 5%. 0.2% Proof stress in the region of $280n/mm^2$ is again similar to that obtainable from the ferritic grades of S.G.

Casting properties are nearer to grey iron than to S.G. with, as might be expected, less tendency to include oxides and dross in the casting. Total shrinkage is only slightly higher than flake irons and provided moulds are hard only nominal feeding should be necessary. Conductivity and thermal stability are also similar to grey iron. It can be produced without alloying with a high ferritic content (giving a Brinell Hardness of 160–180HB) and having good machinability. Notch impact strength is a little lower than S.G.

This combination of properties makes C.G. iron a very attractive proposition for:

automotive castings.
exhaust manifolds.
cylinder heads.
engine blocks.
brake parts.
hydraulic and pneumatic parts

where there is a constant need to save weight and improve efficiency without sacrificing strength, toughness, thermal stability and conductivity.

Methods of Producing Compacted Graphite Iron

1. An addition of cerium-mischmetall to an iron of low sulphur content in a particular proportion to the sulphur. A major disadvantage to this process is that the addition of cerium (a potent carbide stabiliser) must be exact to avoid the retention of carbides.
2. As for the production of spheroidal graphite (q.v) but undertreating with magnesium. The constraint is the very narrow range of residual magnesium that will give acceptable compacted structures.
3. As (2) but including about 0.15% of titanium which widens the working range for residual magnesium. Process costs are increased and scrap must be strictly segregated as titanium is not acceptable in the scrap for subsequent use in grey or S.G. iron. Strength would be reduced in flake irons due to its reaction with nitrogen and degenerate spheroids are likely in S.G.
4. Additions of 0.25—0.4% of calcium with simultaneous small additions of rare earths and magnesium metal in specified proportions to the base iron sulphur content.
5. A combination of magnesium ferro-silicon with a relatively high rare earth content and a significant proportion of calcium allows C.G. structures to be produced over a wide range of residual magnesium values. There are fewer disadvantages attached to this process than any of the others. The alloy is available as VERMICALLOY.

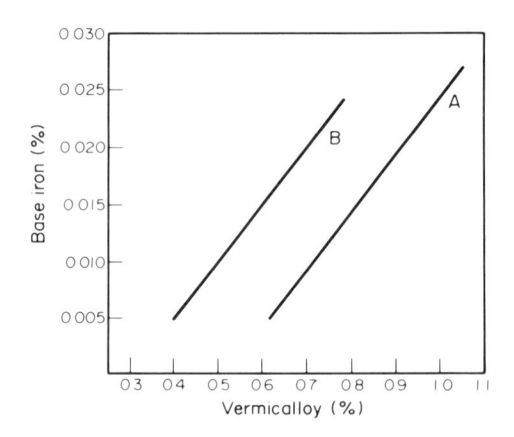

Fig. 6.6. % VERMICALLOY.

The graph shows the VERMICALLOY addition rates related to the base iron sulphur content. Line A was developed using the pour over method without covering and a treatment temperature of 1500°C. Lower treatment temperatures will give higher recovery rates. Sulphur contents below 0.005% are not advised because of a strong tendency for carbides to form. Line B shows the addition rates for an in-the-mould reaction chamber, again in relation to base iron sulphur content.

It is essential that the Mg:S ratio is maintained within the limits 2:1 and 1:1 to ensure correct C.G. structures. The higher ratio is favoured for the heavier sections.

Whichever method of production is used the size and distribution of the graphite in the solidified casting is controlled by a suitable inoculation treatment immediately following the alloy addition stage, exactly as for a spheroidal graphite iron. A fully soluble silicon-based inoculant such as INOCULIN 25 should be used.

At the moment there are no national or international standards for compacted graphite irons as far as is known.

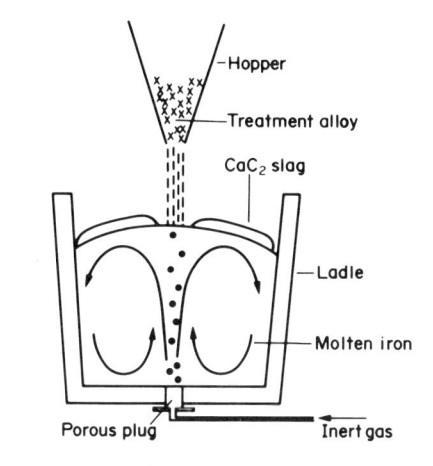

Fig. 6.7. Porous plug treatment technique—schematic.

308 THE FOUNDRYMAN'S HANDBOOK

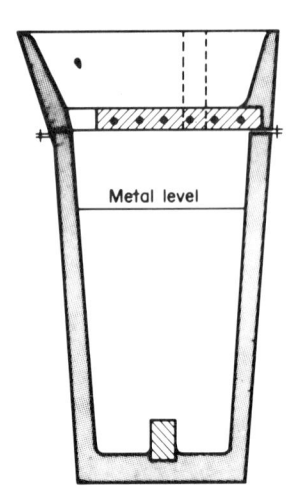

Fig. 6.8. General arrangement of the tundish cover ladle used in the treatment of s.-g. iron.

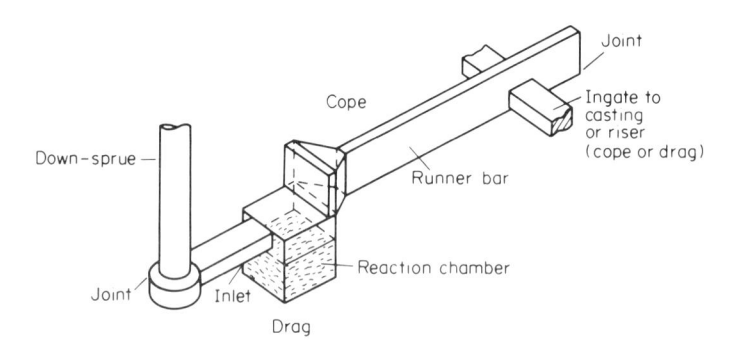

Fig. 6.9. Typical gating system for inmould nodularising technique.

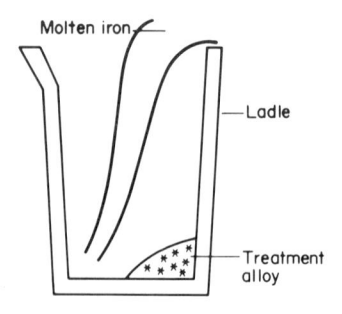

Fig. 6.10. Pour over treatment technique—schematic.

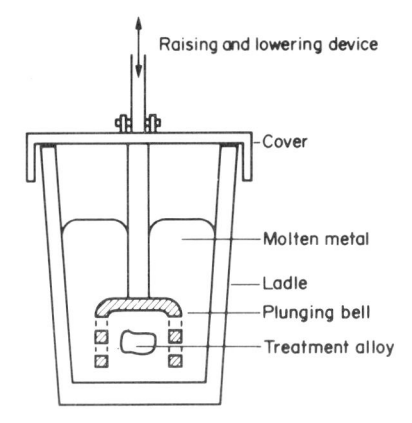

Fig. 6.11. Plunging treatment technique—schematic.

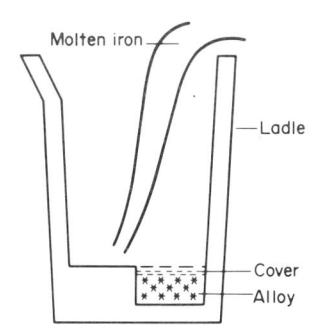

Fig. 6.12. Sandwich treatment—schematic.

SCRAP DIAGNOSIS—ITS CAUSE AND CURE
WITH PARTICULAR APPLICATION TO IRON CASTINGS

Many foundrymen are often in doubt when faced with a scrap casting, and very often the cure is obtained by rule-of-thumb methods, no direct approach to the trouble being made. It is hoped that the following list will help in a quicker diagnosis of scrap and its cure.

Defects	Appearance	Cause	Remedies
MISRUN	May appear as: 1. Holes in the thin sections of a casting. Edges are smooth and well rounded; surface of metal round holes smooth and often shiny. 2. As a line, when two streams of metal have met but not fused together. Fracture may occur along this line. Casting surface may be smooth and shiny.	1. Low pouring temperature. 2. Low fluidity. 3. Hard mould (mainly on very thin casting). 4. Core shift, causing uneven thicknesses. 5. Vent low on sand. 6. Very high moisture content. 7. Pouring practice.	1. Provide hotter metal at cupola spout; reduce heat losses in ladle by using flux coverings. 2. Increase Carbon and Phosphorus. 3. Avoid excessive ramming. 4. Take extra care in positioning. 5. Increase vent by means of vent wire or by adding Silica sand in mixing. 6. Reduce moisture compatible with moulding. 7. Keep runner bush full of metal during pouring.
SHRINKAGE and DRAWS	Rough cavities entering casting on heavy sections, or at the joint of change of sections. Saucer-shaped depressions on heavy sections, usually with rough edges.	1. Incorrect gating and feeding.	1. Use risers to feed heavy sections and ensure that they are filled with hot metal. If using open risers, use feeding flux, if using blind risers use feeding cores. Embody chills where a heavy section of boss cannot be fed directly with a riser.

IRON CASTINGS

SLAG	Similar to above. See Shrinkage and Draws (previous page). Cavities are generally more saucer-shaped and smoother. Slag may be seen before cleaning the castings.	1. Dirty metal. 2. Incorrect gating.	1. Remove all slag from metal before pouring. Thicken slag with sand before skimming. 2. Incorporate skim gates or strainer cores in runner systems. Keep runner bush full whilst pouring.
POROSITY	Castings "weep" under pressure test. Machined surfaces show cavities in thick sections or a series of pin-holes on machined skin.	1. Wrong type of metal. 2. Running and feeding system. 3. Gassy metal.	1. Reduce Silicon or Phosphorus content. 2. See shrinkage. 3. Degas and scavenge well.
HARD METAL	Bright areas on machined faces, often at corners or edges of thinnest sections. May occur as scattered hard spots. Shows, when broken, a white fracture.	1. Wrong type of metal. 2. High moisture content. 3. Pouring practice.	1. Increase Silicon content by: (a) altering metal mixture; (b) introducing Silicon into ladle. 2. Reduce moisture. 3. Avoid splashing metal down runners and risers—"plug" sprues very helpful.
SCABS	Rough "warty" excretions on surface of casting, mainly on heavy sections.	1. Uneven ramming. 2. Incorrect gating. 3. Improperly dried moulds. 4. High clay content in moulding sand.	1. Ram more evenly. 2. Gate so that an even flow of metal is obtained over surface 3. Avoid too rapid drying and allow time for heat to penetrate through the mould. 4. Change moulding sand.

SCRAP DIAGNOSIS—ITS CAUSE AND CURE
WITH PARTICULAR APPLICATION TO IRON CASTINGS—(continued)

Defects	Appearance	Cause	Remedies
DISTORTION	Casting shows swelling on surface.	1. Uneven mould hardness due to insufficient ramming to withstand metal pressure. 2. Poor weighting practice.	1. Ram evenly and firmly. 2. Increase weight on moulds and ensure it is distributed evenly.
ROUGH SURFACE	Casting surface rough.	1. Metal penetration. 2. Low coal dust content. 3. Hard sand surface with low permeability. 4. Moulding sand too "open."	1. Use finer sand or use mould dressing. 2. Make additions of coal dust. 3. Open up the sand. 4. Close-up sand somewhat.
CRACKS	Hair line cracks showing on casting. When broken, discoloration shows that crack was produced while casting was hot. No discoloration shows cold crack.	1. High dry strength of sand. 2. Cores too hard. 3. Casting strains. 4. Mechanical reasons.	1. Ram softer to allow casting to contract. 2. Reduce oil in cores. 3. Gate evenly to avoid these. Break mould to allow of free contraction. 4. Pack casting with wood or old tyres in tumbler. Take care in breaking off risers. See that risers are provided with correctly designed necks.

BLOW-HOLES	Rough shaped holes occurring on the outside of the casting or in the thicker sections. May be found just below surface on machining. In severe cases, section of casting may be hollow. Cavities may be dull or bright depending on conditions under which they have formed.	1. Low vent on moulding or core sand. 2. Hard ramming. 3. High moisture content. 4. Rusty or damp chills and chaplets. 5. Very hard cores. 6. Insufficient venting in cores. 7. Incomplete baking. 8. Damp pouring ladles. 9. Too low a pouring temperature.	1. Increase vent by use of vent wire or open sand with additions of Silica or by the use of a coarser Silica sand. 2. Avoid excess ramming. 3. Reduce moisture to minimum, consistent with workability. 4. Ensure chaplets are dry, and coat chills with oil or proprietary dressing before use. 5. Reduce oil in sand. 6. Ensure vents are clear. 7. Bake until centre is dry and hard. 8. Thoroughly dry all pouring ladles. 9. Increase pouring temperature.
DIRT	Rough cavities and pits in casting surface. If examined before cleaning the sand may often be seen.	1. Strength of sand low. 2. Loose ramming. 3. Direct wash of metal on sand surface, e.g., cods, corners, etc. 4. Poor finish of gating system. 5. Displacement of sand by cores. 6. Disturbed moulds. 7. Insufficient taper on patterns.	1. Increase green bond. 2. Ram evenly. 3. Avoid direct wash with well designed runners. 4. Finish of running system should be as good as mould. Make bushes and runners with good facing sand. 5. Blow out after placing cores. 6. Place weights carefully; avoid knocking moulds. 7. Increase taper to allow clean lift.

REMOVAL OF SAND FROM IRON CASTINGS BY PICKLING

The pickling solution is made up from water and a mixture of equal parts of hydrofluoric and sulphuric acids. The mixed acids being present to the extent of 10%. A lead-lined tank is required for holding the pickling solution. For preference a match board lining is fitted in order to protect the soft lead. The size of the tank is determined by the size of the casting and the output required, allowing about four hours for pickling.

Two other tanks are required, one to contain clean running water, and the second an iron tank heated either by steam coil or gas burner, and filled with a dilute solution of soda.

To start up, No. 1 tank is filled to just below the desired working height with water, and hydrofluoric and sulphuric acids are added at the rate of one litre of each to every 20 litres of water. (Always add acid to water to avoid accident—the concentrated acids are dangerous and skin contact, splashing, etc., must be avoided.)

No. 2 tank is filled with water, and No. 3 tank with a 10% solution of washing soda, and heated to near boiling.

Suspend the castings in No. 1 tank by means of mild steel or Monel metal hooks and allow to remain a sufficient time to remove or soften the sand and core adhering to the casting. The length of time will be judged by inspection, the usual period being between 3 and 4 hours. Remove the castings and swill with a strong jet of water to remove all adhering sand and pass through the water and soda tanks. Leave in the last tank long enough for the casting to get nicely warm, so that penetration of the neutralising soda is obtained and sufficient heat absorbed to dry the castings quickly. Finally rinse with clean water.

The strength of the acid bath is maintained by daily additions of an equal quantity of each acid, the requirements are judged by the evolution of gas from the castings when in pickle, which should resemble the rise of gas from soda water recently drawn from a syphon.

As a rough guide it should be possible to pickle an average of 30 four-cylinder motor engine blocks with 5 litres of each acid.

Mud and insoluble dirt accumulate in the pickling tank and must be cleaned out from time to time. The period varies with amount of castings put through; on an average with a pickling tank going continuously, cleaning out will be needed at the end of every six to eight weeks.

Measurement and handling of the acids are performed by means of a lead jug, and the hands of the operators are protected by gauntlet rubber gloves.

The pickling process described removes all traces of sand and other siliceous material. Not only does it produce a cleaner looking casting but also effects considerable economies in subsequent machining operations by cutting down tool costs.

Waste material from the process including sludge from the bottom of the tanks as well as "killed" acid must be disposed of very carefully as it is often extremely poisonous and corrosive.

CAST IRONS FOR RESISTING HEAT

CLASSIFICATION OF HEAT-RESISTING CAST IRONS

These may be grouped under five main classifications:
1. Certain white and unmachinable cast irons which do not grow under the influence of high temperature.
2. Close-grained cast irons containing low silicon and carbon contents, and small quantities of phosphorus, corresponding with good engineering properties, and sometimes having alloy additions of the order of about 3%.
3. Irons of the Silal type with silicon in excess of 4%.
4. Austenitic cast irons.
5. High chromium Ferritic cast irons.

1. White Cast Irons

The carbon in white cast irons is all in the combined form and there are no graphite flakes. When the composition is such that the combined carbon is stable at high temperatures, no growth results when the iron is repeatedly heated and cooled, but there is a tendency to develop cracks because of its brittleness. An iron with the following composition was found to be immune from growth at 900°C.: Total carbon 2.66%, silicon 0.59%, and manganese 1.64%. Although white irons do not grow, they scale rapidly under oxidising conditions, and their application is limited because of their brittleness and unmachinability.

2. Low Silicon Irons

Such irons, with about 1.2% silicon, and about 3.2% total carbon, have a fine, close-grained structure. They are also strong and possess some ductility at normal temperature. Owing to the "critical" or phase change demarcation which takes place between 700° and 750°C., it is apparent this class of iron should not be utilised for components whose normal service life necessitates working at temperatures above a maximum of 700°C. Service, at temperatures up to that mentioned, from this grade of cast iron can be improved by the addition of chromium up to 0.5% and nickel at 1.5 to 2.0%.

Irons under this heading have the advantage over those in Groups 3 and 4 in that they are easier to make into sound castings which will withstand rough usage. On the other hand, they oxidise readily, so where oxidising conditions are severe, they fail quickly by scaling.

Small changes in the amounts of sulphur, manganese, and phosphorus, have no appreciable effect on the tendency of these irons to grow at high temperatures. The amount of phosphorus which can be allowed depends upon the conditions. Below 900°C. phosphorus helps to maintain the iron rigid. Above this temperature it rapidly becomes plastic and the iron may collapse. The lower the phosphorus the greater the ductility and hence the less liability to cracking through repeated heating. Thus for use at temperatures below 900°C., up to 0.75% phosphorus may be present, but for castings subjected to more than 900°C., the phosphorus content should be about 0.3% maximum.

3. High Silicon Irons

Developed by the B.C.I.R.A. and covered by British Patent No. 323,076, known by the registered trade mark "Silal". The group carries silicon contents ranging between 4% and 10% and consequently possesses a structure of finely distributed graphite in a ferritic matrix. Their resistance to high temperature deterioration is due to the following reasons:

1. Low total carbon content, yielding a small graphite quantity and structure, which ensures the maximum resistance to gas penetration.
2. Resistance to oxidation by reason of high silicon content.
3. The volume change points can be raised beyond the maximum service temperature by the use of sufficient silicon.

In any application, the actual silicon content is determined by the temperature; the higher the temperature, the higher the silicon content required. As the silicon is increased, the irons become weaker and more brittle at normal temperatures and it follows therefore that the minimum silicon content necessary to prevent growth should always be used. In the ordinary way, the lowest silicon contents vary between 5% and 6%. This grade, however, is not recommended for service temperatures above 900°C. Above that temperature, scaling and growth take place, but will not occur if irons of higher silicon content, up to 10%, are used. Total carbon should be about 2.5% and phosphorus at less than 0.5%. Manganese should not exceed the quantity necessary to neutralise the sulphur present, that is, with 0.1% sulphur, the manganese should be 0.5%. An iron such as this can be made readily in the cupola, from a charge composed of ferro-silicon and steel scrap, because the solubility of carbon in molten cast iron is lowered by 0.25% by each 1% of silicon present.

4. Austenitic Irons

To overcome the brittleness of "Silal" attempts were made to impart some ductility by adding nickel and chromium. The resulting grey, non-magnetic iron called "Nicrosilal" was tough and ductile and resisted growth up to 1000°C. Its nominal composition was:

T.C. 2.0%
Si 5.0%
Mn 1.0%
Ni 18.0%
Cr 2.5%

Many more compositions based on the same approach have become accepted the more well-known now being the Ni-resist series. These include both flake and spheroidal graphite irons as indicated in the following tables which represent the major part of European practice.

5. High Chromium Cast Irons

These also have useful application in the high temperature range where resistance to scaling is a major requirement. The structures of these irons consist of primary iron-chromium carbides in a matrix of iron-chromium solid solution and secondary iron-chromium carbides associated with ferrite. There are two main grades. The first contains 16% of chromium with about 1.9% each of carbon and silicon. It is produced under the registered name "Enduron" by the Sheepbridge Stokes Company. The Vickers Diamond hardness is between 280 and 320, which makes it more difficult to machine than the heat-resisting irons mentioned previously. Also it is more susceptible to cracking. However, it resists scaling at temperatures from 900° to 950°C.

An iron containing chromium at 33% with total carbon 1.2% and silicon 2.0% has the best scale resistance of any of the heat-resisting irons and can be used at temperatures as high as 1050°C without marked deterioration. As in the case of the 16% chromium iron it is not easily machined and design of parts must be considered in relation to its increased brittleness and greater liability to cracking.

Summary

A general summation of the applicability of different kinds of cast iron compositions to service at high temperatures can be drawn up as follows:

Temperatures up to 600°C. Straight unalloyed grey irons of the high duty type and possessing sound, dense, fine-grained structures will give satisfac-

ISO AND BRITISH STANDARDS FOR FLAKE GRAPHITE AUSTENITIC CAST IRONS—CHEMICAL COMPOSITION AND MECHANICAL PROPERTIES.

ISO 2892–1973 British Standard BS 3468–1974	Chemical composition							Tensile strength, (R_m) min			International trade name
	C% max.	Si%	Mn%	Ni%	Cr%	Cu%	Kgf/mm²	N/mm²	tonf/in²	
L–Ni Mn 137	3.0	1.5–3.0	6.0–7.0	12.0–14.0	0.2 max.	0.5 max.	*14.3*	140	*9.1*	—
L–Ni Cu Cr 15 6 2	3.0	1.0–2.8	0.5–1.5	13.5–17.5	1.0–2.5	5.5–7.5	*17.3*	170	*11.0*	Ni-Resist 1
L–Ni Cu Cr 15 6 3	3.0	1.0–2.8	0.5–1.5	13.5–17.5	2.5–3.5	5.5–7.5	*19.4*	190	*12.3*	Ni-Resist 1b
L–Ni Cr 202	3.0	1.0–2.8	0.5–1.5	18.0–22.0	1.0–2.5	0.5 max.	*17.3*	170	*11.0*	Ni-Resist 2
L–Ni Cr 203	3.0	1.0–2.8	0.5–1.5	18.0–22.0	2.5–3.5	0.5 max.	*19.4*	190	*12.3*	Ni-Resist 2b
L–Ni Si Cr 20 5 3	2.5	4.5–5.5	0.5–1.5	18.0–22.0	1.5–4.5	0.5 max.	*19.4*	190	*12.3*	Nicrosilal
L–Ni Cr 303	2.5	1.0–2.0	0.5–1.5	28.0–32.0	2.5–3.5	0.5 max.	*19.4*	190	*12.3*	Ni-Resist 3
L–Ni Si Cr 30 5 5	2.5	5.0–6.0	0.5–1.5	29.0–32.0	4.5–5.5	0.5 max.	*17.3*	170	*11.0*	Ni-Resist 4
L–Ni 35	2.4	1.0–2.0	0.5–1.5	34.0–36.0	0.2 max.	0.5 max.	*12.2*	120	*7.8*	—

IRON CASTINGS

ISO AND BRITISH STANDARDS FOR NODULAR GRAPHITE AUSTENITIC CAST IRONS—CHEMICAL COMPOSITION AND MECHANICAL PROPERTIES

ISO 2892–1973 British Standard BS 3468–1974	C% max.	Si%	Mn%	Ni%	Cr%	P% max.	Cu% max.	Tensile strength, (R_m) min kgf/mm²	N/mm²	tonf/in²	0.2% proof stress, ($R_{p0.2}$) kgf/mm²	N/mm²	tonf/in²	Elongation, (A) min.	International trade name
S-NiMn 13 7	3.0	2.0–3.0	6.0–7.0	12.0–14.0	0.2 max.	0.080	0.5	39.8	390	25.2	21.4	210	13.6	15	Ni-Resist D-2
S-NiCr 20 2	3.0	1.5–3.0	0.5–1.5	18.0–22.0	1.0–2.5	0.080	0.5	37.7	370	24.0	21.4	210	13.6	7	Ni-Resist D-2B
S-NiCr 20 3	3.0	1.5–3.0	0.5–1.5	18.0–22.0	2.5–3.5	0.080	0.5	39.8	390	25.2	21.4	210	13.6	7	Nicrosilal
S-NiSiCr 20 5 2	3.0	4.5–5.5	0.5–1.5	18.0–22.0	1.0–2.5	0.080	0.5	37.7	370	24.0	21.4	210	13.6	10	Spheronic
S-Ni 22	3.0	1.0–3.0	1.5–2.5	21.0–24.0	0.5 max.	0.080	0.5	37.7	370	24.0	17.3	170	11.0	20	Ni-Resist D-2C
S-NiMn 23 4	2.6	1.5–2.5	4.0–4.5	22.0–24.0	0.2 max.	0.080	0.5	44.9	440	28.5	21.4	210	13.6	25	Ni-Resist D-2M
S-NiCr 30 1	2.6	1.5–3.0	0.5–1.5	28.0–32.0	1.0–1.5	0.080	0.5	37.7	370	24.0	21.4	210	13.6	13	Ni-Resist D-3A
S-NiCr 30 3	2.6	1.5–3.0	0.5–1.5	28.0–32.0	2.5–3.5	0.080	0.5	37.7	370	24.0	21.4	210	13.6	7	Ni-Resist D-3
S-NiSiCr 30 5 5	2.6	5.0–6.0	0.5–1.5	28.0–32.0	4.5–5.5	0.080	0.5	39.8	390	25.2	24.5	240	15.5	—	Ni-Resist D-4
S-Ni 35	2.4	1.5–3.0	0.5–1.5	34.0–36.0	0.2 max.	0.080	0.5	37.7	370	24.0	21.4	210	13.6	20	Ni-Resist D-5
S-NiCr 35 3	2.4	1.5–3.0	0.5–1.5	34.0–36.0	2.0–3.0	0.080	0.5	37.7	370	24.0	21.4	210	13.6	7	Ni-Resist D-5B

tory service up to about 600°C. In the lower temperature ranges phosphorus is not a harmful element so far as growth and scaling are concerned. Actually, in the amounts usually present in light casting mixtures it tends to improve scaling resistance under conditions of heat and oxidation within the limits of temperature stated. Phosphorus also imparts increased stiffness and rigidity, provided these temperatures are not exceeded.

Temperatures up to 700°C. Here again there is no real need to depart from grey iron compositions within the category encompassed by the close-grained alloys described in the group two classification. For this increased temperature range a simple ladle addition of chromium will confer the required stability of the pearlite and decrease the size of the graphite flakes.

Chromium is effective in retarding decomposition of the carbides at a ratio of about 0.6%. Above this value there is some risk of obtaining too great a hardness for easy machinability, especially in thinner sections. However, undue development of hardness can be guarded against and higher percentages of chromium used if a suitable proportion of silicon is introduced simultaneously with the chromium.

Where conditions of heat shock have to be withstood it is wise to use a relatively high total carbon content so as to exploit to the full the buffer or cushioning effect of the higher graphite precipitation in the presence of large temperature differentials. An example of a typical composition follows:

Total carbon 3.40–3.70%.
Silicon according to section of casting.
Manganese about 1%.
Sulphur less than 0.12%.
Phosphorus 0.3% maximum.
Chromium 0.6%.

The silicon content will vary according to the average sectional thickness of the casting:

From 9 mm to 18 mm average section thickness 2.3% silicon.
From 18 mm to 38 mm average section thickness 2.1% silicon.
From 38 mm to 60 mm average section thickness 1.9% silicon.

Temperatures up to 850°C. Still keeping within the group two range of composition, a 1% addition of aluminium will reduce oxidation and scaling losses up to 850°C. At this temperature the aluminium increases the life to about five times that of unalloyed grey cast iron. A somewhat greater resistance to heat and scaling can be imparted by raising the aluminium from 2% to 4%. However, ratios between 1% and 2% must be avoided. With the higher aluminium contents, steps must be taken to guard against the inclusion of defects caused by the entrapment of aluminium oxide in the castings.

Temperatures up to 900°C. For this relatively high temperature resort must be made to irons of the high silicon silal type described in group three. A composition which can be recommended is:

Total carbon	2.2%.
Silicon	6.25%.
Manganese	0.50%.
Sulphur	lowest.
Phosphorus	0.50% maximum.
Copper	1.50%.
Chromium	0.50%.

The cupola charge for this is based on silvery pig iron or ferro-silicon and arranged to yield a silicon content of 6.5%. Copper and chromium are added to the molten metal in the ladle.

Temperatures above 900° C. Here it is necessary to make use of the highly alloyed austenitic compositions indicated in group four. Some of these have well known proprietary names such as Nicrosilal, Ni-Resist and Causal, which are the subject of patents.

Useful alternatives to the austenitic series are the class 5 Irons of 16% and 33% chromium content. Their resistance to scaling is very good and chromium is a less expensive alloying element than nickel.

CAST IRONS FOR RESISTING CORROSION

When considering corrosive attack the position is rendered much more difficult than for heat resistance because of the almost infinite combination of corrosive agents, their concentrations, temperature and other conditions. Therefore, every case must be considered on its own as an individual problem. It is possible, however, to give general guidance as to what type of composition of iron will be most likely to give satisfactory service in particular cases.

No attempt can be made to cover the action of all possible corrosive media but perhaps it is true to say that cast irons which can withstand attack by the three principal mineral acids, hydrochloric, nitric and sulphuric are likely to have a useful life under many other conditions of acid corrosion.

1. Low Alloy Cast Irons

These cannot be expected to show any outstanding improvement over the range of unalloyed cast irons, except in so far as small additions of alloying elements contribute towards stronger, sounder structures and freedom from non-metallic inclusions. An outstanding instance is the resistance against caustic alkali attack which is imparted to cast iron containing up to 2% of nickel and 0.5% of chromium.

2. High Nickel Cast Irons

The high nickel austenitic cast irons have been described in group 4 of the heat-resisting irons, page 317. Of these, the best known is Ni-Resist which has satisfactory founding properties, is readily machinable and considerably tougher than ordinary grey cast iron. Its tensile strength is not very high, about 200 N/mm^2.

Ni-Resist compositions withstand well the action of sulphuric acid at atmospheric and moderately raised temperatures in all strengths from about 20% to 80% and up to 60°C. Actually it performs better than 18/8 stainless steel and is probably the most economical material for standing up against sulphuric acid corrosion.

With hydrochloric acid Ni-Resist iron is not so good. In the case of cold dilute acid it has a field of usefulness, but is unsuitable in the case of hot and strong hydrochloric acid solutions.

For nitric acid Ni-Resist is not suitable at all, but it offers an excellent resistance to caustic soda over a wide range of strength and temperatures. Under conditions of attack such as with industrial atmospheres, sea and tap water, surface rusting occurs but the rate of corrosion falls off with time to about one-fifth to one-eighth that of ordinary grey cast iron.

Copper is normally present in Ni-Resist compositions and produces a blue corrosion product which cannot be tolerated in the food, soap and cosmetic industries. Austenitic irons without copper, and containing about 21% nickel of Nicro-Silal are then employed. Cast irons of this type are non-magnetic and have a higher coefficient of thermal expansion than that of ordinary iron.

3. High Chromium Cast Irons

Item 5 in the section dealing with heat resisting cast irons on page 317 gives a general description of these. Although they have a much higher tensile strength than the austenitic irons (550 N/mm^2 with 30 to 33% chromium) they are not so tough. Also their hardness of approximately 300 Brinell makes machining more difficult.

Because chromium has the property of forming continuous oxide films of high stability it is able to confer a high resistance to corrosion under strongly oxidising conditions such as are met in resisting the attack of nitric acid. To exert its maximum effect the chromium must be present in solid solution and not in the form of carbide. Thus with chromium at 33% and just over 1% of carbon there remains enough chromium in solid solution to ensure good resistance to oxidising acids. So for hot 10% nitric acid the high chrome iron has a resistance equal to 18% stainless steel. In fact it has a satisfactory resistance to nitric acid at all strengths and all temperatures except for strong boiling acid when its behaviour becomes irregular.

High chromium cast irons offer little resistance to sulphuric and hydrochloric acids. In this respect they are complementary to the high nickel irons giving good service under conditions where the latter fail and vice versa. Very good resistance is offered by the high chromium irons to sea and tap water attack. Exposures in industrial atmospheres of up to 2 years have left the surface still bright.

As distinct from the austenitic cast irons the 33% of chrome iron is strongly magnetic and its co-efficient of expansion is similar to that of unalloyed iron and steel at about $11-12 \times 10^6$ per degree C. Most of these details are from information published by Sheepbridge Engineering Ltd.

4. High Silicon Cast Irons

This material containing silicon at an average 15% is known under various trade names such as:

Ironal, Tant Iron, Duriron, Durichlor, Anaciron, Corrosion Acidur, Antiacid, Duracid, Eliante and others. It is unmachinable, except by grinding, and very brittle.

Tests which have been made indicate that the silicon content best able to resist different mineral acids should be as follows:

10% Sulphuric acid	14.8% to 17.3%.
10% Nitric acid	16.0%.
10% Hydrochloric acid	18.5%.

Thus in practice, the limits of chemical composition recommended are:

Carbon	0.2–1.0%.
Silicon	13.0 to 18.0%.
Manganese	0.25 to 2.0%.
Sulphur	0.05% maximum.
Phosphorus	0.20% maximum.

In this range of composition the following approximate physical properties are obtained:

Density	6.8
Tensile Strength	100 N/mm^2
Compression Strength	530 N/mm^2
Brinell Hardness	450 to 500.
Melting point	1200°C.
Solid contraction	3%.

At 13% silicon, the carbon solubility is about 1.75% at the point of solidification. With silicon at 18% carbon solubility is reduced to about 0.8%. Greater quantities of carbon can be held in solution in the molten alloy, dependent upon temperature and melting conditions. However, in view of the graphitising action of silicon, all the carbon is present as graphite in the solidified alloy at atmospheric temperature.

Unless care is taken in the selection of raw materials, in melting and in pouring, trouble is likely to arise through the separation of free graphite or "kish" in the mould, together with the presence of flake form graphite in the structure. This further embrittles an already weak material. In this respect it has been found that graphite in the eutectic form is best.

The B.C.I.R.A. has evolved a method for controlling the form of the graphite in high silicon cast irons. It is accomplished by adding from 0.02% to 1.0% of cerium to the molten metal, immediately prior to pouring. On cooling the graphite separates out as finely divided under-cooled and any excess over the eutectic quantity takes on the nodular formation. High silicon cast iron given the cerium treatment shows improved mechanical properties including less brittleness and a marked freedom from porosity and cavities.

For the cerium to be fully effective the base iron must be low in sulphur. An example taken from British and American patents covering the use of cerium in this way was melted in an indirect arc furnace. The charge was:

8 kg pig iron.
15 kg Ferro-Silicon (46% silicon).
24 kg Mild Steel Scrap.

The pig iron had the chemical analysis T.C. 3.94%; Si 2.78%; Mn 0.57%; S 0.017%; P 0.029%. 25 kg of iron from this charge was tapped into a ladle and 60 g of cerium mischmetall added. Final analysis obtained was:

Total carbon	1.19%.
Silicon	15.0%.
Manganese	0.59%.
Sulphur	0.015%.
Phosphorus	0.038%.
Cerium	0.118%.

High silicon irons of this type cannot be machined and any finishing has to be done by grinding. Generally, if the castings can be machined with a cutting tool, it is an indication of unsuitability for withstanding severely corrosive conditions.

The high shrinkage of these alloys causes trouble in founding and recent investigations indicate that this property is related, in some degree, to the state of the carbon. It has been suggested that shrinkage increases as more carbon separates as graphite which is the converse of what happens with ordinary grey cast iron alloys. On the whole, it appears that the higher the purity of the iron-silicon alloy the less is the likelihood of cooling trouble.

High silicon cast irons may be melted in the cupola, reverberatory, rotary, crucible, electric furnace (arc or high frequency). Of these, the cupola must be considered the least suitable in view of keeping the carbon within the desired limits.

The pouring temperature will depend on the size and section of the castings in production. It should be kept as low as possible, the normal range is between 1220° and 1280°C. High casting temperature favours the formation of large columnar crystals. On the other hand, a low temperature produces a fine grain. Liquid shrinkage is not excessive, the real difficulties result from the high solid contraction together with the low strength. Moulds must be loosely rammed and cores made to collapse readily. Any parts of the mould likely to cause resistance to free contraction must be eased after casting. Care must be taken to prevent the formation of fins which encourage cracking. Avoid use of chaplets where possible; if used they must be of the same composition as the casting.

Correct design is a fundamental requirement. Sharp angles must be avoided and also sudden changes of section. Use generous fillets. Vessels such as boiling pans, receivers and stills should be made with a curved bottom, as large flat surfaces set up contraction difficulties. All joints in pipe work must be formed by means of flanges or spigot and socket since the alloy is too hard to machine or thread.

One of the troubles is to ensure a sufficient freedom from dissolved gases to produce sound castings. Precautions need to be taken at every stage of manufacture to guard against the introduction of hydrogen to the molten metal. It has been established that the presence of hydrogen and carbon monoxide plays a most important part in causing porosity and unsoundness.

As well as resisting acids, the high silicon cast irons can be used against other corroding media and have stood up well also under conditions of erosion and abrasion, such as for example pipes for handling wet sand. The material cannot be recommended for vessels working under internal pressures of more than 3.5 kg/cm^2. It is also unsatisfactory for withstanding rapidly fluctuating temperatures although the danger of cracking is lessened by the provision of suitable facilities for expansion and contraction. For pumps and valves high silicon cast iron is not recommended if the temperature changes are likely to exceed 40°C. In the case of large tower sections subject to appreciable temperature change, the height must not exceed 1½ times the diameter.

CAST IRONS FOR RESISTING WEAR

Unalloyed Cast Irons

Any grade of cast iron possessing an all white or chilled structure will resist wear and abrasion better than grey cast iron. For the fully white fracture the Brinell hardness will be about 450 to 500. Hardness will vary according to whether the iron is cast in sand or in a chill mould. Also if the cooling rate is not sufficient to retain all the carbon as carbide then the hardness will fall as the structure becomes more mottled and approaches full greyness. Generally, for sand castings to possess an all white fracture under conditions of normal cooling the silicon content will need to be less than 1.0%. When cast into chill moulds the silicon ratio required to produce graphite free structure will depend on the speed of cooling.

Cast iron with an all chilled structure is not machinable in the ordinarily accepted sense of the term. This fact must be borne in mind when considering its suitability for any given purpose. If shaping or boring has to be done then nothing much beyond an all pearlitic matrix cast iron can be employed in which the combined carbon is just below 1.0%. Such cast irons are not wear resistant in the sense of possessing specially high resistance which is being considered under this heading.

High Chrome-Carbon Irons

Some wear resisting alloys develop their full properties only by becoming work hardened. There are many instances involving wear by abrasion in which the scouring action of the abrasive is not sufficient to induce work hardening. In such cases, the high chromium-high-carbon, wear hardening cast irons find useful application. Their range of composition is as follows:

Total carbon   2.25–2.83%.
Silicon   0.50–1.50%.
Manganese   0.50–1.25%.
Chromium   24.0–30.0%.
Nickel   minimum.
Iron   balance.

They have a Brinell hardness of from 500 to 550 as cast but can be hardened by heat treatment up to 600 Brinell. The heat treatment, which gives greatly improved wear resistance, consists of heating to a temperature of 1100°C. When this temperature is attained, hold it for an hour. Then the castings are allowed to air cool. This heat treatment causes the formation of very unstable austenite, so unstable that it will transform to harder martensite under mild abrasive or rubbing action.

These high-chromium, high-carbon irons can be rendered softer for grinding or simple machining by annealing at 760 to 790°C for 12 to 24 hours, followed by air cooling. By this treatment the hardness is lowered to between 350 and 450 Brinell.

Heat Treated Low Alloy Cast Irons

Whilst heat treatment of iron castings has limited application and in general is used for simple shapes only, it may offer a solution to some problems of wear. A composition which can be used is:

Total carbon 3.3%.
Silicon   1.8%.
Nickel   1.5%.
Chromium   0.4%.

After quenching in oil from 850°C. and tempering at 300°/350°C, a hardness of 400 to 450 Brinell is developed.

Ni-Hard

A completely white cast iron which has the property of air hardening by reason of its relatively high nickel and chromium contents. It is a development of the International Nickel Company and can be made in two grades. High carbon imparts maximum hardness whilst the lower carbon grade is used for greater strength properties. Chill castings possess more strength and are harder than sand castings.

Machining is possible only by the use of special tools, otherwise resort to grinding must be made. Inserts of grey cast iron can be "cast in" where it is required to drill and thread.

All Ni-Hard castings benefit from stress-relieving by heating to between 250° and 300°C. No loss of hardness results but strength is improved. Composition and properties as given by the International Nickel Company are:

Composition per cent.	High Hardness	High Strength
Total carbon	3.2–3.6%	2.8–3.2%
Silicon	0.4–0.7%	0.4–0.7%
Manganese	0.3–1.2%	0.3–1.2%
Sulphur	0.15% max.	0.15% max.
Phosphorus	0.3% max.	0.3% max.
Nickel	3.3–4.8%	3.5–5.0%
Chromium	1.2–1.7%	1.2–1.7%
Molybdenum	0.0–0.4%	0.0–0.4%

Specific Gravity 7.6–7.8
Weight lb cu. in. 0.275–0.28
Pattern Makers' shrinkage in/ft. $\tfrac{5}{32}-\tfrac{1}{4}$

Hardness:	Sand Cast	Chill Cast	Sand Cast	Chill Cast
Brinell (Tungsten Carbide Ball)	550–650	500–725	500–600	550–650
Vickers D.P.N. (average)	590–700	640–800	540–640	590–700
Rockwell C (average)	54–60	57–64	52–57	54–60
Scleroscope (average)	72–82	75–90	68–75	72–80
Tensile tons/in^2	16–19	18–26	18–22	22–27

Tellurium Additions

Tellurium introduced to molten cast iron promotes the formation of a hard chill either with or without the aid of metal chillers. It increases the amount of massive carbide in the chill so that a higher hardness and better wearing properties are imparted.

The chilling effect of tellurium does not persist in the re-melted scrap. Thus segregation of scrap, essential with chromium additions, is quite unnecessary. The hardness developed in a tellurium chill is about 450 Brinell.

There is, however, a warning based on evidence put forward by the B.C.I.R.A. that the use of scrap cast iron, containing tellurium, in furnace charges, may cause trouble. It appears that traces of tellurium may persist after remelting in ratios as small as 0.0007%. Such residual quantities of tellurium may give rise to a detrimental arrangement of the graphite which causes very low strength, particularly in grey iron castings of relatively heavy section which cool slowly.

For direct introduction of tellurium into molten cast iron, the best method is to enclose the element in copper tubes or to employ a tellurium copper alloy. Copper tubes containing weighed quantities of tellurium are available. When treating small volumes of iron the required number of TELLURIUM TUBES can be placed in the bottom of a hand shank and the iron tapped on to them.

In view of the relatively low boiling point of tellurium and its great affinity for oxygen, it is advisable not to allow the tubes to float on the surface. If they do the effect may be lost and the fumes evolved will be a nuisance. The most effective and trouble-free method of making the addition, especially in crane ladles, is to wire or wedge the tubes to a stirring rod and plunge them well below the molten surface.

Successful use of tellurium as a chilling medium is dependent on control, especially with respect to:

1. Grade of cast iron to be treated.
2. Quantity of tellurium used.
3. Effectiveness of the method of making the addition.

The degree of chill depth for any given tellurium treatment would depend on the character of the iron. Thus a low silicon iron will chill much deeper or in much heavier sections than a soft grey iron used for ordinary light engineering castings. Small percentage additions to tellurium only are needed and 0.02% is about the maximum ever required.

Besides direct addition of tellurium to molten cast iron it is possible to impart a chilled surface to iron castings by painting the mould or core, adjacent to the area desired hard, with a dressing containing tellurium. TELLURIT is a mould paint of this kind and is available in both water and spirit-based versions.

INOCULATED GREY CAST IRON

Cast iron is sensitive to section thickness; that is to say, the physical and mechanical properties will vary with section or, more correctly, with cooling rate. Arising from this characteristic, it is difficult to predict accurately the mechanical properties of a casting and variations in surface hardness, even hard spots will certainly occur.

The classical means of achieving the desired physical properties and structure of cast iron, in relation to section, has been, and to a large extent still is, to change its chemical composition. For practical reasons, a compromise must be expected because it is impossible to adjust the chemical composition to give optimum properties in each individual casting.

Difficulties also arise when a foundry is attempting to make various types of casting of different section thickness from one type of cast iron alloy.

Inoculation

Briefly, inoculation of grey iron is a means of reducing sensitivity to section, thereby enabling the foundryman to produce castings of more consistent quality. While inoculation does not compensate for an unsuitable iron composition, it does greatly reduce the effect of small variations in chemical composition and hence simplifies the function of metal production and control.

The physical properties of grey iron are largely determined by its micro structure. Since variations in micro structure are reflected in physical property variations, the problem of improving physical properties and achieving greater consistency of microstructure must be tackled by some method of improving the microstructure. This is essentially what inoculation does.

Desirable physical properties include good machinability, high tensile strength coupled with toughness, pressure tightness, resistance to wear and useful hardness values. Above all, the effect of casting section on these properties must be minimised, and consequently the micro structure must be as uniform as possible throughout the casting, whatever the variations in section.

Castings produced from iron which has not been inoculated are likely to reveal a white or mottled structure, particularly in thin sections. Such irons frequently show an interdendritic distribution of the graphite flakes coupled with a coarse "eutectic cell" structure, in heavier sections. It is the chill structure which largely accounts for complaints regarding inconsistent machineability.

The reason for these variations lies in the difference in the rate of cooling between the outside of the casting and the interior. This is unavoidable but inoculation does minimise the influence of cooling rate on the micro structure; thereby producing more or less constant properties from centre to edge of a given casting.

What Inoculation Does

In the metallurgical sense, grey cast iron is a brittle material, and even those alloys which come under the category of high duty iron, possessing high tensile strength, may not have increased toughness. In thin sections, the latter possesses hyper-section sensitivity and exhibits under-cooled graphite structures at the surface. The resulting structures are particularly brittle to impact loads, which under service conditions, can off-set any improvement obtained in the bulk of the material.

Inoculation has a marked improvement on toughness, though this property is difficult to define and even more difficult to measure quantitatively.

It is known that a relationship exists between impact strength and the "eutectic cell count" of a grey cast iron. The greater the number of eutectic cells per unit area, the higher is the impact resistance of the cast section. Increasing the number of eutectic cells for a given area is a major outcome of inoculation and for this reason therefore inoculation always increases transverse strength and can effect a marked increase in deflection under load. Since inoculation refines the "grain size" (i.e., increases the eutectic cell count) and also promotes the formation of Type A graphite, the tensile strength is also improved. A very important feature is the reduction in chilling characteristics, thereby greatly reducing surface hardness and producing thin sections free from iron carbide.

While a great deal is known about the theory of inoculation, it is not yet fully understood. Some of the evidence is conflicting regarding the effects of different types of inoculant on various types of iron. The origin of these differences in response to a given inoculant originates in variations of chemical composition, charge composition and melting conditions existing in the cupola. With electric furnace melting, such factors as bath temperature, holding time and general melting routine can all affect the response of the iron to a given inoculant.

A considerable amount of work has already been carried out in order to evaluate these inter-related factors and this has led to the introduction of INOCULIN 25. This product has been shown to give more consistent results over a wider range of iron alloys than any other inoculating material currently available.

Many currently available inoculating agents suffer from practical disadvantages; calcium silicide is very good for high steel mixes but is not particularly potent in inhibiting chill in thin sections. Its major disadvantage, however, is that it produces a viscous slag on the surface of the ladle which is difficult to remove and all too frequently results in slag inclusions in a casting.

Ferro-silicon is a good inoculant for inhibiting chill but it is not very effective in heavy sections. Quite large additions have to be made which tend to lower the metal temperature, and this also produces a slag on the metal in the ladle. A serious disadvantage of ferro-silicon from the practical point of view is that the inoculation effect appears to be controlled by the amounts of aluminium and/or calcium present as impurities in the ferro-alloy. Indeed, it has been suggested that without the impurities ferro-silicon would be virtually worthless as an inoculant. The amounts of these elements present can vary widely depending on the origin of the ferro-alloy but most varieties used for inoculating purposes will contain between 1.0% and 1.5% of aluminium and up to 0.5% of calcium. Some grades can have up to or even beyond 2.0% of aluminium but in these quantities the possibility of "pin-holing" exists due to promotion of a metal-moisture reaction from retained aluminium in the iron. For consistent inoculating effect therefore it would appear desirable to obtain a consistent ferro-alloy: the inoculating effect due to ferrosilicon is also reported to fade quite rapidly, which can be a marked disadvantage if using large ladles or casting heavy sections.

One of the advantages of INOCULIN 25 is that it is rapidly and fully soluble and so makes minimum contribution to any slag problem. Application can therefore be at the last possible moment before the metal enters the mould.

The Application of an Inoculant

The method of adding an inoculant, as distinct from the quantity added, can be very important. As described above the simplest system probably consists of some form of trickle dispenser discharging into the cupola or furnace spout in a region of some metal turbulence. Treatment should start when the base of the ladle is covered and should continue for at least 2/3 of the tapping time. The onset of "fading" is, however, almost immediate and minimum delay should occur before casting.

A number of other spout application methods have been proposed including mechanically fed alloy wire or filled tube, etc. but will suffer from the same factor of fade with time before casting. To take account of this, particularly in mass production foundries where constant in-the-mould conditions are vitally important, inoculant application has switched to the mould itself. Where very large numbers of similar moulds are constantly passing the pouring station it obviously makes sense to inoculate the individual mould to avoid variations in metal quality and sensitivity to section. Indeed if automatic pouring is employed it is the only way of ensuring consistent metal treatment. Several methods of inoculating in this situation are in existence.

Where moulds are all the same size and indexed into position it is possible to install an electronic automatic inoculant dispenser such as the MSI System 90 equipment developed by British Cast Iron Research Association and Foseco in co-operation. The principles involved are illustrated in Fig. 6.13: note that a specially graded inoculant—INOCULIN 90—must be used to remain compatible with the mechanics and deliver a constant pre-set amount. The MSI.90 System is operated by compressed air and electricity and can be fully automated and actuated by either mechanical or visual signals.

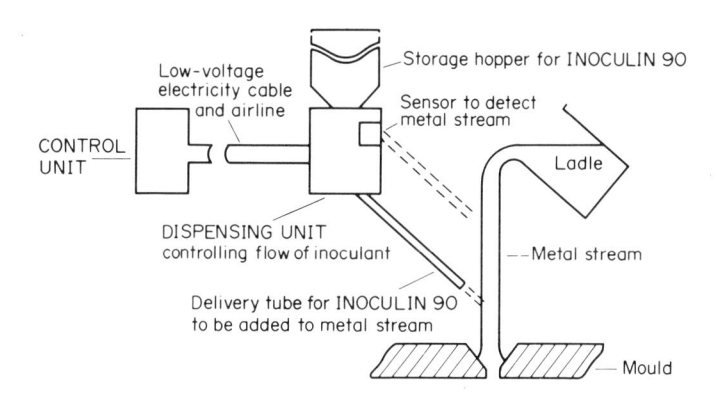

Fig. 6.13. The principles involved in MSI System 90.

Simpler, less sophisticated methods can be based on INOTAB or INOPAK types of product. Both are based on a fully soluble ferro-silicon of carefully selected grading to ensure rapid and complete solubility. INOTAB, as the name suggests, is a tabletted material available in a range of sizes to suit the weight of metal in a particular box. The tablet may be located in the runner bush, on the upstream side of a strainer core, filter cloth or ceramic filter or it may be pressed into a moulded recess on the top side of the runner bar. INOPAK is a similar closely graded silicon-based inoculant in fine powder form sealed into a paper sachet rather like a tea bag. Again it is located in the runner bush or upstream of a filter. Since both INOTAB and INOPAK are rapidly and fully soluble some users simply drop them down the downsprue into a well or sump at the bottom and pour metal as usual. Without an ingate filter however some possibility of penetration of unabsorbed inoculant into the mould cavity must always exist.

A similar in-the-mould technique can be utilised with bulk granular inoculants provided a special design of reaction chamber is incorporated into the running system—rather similar to the reaction chamber used for the in-mould method of making S.G. iron. Again a filter would be a desirable feature to prevent possible carry-over into the mould cavity.

The Influence of Base Iron Composition

A major factor which influences the response of an iron to a given inoculant is the composition of the base charge. For example, a soft grey iron (produced from 40% pig iron, remainder returns and bought scrap) will, after treatment with INOCULIN 25, be less likely to chill and form undercooled graphite structures. As a result, castings will be tougher and have better and more consistent machining characteristics. In castings of up to 25 mm section, inoculation will serve to increase the eutectic cell count, with the result that the main body of the casting will be tougher. Above 25 mm section, the structure begins to coarsen rapidly and physical properties will fall off. Inoculation will not prevent this happening with a charge composition of this type.

By progressively increasing the steel content of the charge the effect of increasing section on coarsening of the structure is reduced, provided that the iron is inoculated. INOCULIN 25 has a distinctly marked effect in promoting this characteristic, while at the same time, the hardness of castings does not noticeably increase. The iron is also less sensitive to chilling in thin sections than is the case with uninoculated iron.

With increasing steel content, the total carbon content is progressively reduced, thus the volume of graphite formed is less and, as a consequence, the iron is potentially stronger and tougher. Again, inoculation of the charge with INOCULIN 25 ensures that this potential is realised.

Thus, whatever the basic composition of the iron charge, inoculation enables the foundryman to produce a more consistent product showing an all round improvement in mechanical properties and in chill reduction.

THE WEDGE CHILL TEST

Importance of Control

In jobbing foundries, where the value of individual castings is often great, and in mechanised foundries, where a considerable value of castings can be produced in a very short time, the use of an unsuitable iron can in either case result in very substantial losses. Some means, therefore, of testing the quality and so suitability of an iron as it comes from the cupola is thus very desirable. A technical committee (T.S.6) of the Institute of British Foundrymen was appointed in 1944 to report on suitable testing methods, and one of their several excellent recommendations was the Wedge Chill Test. Much of the information given here is abstracted from the report of this sub-committee.

Chilling Indications

Different types and compositions of cast iron usually exhibit well-marked differences in chilling tendencies when quickly cooled from the molten state. This chilling tendency is related to melting conditions, chemical composition, particularly as regards carbon and silicon contents, and also to the section thickness cast. For example, in a casting of normal thickness, one iron may exhibit no chilling, another grade might produce a casting having a completely white fracture, while an intermediate grade could produce a casting having a fracture partly grey and partly white. Thus, by casting an iron under carefully controlled conditions of section thickness, etc., a rapid means of assessing its suitability or otherwise for a certain type of casting is provided. The wedge test will also indicate the degree to which depth of chill is reduced after inoculation of the iron.

Wedge Dimensions

In order to minimise the time required for carrying out the test and also the cost involved, the wedge chosen should be the smallest possible consistent with giving a reasonable proportion of white to grey fracture with the particular type of iron in use. There are eight standard sizes of wedge from A, the smallest, to H, the largest, as detailed in Fig. 6.14 and the table. The table also shows the time required to complete the test.

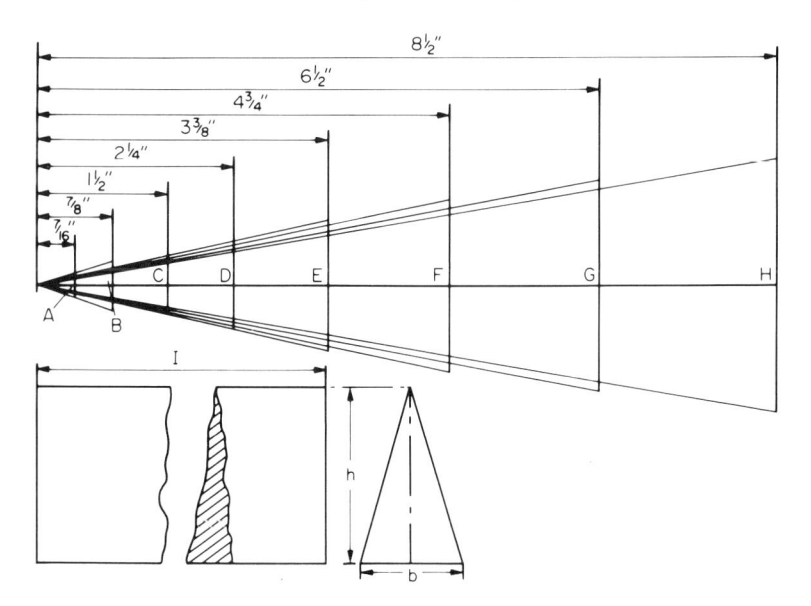

Fig. 6.14.

An iron required for thin sectioned castings will usually have high total carbon and high silicon contents, in order to avoid chilled sections in the castings themselves. Such irons would, therefore, show little, if any, chill in the test pieces, and in testing, one of the smaller wedges A, B or C, would be chosen. On the other hand, in testing an iron required for heavier section castings, requiring high strength or high hardness, and which would show a completely white fracture in small test pieces, one of the larger wedges, D, E, F, etc., would be used.

Sand for Mould

To minimise variations through sand conditions, it is recommended that wedges be cast in moulds made of moisture-resistant oil sand. For casting, the moulds can be set in a moulding box, and backed up with loosely-packed moulding sand. Stocks of moulds should always be available.

Testing Procedure

A sample of the metal to be tested is cast into an appropriate size of wedge mould. The pouring temperatures should be standardised, and preferably checked by means of a pyrometer. When the metal has solidified, the wedge should be removed from the mould at a dull red heat—about 600°C (1100°F.)—and quenched in water. The base side of the test piece should be the first to enter the water, and piece moved quickly about the water to avoid spurting by the steam. On removing from the water, the wedge is broken up by a breaking block, as shown in Fig. 6.15.

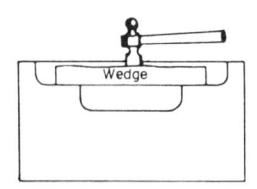

Fig. 6.15.

The fractured section will then clearly indicate a white area at the tip, as indicated (Fig. 6.16). It is the measurement of the base dimensions W, in millimetres, which is taken as the wedge reading. Figure 6.17 shows a useful device for making quick measurements.

Fig. 6.16.   Fig. 6.17.

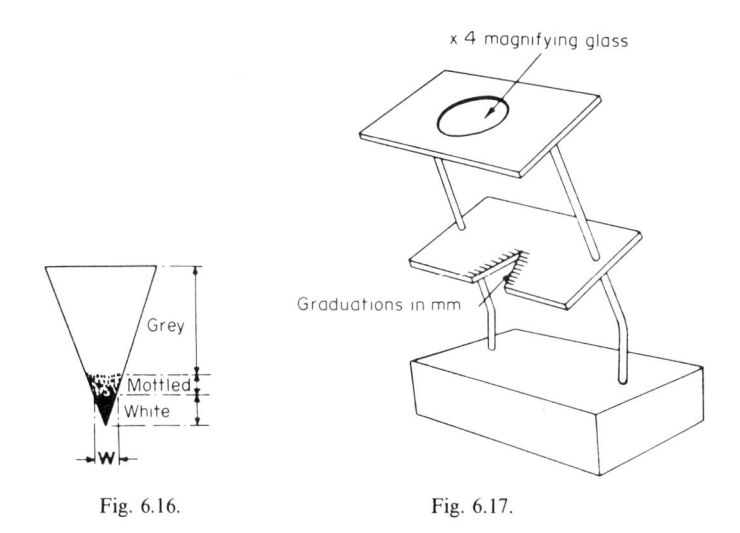

Applications of the Test

The wedge chill test is of the greatest possible value in controlling the quality of iron to meet required specifications. If, in running a routine test, a variation in the depth of chill appears, it indicates that a change in the composition of the iron has occurred; provided that cupola operation, pouring temperature and mould conditions have remained constant. The test also indicates the efficiency of an inoculating treatment; or whether the effect of inoculation has faded.

Apart from its usefulness as a means of checking the consistency of metal as it comes from the cupola, the test may also be used to gauge the suitability of an iron for any specific purpose. The depth of chill obtained can be compared with the depth of chill in wedges cast previously from an iron of a composition known to be suitable for the job in question. For example, if on comparison the depth of chill produced by the iron being tested is greater than that in the control wedges, its composition can be adjusted by making suitable additions of silicon to the ladle.

In the manufacture of high duty types of cast iron with charges containing steel scrap, the wedge test provides an almost indispensable means of control. These types of iron require inoculation treatment at the cupola spout, and the difference in degree of chill shown by wedge tests taken before and after inoculation is a measure of the effectiveness of this treatment.

INOCULIN 25 will reveal its efficiency very well by taking wedge chill test samples before and after inoculation. A guide to the suitability of the inoculated iron, as judged by the wedge test, is that the chill width should not exceed one-third of the average cross sectional thickness of the castings which are to be made. For example, iron which gives a chill measurement of 6 mm in the wedge should not be poured into castings which have a section thickness of less than 18 mm. This ruling applies where maximum tensile strength is required but for general purposes, a wedge value of one-quarter of the average cast section thickness can be adopted.

Wedge	Base (b) mm	in	Height (h) mm	in	Length (l) mm	in	Time taken to complete test
A	6	$\frac{1}{4}$	11.1	$7/_{16}$	57.1	$2\frac{1}{4}$	35 seconds
B	12.7	$\frac{1}{2}$	22.2	$7/_8$	101	4	40 seconds
C	19	$\frac{3}{4}$	38	$1\frac{1}{2}$	127	5	1 minute
D	25.4	1	57.1	$2\frac{1}{4}$	127	5	$1\frac{1}{2}$ minutes
E	38	$1\frac{1}{2}$	85.7	$3^3/_8$	152	6	3 minutes
F	51	2	120.6	$4\frac{3}{4}$	152	6	10 minutes
G	63.5	$2\frac{1}{2}$	165.1	$6\frac{1}{2}$	203	8	15 minutes
H	76.2	3	215.9	$8\frac{1}{2}$	203	8	20 minutes

Ramming Mix or Patching for Cupola Lining

Ganister rock $-\frac{5}{16} + 20$ mesh   40%
Fine silica sand   40%
Fireclay of maximum refractoriness and plasticity   20%
Moisture   7% approximately

When used as a patching or daubing material for cupolas, all the slag from the previous heat must be removed from the brickwork or lining, otherwise when the furnace is in blast the old slag will melt and allow the new patching material to come away.

SECTION VII
DIE-CASTINGS

GRAVITY DIE OR PERMANENT MOULD CASTINGS

General Notes

In this method of casting, molten metal is poured (under "gravity") into a metal ("permanent") mould or die. The metals usually cast are primarily those aluminium alloys that are suitable for this technique, some copper base alloys, e.g. gravity diecasting brass and aluminium bronze, and occasionally zinc, tin, or lead (battery plates) alloys: some ferrous diecasting is also carried out. The die itself can be made of any metal with a similar or higher melting point than the metal being cast, but for a number of reasons a good, close-grained haematite type cast iron is most commonly used. Cores are usually made from heat-resisting steel, but for some castings, e.g. cocks, etc., increasing use is being made of non-reusable cores made from resin or silicate-bonded sand. This method introduces its own practical difficulties in that the rates of heat extraction from the cast metal as between the metal die and the sand core are vastly different, and unfed shrinkage associated with the sand cored face is common.

Cast Iron Composition suitable for Gravity Dies

A chemical analysis which works well in practice lies within the following limits:

Total carbon   3.0–3.3%.
Silicon   1.2–1.6% according to section thickness.
Manganese   0.8–1.2%.
Phosphorus   0.3%.
Sulphur   0.12%.

The heat resisting properties of this material can be improved by making an addition of 1% Simcronite 2 to the molten iron. Dies should be cast with metal at as high a temperature as possible, and it would be an advantage to cast into a heated mould, so as to get more uniform cooling. Before using it is advisable to stress anneal the castings by soaking for at least 1 hour at 790°C and cooling down slowly in the furnace.

Removable Steel Cores

Cores are constructed from heat resisting steel of which there is a variety of grades. A suggested composition is:

Total carbon	0.3–0.4%.
Silicon	0.3–0.4%.
Manganese	0.2–0.3%.
Chromium	3.0–3.75%.
Vanadium	0.45%.
Tungsten	8.0–9.0%.
Cobalt	1.0–1.5%.

The heat treatment for this steel is: heat slowly to 890°C and quench in oil. Temper at 550–580°C by heating slowly and cooling in still air.

Coating the die

It will be appreciated that when molten metal is cast into a relatively cool die the rate of heat extraction, and hence the speed of solidification, can be very fast indeed. So much so in fact that unless special precautions are taken it would be almost impossible to cast any but the most simple shapes. Control of transfer of heat from metal to die is achieved, firstly, by dressing the die with an insulating refractory coating and, secondly, by control of metal and die temperatures. In practice the die-dressing assumes the greatest importance, and temperature control centres mainly around control of pouring temperature. A smoothly cycling die will normally reach a satisfactory temperature balance with its surroundings, and this should be the aim of the operator.

Before casting begins the die is heated to a temperature of about 110–130°C for the application of the refractory dressing by spray gun. If the die is much hotter than this the coating will "frizzle" and leave a rough dressing, while if it is cooler the dressing may "run". Coating a die properly is very much an art. Obviously where maximum insulation is required (runners and risers particularly) the dressing must be applied more thickly, and where maximum chilling is necessary the coating may be only very thinly applied or even subsequently removed entirely. The graduation between maximum chilling and maximum insulation is usually the product of experience both with the die and the dressing being used and the general principles of the process. A "coarse" dressing will give maximum insulation while a "smooth" dressing, while giving a smoother casting finish, will not allow the metal to "run" so easily and may not therefore be so suitable for thin-sectioned castings. Only the die cavity is dressed, and any coating falling on the mating faces or on core prints, etc., is removed. These faces can be treated with a graphite in water dispersion if necessary to improve die operation.

After applying the refractory coating the die is re-heated to about 300–400°C to frit the dressing into position and provide sufficient die superheat to begin casting. Dies should always be above equilibrium temperature at the commencement of a run: if attempts are made to work the temperature up by casting metal into them, the usual results are "stickers" and rapid removal of the die dressing. Small removable cores can rapidly overheat, and it is usual either to use duplicate cores (one set cooling while the other is in the casting) or to dip the entire core into a dilute suspension of graphite in water. A small amount of graphite is picked up each time and will help in providing easier extraction.

Dies for copper base diecastings may not be coated with a refractory dressing as discussed above—a lot will depend on the complexity of the casting. Because of the higher casting temperature, however, die temperature fluctuation will be greater, and it is usually necessary to quench the whole die in an aqueous graphitic dispersion as described for cores. Again, some graphite is picked up each time and forms an essential part of the technique.

In ferrous diecasting the dies may be coated with an insulating—lubricating dressing and, in addition, may be "sooted" with an acetylene flame.

Die Coatings

(a) White insulating dressings:
  DYCOTE 39 (smooth), 34 (coarse), and 140 (medium).
(b) Graphite in water dispersions:
  DYCOTE 11, 38.
(c) Graphite in oil dispersion:
  DYCOTE 40, 410.
(d) Combined insulating/lubricating dressing:
  DYCOTE 36.

Melting Procedure for Aluminium Alloys

Because of the nature of the diecasting process, complex and sophisticated metal techniques are not usually required. Only occasionally is it necessary to obtain complete freedom from dissolved gas in the metal or a high level of grain refinement. The rapid rate of solidification usually ensures that any gas porosity is fine and adequately dispersed and also that grain structures are small and equiaxed. Thus the only treatment accorded the usual bale-out type of furnace is to dross off periodically with COVERAL 11 using about 0.5% of the weight of metal. Indeed, it is frequently advantageous to make deliberate additions of gas to the metal so that isolated shrinkage unsoundness in the casting is offset by the dispersed gas porosity. To accomplish this, DYCASTAL is added either to the furnace as a plunged tablet or as a "pinch" in the ladle before it is filled.

When, for special quality reasons, it is necessary to provide a full degassing and grain-refining treatment, the technique outlined for sand castings should be used.

The rapid solidification rate in a metal die will also ensure a fair degree of modification of the eutectic in the aluminium–silicon alloys such as LM6, especially in the thinner sections. Where extra modifying potential is required, metallic sodium may be added as NAVAC or NAMETAL or a modifying flux such as COVERAL 36A or 29A may be used.

Special Aluminium Alloys

Under this heading come alloys such a Birmabright (LM5) and "Y" Alloy (L35) for which the technique is as follows:

Melt down the ingot and process scrap under a layer of COVERAL 65, superheat to a maximum temperature of 750°C, and skim. Add a further 2% of the charge weight of COVERAL 65. Allow the flux to fuse on the surface, followed by a thorough puddling of the "wet" slag for 3 to 4 minutes, employing for the purpose a skimmer tool with a rotary motion. Any remaining "islands" of slag or dross should be skimmed off and the sides of the crucible at the metal level gently scaped. Ladling out then proceeds in the usual manner.

In the case of a bale-out furnace working continuously with more ingot metal and scrap being fed in from time to time, the routine outlined would be repeated at intervals dependent upon the rate at which the metal is consumed.

These alloys are susceptible to grain growth, even when cooled relatively quickly as in gravity die-casting. To ensure the production of a fine grain size, NUCLEANT 2 tablets are plunged on a basis of 0.25% by weight.

Although the technique outlined for these special alloys is also suitable for Lo-Ex (LM13), there is an alternative procedure which may be considered simpler. COVERAL 11 is used in the bale-out furnace periodically, purely for cleansing and drossing off. Degassing with DEGASER 190 tablets is carried out at prescribed intervals dependent upon local conditions.

Slush Casting

This is generally carried out in pure aluminium and applied to such components as spouts for kettles and teapots. Common troubles experienced are cracking and sticking to the die.

Pick up of iron is a fruitful cause of cracking so it is best not to use a steel crucible for melting. In addition, all plungers and other steel tools introduced into the molten metal should be well-coated with DYCOTE 36 and dried before use.

Melt down and dross off the pure aluminium with COVERAL 11. NUCLEANT 2 tablets should be plunged at intervals on a basis of 0.25% by weight to counteract excessive grain growth to which pure aluminium is prone. Suitable intervals for applying the NUCLEANT and drossing off will depend on the rate at which the molten metal is used and can coincide with the make up of the bath by additions of fresh ingot or returned scrap. A suitable pouring temperature will be something of the order of 720°C.

Aluminium Bronze

For gravity die-casting in this alloy the 88/9/3, copper–aluminium–iron composition is widely used. Melting is done with the aid of ALBRAL 3 which exerts a powerful solvent action on oxides, including alumina. The latter is formed continuously at the high temperatures involved in the casting of an aluminium bronze alloy, but a protective cover will inhibit is formation. It is advisable to work in small quantities of ALBRAL from time to time, a plunger or similar tool being employed for the purpose. The metal may be further deoxidised by plunging one DEOXIDISING TUBE E3 to each 50 kg of metal.

Under conditions of working, fresh metal has to be added as casting proceeds. Thus, the treatment with ALBRAL and DEOXIDISING TUBES must be repeated at suitable intevals. The exact nature of the technique employed will depend on whether a bale-out furnace is used to hold a reservoir of metal which is replenished from a separate melting furnace.

An average pouring temperature is around 1200°C, which makes it necessary to use special hand ladles of heat-resisting alloy steel. For the same reason the dies are sometimes sunk in special steel although close-grained cast iron is employed satisfactorily.

For coating the gravity die blocks DYCOTE 36 is suggested. The steel cores, especially if of thin section, need cooling down in between each cast. This is conveniently done by dipping them into an aqueous suspension of DYCOTE 11. The graphite coating thus applied imparts the desired lubricating properties, and assists in the extraction of the cores.

Turbulence must be guarded against when pouring the die, bearing in mind that aluminium-bronze is a dross forming alloy. The relatively high solidification shrinkage must also be taken into consideration in allocating the dimensions of feeding risers. In the effort to avoid turbulent flow and agitation of the molten alloy in the die cavity, full advantage should be taken of dies which lend themselves to tilting during pouring, the die being brought back to the vertical position as metal enters the feeder.

A precaution against cracking is to rotate cores and pins whilst the metal is being poured. These should be withdrawn as soon after casting as possible. In fact, the whole die assembly should be released as quickly as is practicable after the component has set.

DEOXIDISING TUBES L have been applied successfully to the gravity die-casting of aluminium-bronze, particularly in cases where cracking has been experienced due to the presence of impurities, especially bismuth. Aluminium-bronze is very hot short at its critical temperature, and the presence of impurities tends to interfere with the normal mechanism of solidification, leading to minor or major cracking. DEOXIDISING TUBES L also influence grain refinement and act as a degasser. A DEOXIDISING TUBE LI containing 5.5 g lithium, is normally sufficient to treat 75 to 100 kg charge. The tubes must be completely immersed.

Gravity Die-casting of Brass

The chemical specification for die-casting brass is substantially 60/40 copper–zinc alloy, containing also up to 0.5% aluminium and from 0.5–1.0% of lead.

Where multiple piece dies are used they are taken apart and dipped in a water suspension of DYCOTE 11 after each pour. In this way the dies are kept cool and the tendency for build-up to occur on the face of the dies is considerably reduced. The steel cores are also dipped in the suspension of DYCOTE 11 after each cast. An alternative coating for the die blocks is DYCOTE 36, or DYCOTE 61 which is a graphite-based dressing developed to reduce zinc oxide build-up on the die walls.

Melt the brass under a protective flux cover of 1% by weight of ALBRAL 3. Thereafter about 0.5% of the charge weight is stirred into the melt at intervals, preferably at the time of adding fresh alloy to the holding furnace. The dross which forms on the surface of the melt can be skimmed off following each application, or pushed aside while ladling out. Superheat the bath to a temperature of about 1100°C and pour the dies with metal at about 1050°C.

Gravity Die-casting of Gunmetal

Gunmetal alloys have not been employed very extensively for gravity die-casting in Great Britain. There are a number of reasons which include short die life, extraction difficulties, and sub-surface porosity, etc. Even so, satisfactory gunmetal die-castings can be produced provided the component is small and of simple design.

In melting down the initial bale-out charge, CUPREX 1 blocks should be employed on a basis of 1% of the charge weight. Thereafter melting will be continuous with fresh ingot metal and returned scrap being added from time to time to maintain the bath of molten alloy. Subsequently REGENEX may be plunged at appropriate intervals to keep the liquid metal clean and degassed.

Following the degassing operation, strong deoxidation is carried out by plunging one DEOXIDISING TUBE DS4 to each 25 kg of melt. This will also have the effect of promoting fluidity to a maximum degree.

Pouring temperatures will range from 1150° to 1200°C, and the die should be heat-resisting cast iron or steel to stand up successfully.

DYCOTE 36 is recommended for spraying the die blocks, which should be sunk with as much taper as dimensional tolerance will allow, otherwise extraction difficulties will become a major problem.

Eliminate any tendency to build up back pressure by cutting venting grooves in the die at as many points as design permits.

Before casting commences always pre-heat the dies to a working temperature of 350–400°C with the idea of limiting thermal shock and blowing.

Gravity Die-casting of Lead and Lead Base Alloys

Melt down under a cover of 1% of the charge weight of PLUMBREX 2 which is also used for drossing off purposes prior to pouring. A further small quantity may be plunged occasionally for thorough cleansing.

The die face may be coated with an insulating dressing as used for diecasting of aluminium alloys (see appropriate section) if feeding problems are anticipated and directional solidification is required. For simple castings, however, it is often sufficient simply to apply a coating of silicone (most conveniently from an aerosol) to the bare die surface.

Battery plate manufacturers typically employ a highly insulating coating because of the very thin grids they are making, and this is usually based on cork flour with a silicate binder. Individual casters frequently add additional materials, cellulose binder derivatives being most common. Battery grid casting is a very specialised section of gravity diecasting of lead, and most producers have evolved their own individual techniques.

Gravity Die-casting of Magnesium Base Alloys

It is best not to attempt the production of complicated components containing many changes of section. Dies normally used for aluminium alloys need to have the runner-riser system enlarged. Magnesium alloys are more prone to shrinkage cavitation.

The following observations apply to the bale-out furnaces fitted with steel crucibles but where no protection is afforded the molten alloy by hooding and a continuous atmosphere of sulphur dioxide. Under such conditions it is essential that the surface of the melt be covered by a thin "barrier" of INERTEX, thus reducing burning to negligible levels.

Into the pre-heated crucible place MAGREX 60 on a basis of about 3% by weight of the final charge. Add pre-heated magnesium ingot and/or scrap, in stages, until the melt is complete. Keep the flux cover intact and make good any exposed areas by a further addition of MAGREX 60. Keep the melt to a working temperature of 730–750°C for the refining stage.

For refining, the MAGREX 60 fluid slag is thoroughly stirred into the bulk of the metal over a period of 3 to 4 minutes. If properly conducted the surface of the magnesium assumes a bright lustre. A "frothy" appearance indicates the need of continued refining with a fresh introduction of fluxing compound.

The slag loaded with entrained oxides and impurities gradually sinks although some will cling to the walls of the pot. Atmospheric attack upon the surface of the melt should be curtailed by adding a little fresh flux.

Grain refinement is carried out by plunging NUCLEANT 200 tablets at a rate of approximately 125 g per 50 kg of magnesium alloy. In most alloys these tablets have a sufficiently high density to sink of their own accord. Simply displace the flux on the surface of the melt and place the tablets on the exposed surface. Allow the reaction to take place steadily over a period of 2 to 3 minutes, after which the charge is allowed to settle for 10 minutes under its protecting cover of flux compound.

Finally, skim off the surface flux in stages. As the metallic surface is exposed, cover the area immediately with INERTEX, so preventing any tendency to burn. Ladling out can then commence. The die cavity is flushed with sulphur dioxide, delivered under pressure from a cylinder, prior to each pour. The die blocks are dressed with a strongly adhering coating such as DYCOTE 140.

Gravity Die-casting of Tin

As a covering flux during melting use STANNEX 101 which can also be applied at intervals for cleaning up the melt when fresh ingot or scrap is added. This should be placed on the surface of the molten tin and plunged slowly as directed.

The temperature of the molten tin bath should not rise above 280°C, and a temperature of between 260° and 270°C is suitable for pouring into the die. DYCOTE 39 is used for coating the die or silcone, as described for lead alloys, may be used.

PRESSURE DIE-CASTING

General Notes

The following points are of importance for the production of sound pressure die-castings in whatever metal or alloy.

1. Use as low an injection temperature as will ensure complete filling of the die cavity.
2. Maintain a suitable average die temperature free from sharp local temperature gradients.
3. Position the gate so that the injected metal does not impinge directly upon die or core surfaces.
4. Employ an injection pressure high enough for a strong pressure pulse to be transmitted through the solidifying casting at the end of the injection stroke. Where these conditions are observed the incidence of porosity, shrinkage voids and surface imperfections is reduced to negligible proportions.

Composition of Die Steels

It is usual to employ steels of heat-resisting composition for the construction of pressure dies and cores where required.

Die-casting die steels are usually machined in the soft annealed condition and heat treated after machining. The temperature and length of time required for the various steps in heat treatment depend on the type of steel and the size and intricacy of the die being treated. It is best to obtain the advice of the makers of the steel and their instructions for suitable heat treatments.

Some typical chemical analyses of tool steel for die-blocks are given in the following tables taken from the book on die-casting by Charles O. Herb.

DIE-CASTINGS

COMPOSITION OF STEELS COMMONLY USED FOR DIE-CASTING DIES

Chemical element	Letters designating different types of steel							
	A	B	C	D	E	F	G	H
Carbon	0.20–0.50–	0.40–0.50–	0.40–0.50–	0.30–0.40–	0.30–0.40–	1.40–1.50–	0.35–0.45–	0.25–0.35–
Manganese	—	0.65–0.80–	0.40–0.80–	0.20–0.35–	0.20–0.35–	—	0.20–0.35–	0.20–0.35–
Silcon	—	0.30–	0.30–	0.8–1.0	0.8–1.0	—	0.35–	0.35–
Chromium	—	0.75–0.90–	2–2.50	4.75–5.75	4.5–5	12–13	2.5–3.5	2.5–3.25
Vanadium	—	—	0.15–0.30	—	—	—	0.30–0.60–	0.30–0.60–
Nickel	—	1–1.50	—	—	—	—	—	—
Tungsten	—	—	—	4–5	0.75–1.25–	—	8–10	12–16
Molybdenum	—	—	—	—	1–1.50	0.50–0.60	—	—
Cobalt	—	—	—	0.50–	—	3–4	—	—

BRINELL HARDNESS OF DIE STEELS LISTED IN TABLE ABOVE

Base alloys for which the listed die steels are used	Letters designating different types of steel							
	A	B	C	D	E	F	G	H
Tin	220	170–220	—	—	—	—	—	—
Lead	220	170–220	—	—	—	—	—	—
Zinc	170–250	170–300	300–402	—	360–402	—	—	—
Magnesium	—	—	387–430	402–444	402–444	—	—	—
Aluminium	—	—	402–444	420–460	420–460	430–517	444–517	444–517
Copper	—	—	—	—	—	430–517	444–517	444–517

Metal Treatment for Aluminium Base Alloys

The popular British aluminium base alloys for pressure die-casting include: LM1, LM2, LM4, LM6, LM25.

The melting and fluxing techniques for these alloys have already been dealt with (see pages 139 to 156) and in the main these instructions apply here. In some cases modifications are advisable and these are included in the notes which follow.

Aluminium Alloy LM6

As melting down and drossing off flux the low melting point COVERAL 36A is recommended. This can be used for the initial melting down in the bale-out furnace and applied for cleaning up the surface of the change at appropriate intervals and when fresh ingot metal or scrap are added.

Aluminium Alloy LM2

The instructions given for LM6 apply equally in this case also. Sometimes trouble is experienced due to the casting welding, or sticking, to the die. So far as aluminium base alloys are concerned an opinion has been expressed that up to about 0.8% of iron in the alloy helps to prevent this taking place. Suitable die dressings are, of course, a factor in overcoming this trouble and some experimentation is often necessary.

Aluminium Alloy LM4 and LM25

The melting and fluxing of this alloy consist in the plunging of DEGASER 190 at pre-determined intervals prior to drossing-off with COVERAL 11. The freezing range of this alloy is between 525 and 625°C. A suitable temperature range of injection is 650–670°C.

Aluminium Alloy LM1

Instructions as for LM4 apply.

Magnesium Base Alloys

If the magnesium alloy is in the form of scrap and is melted down and refined prior to treatment in the bale-out furnace, it should be melted and refined, using MAGREX 60. The technique to be followed here has already been given on page 347.

As the slag is comparatively heavy, much of it sinks to the bottom of the pot, leaving behind a thin but tenacious protective film. About 1% of the charge weight of flux is used. It is customary to apply the flux at intervals which are dictated, mainly, by the rate at which the metal is replenished. From time to time it is necessary to dredge out the spent sludge from the bottom of the crucible.

One big advantage in the case of pressure die-casting magnesium alloys is the fact that the metal is worked at a relatively low temperature, thus curtailing oxidation. The temperature of the magnesium alloy in the bale-out furnace prior to transference to the cold-chamber die-casting machine should be about 680°C. The steel crucible of the bale-out furnace for magnesium base alloys should be of pressed or cast construction. There is no need to employ inhibitors in the form of SO_2 gas for flushing the dies as applies to the gravity die-casting of magnesium base alloys.

As some form of "self healing" surface flux is called for during ladling out, INERTEX is very effective and a light dusting at regular intervals is sufficient to maintain the protective skin over the metal.

Zinc Pressure Die-casting

The zinc base casting alloy used in the majority of cases is Mazak No. 3 or No. 5 to British Standard Specification 1004, alloys A and B. These contain about 4% of aluminium and some copper and very carefully controlled quantities of magnesium, lead, iron, cadmium, and tin, the balance being zinc.

A recommended procedure for the melting of zinc base alloys such as these is the use of ZINCREX flux in a ratio of about 0.5% of the charge weight. The method of application will depend to some extent on the type of pressure die-casting machine in use. Some machines have a self-contained reservoir and work upon the goose-neck principle. In this the capacity is limited and it is not convenient to apply a fluxing compound to the metal in the reservoir. On the other hand, where the die-casting machines are fed from a separate bale-out furnace ZINCREX can be used regularly at intervals with distinct advantage. Suitable intervals would be when making up the bath with fresh ingot metal or returned scrap. A little flux worked into the surface of the metal at these stated intervals will keep the metal bath clean and free from dross and suspended non-metallics. After working in the flux as described the surface dross is removed.

Injection temperatures should be kept as low as possible consistent with soundness and good surface appearance. A suggested range would be from 400–425°C for Mazak alloys, and the temperature of the dies should be kept somewhere in the limits of 180–220°C.

Pressure Die-casting of Brass

The alloy most used is a 60/40 copper–zinc brass containing up to 0.2% of aluminium. This alloy melts at a temperature of about 900°C, and die-casting machines for brass are designed on the principle of casting the metal while it is in a semi-liquid or plastic condition rather than in the completely molten state. In this way, the life of the dies is increased and, in addition, cold water is constantly circulated through plates behind the die face to prevent them from becoming heated beyond 350°C. The alloy is injected into the die at a temperature of about 950°C.

If the alloy containing aluminium is employed it is melted under a cover of ALBRAL 3 flux and treated with DEOXIDISING TUBES DS to give additional fluidity. Use 1% of ALBRAL 3 during the initial melt-down period and then stir in 0.5% at regular intervals such as when the bath is replenished by fresh ingot metal or scrap. Two DEOXIDISING TUBES DS1 can be added per 50 kg of melt and pro rata according to the rate of replenishment of the bath.

Small machines which will cast parts with an area up to 45 cm^2 and a weight up to about 300 g apply a pressure of about 350 kg/cm^2 on the metal being cast. Medium-sized machines will deal with castings up to 90 cm^2 in area, and 600 g in weight. Large machines develop a pressure of 700 kg/cm^2 or even higher.

The largest die-casting machine can be operated at the average rate of 100 shots per hour, while the other machines can be operated at from 1700 to 1800 shots per 8-hour day.

Pressure Die-casting of Lead

Lead base alloys are among the easiest to pressure die-cast, and their low melting point allows a long die life. Usually the lead is alloyed with antimony to increase hardness. The percentage of antimony added is normally between 10% and 12%. Sometimes 4–5% of tin is also added to improve the fluidity.

For melting the alloy we recommend the use of PLUMBREX 2 as a flux used in a ratio of from 0.5% to 1% of the charge weight.

Pressure Die-casting of Tin and Tin Base Alloys

Tin, and alloys based on tin, are also easy to pressure die-cast, their low melting point causing little deterioration of the dies.

When commencing a melt, the initial charge should be melted down under a protective cover of STANNEX 101, using sufficient of it to cover the surface area of the melt. Thereafter the application of the flux will depend upon operating conditions and the type of pressure die-casting machine in use. Some machines have a self-contained reservoir and work upon the goose-neck principle. In this the capacity is limited and it is not convenient to apply a fluxing compound to the metal in the reservoir. On the other hand, where the die-casting machines are fed from a separate bale-out furnace, STANNEX 101 can be used regularly at intervals with distinct benefit. Suitable intervals would be when making up the bath with fresh ingot metal or returned scrap. At these times, place STANNEX 101 to the extent of about 0.25% of the charge weight on the surface of the molten metal and press it slowly to the bottom with the aid of a perforated plunging tool. Hold the plunger down until the reaction ceases, stir well, withdraw the plunger, and skim off any dross or residue.

Dressings for Pressure Die-casting

All the metals mentioned in the previous section can, with the exception only of the copper-based alloys, be most conveniently cast using the new water-dispersible lubricants. In average practice about 1 part of lubricant is dispersed in around 40 parts by volume of water to form a creamy emulsion, stable under most conditions. Fairly generous quantities of this dispersion are sprayed on to the die face compared with the quantity of oil-based lubricant previously used. There is, therefore, a marked cooling effect from this, and it may be possible to reduce the flow of cooling water through a water-cooled die or, alternatively, increase the cycling rate of a non-water-cooled die. In addition to the technical and practical advantages of these dressings they are, compared with the oily types, very much cleaner in use since little oil smoke is given off and there is no fire hazard. They are easily removed by conventional degreasing processes and may therefore be used satisfactorily for castings subsequently painted or plated.

For a particularly difficult casting inclined to "stick" in one or more spots or if a local hot spot exists in an otherwise satisfactory die, the effect of these water dispersible dressings can be reinforced by touch swabbing of the critical area(s) with a filled lubricant such as DYCOTE 670 or DYCOTE 40. The fillers are finely divided aluminium flake and colloidal graphite respectively. These filled dressings can, if required, be diluted with paraffin or white spirit and used conventionally as die lubricants using similar swabbing or spraying techniques. Graphite containing dressings are also frequently used on the back of the moving half of the die to provide lubrication for cores, ejector pins, etc.

The pressure die-casting of brass and other copper-based alloys normally means, because of the higher temperatures involved, that graphite in some form is essential on the face of the die. DYCOTE 40 is therefore used either as supplied or diluted with a little light mineral oil. It is best applied by fine spray to ensure an even thin coating on the die face. If a water-dispersible graphite containing coating is required, DYCOTE 11 or 38 should be used, a suitable dilution being found by trial and error.

SECTION VIII
STEEL CASTING SPECIFICATIONS

AMERICAN SPECIFICATIONS FOR STEEL CASTINGS

Carbon and Low Alloy Cast Steels

For details of the following specifications see pages 358 to 367.

ASTM A 27-62 Mild to Medium-Strength Carbon-Steel Castings for General Application.
ASTM A 148-60 High-Strength Steel Castings for Structural Purposes.
ASTM A 216-63T Carbon Steel Castings Suitable for Fusion Welding for High Temperature Service.
ASTM A 217-60T Alloy Steel Castings for Pressure Containing Parts Suitable for High Temperature Service.
ASTM A 352-60T Ferritic Steel Castings for Pressure Containing Parts Suitable for Low Temperature Service.
ASTM A 356-60T Heavy-Walled Carbon and Low Alloy Steel Castings for Steam Turbines.
ASTM 389-60T Alloy Steel Castings Specially Heat Treated for Pressure Containing Parts Suitable for High Temperature Service.
ASTM A 486-63T Steel Castings for Highway Bridges.
ASTM A 487-63T Low Alloy Steel Castings Suitable for Pressure Service.
SAE 1962 Automotive Steel Castings.
AAR M 201-62 Steel Castings.
ABS Am. Bur. Shipping Steel Castings—1964 Rules Edition—Machinery and Hull Castings.

Lloyds' Register of Shipping 1963—Steel Castings.

High Alloy Cast Steels

For details of the following specifications see pages 368 to 373.

ASTM A128-60 Austenitic Manganese Steel Castings.
ASTM A296-63T Corrosion-Resistant Iron-Chromium and Iron-Chromium-Nickel Alloy Castings for General Application.
ASTM A297-63 Heat Resistant Iron-Chromium and Iron-Chromium-Nickel Alloy Castings for General Application.

STEEL CASTING SPECIFICATIONS

ASTM A351-63T   Ferritic and Austenitic Steel Castings for High Temp. Service.
ASTM A447-50   Chromium-Nickel-Iron Alloy Castings (25-12 Class) for High Temp. Service.
ASTM 448-50   Nickel-Chromium-Iron Alloy Castings (35-15 Class) for High Temp. Service.
MILITARY   MIL-S-16993A December 1954 Steel Castings (12 per cent Chromium).
                MIL-S-867A December 1951 Steel Castings Corrosion Resisting Austenitic.

Pages 356 to 374 are reprinted from the *Summary of Steel Castings Specifications* by permission of The Steel Founders Society of America. Explanation of symbols used see pages 366, 367 and 374.

CARBON AND LOW ALLOY CAST STEELS
PHYSICAL PROPERTIES

Specification	Class	Heat treatment#	Tensile strength p.s.i.	Yield point p.s.i.	Elong. in 2" %	Red. of area %	Other tests: bend, impact hardness†
ASTM A 27-62	N-1	A or N or NT or QT	—	—	—	—	—
	N-2	—	—	—	—	—	—
	U-60-30	—	60,000	30,000	22	30	—
	60-30	A or N or NT or QT	60,000	30,000	24	35	—
	65-35	A or N or NT or QT	65,000	35,000	24	35	—
	70-36	A or N or NT or QT	70,000	36,000	22	30	—
	70-40	A or N or NT or QT	70,000	40,000	22	30	—
ASTM A 148-60	80-40	A or N or NT or QT	80,000	40,000	18	30	—
	80-50	A or N or NT or QT	80,000	50,000	22	35	—
	90-60	A or N or NT or QT	90,000	60,000	20	40	—
	105-85	A or N or NT or QT	105,000	85,000	17	35	—
	120-95	A or N or NT or QT	120,000	95,000	14	30	—
	150-125	A or N or NT or QT	150,000	125,000	9	22	—
	175-145	A or N or NT or QT	175,000	145,000	6	12	—
ASTM A 216-63T	WCA	A or N or NT	60,000	30,000	24	35	Bend—Degrees [90]
	WCB	A or N or NT	70,000	36,000	22	35	[90]
ASTM A 217-60T	WC1	A or NT	65,000	35,000	24	35	[90]
	WC4	A or NT	70,000	40,000	20	35	[90]
	WC5	A or NT	70,000	40,000	20	35	[90]
	WC6	A or NT	70,000	40,000	20	35	[90]
	WC9	A or NT	90,000	60,000	20	35	[90]
	C5	A or NT	70,000	40,000	18	35	[90]
	C12	A or NT	90,000	60,000	18	35	[90]

STEEL CASTING SPECIFICATIONS 359

COMPOSITION

CHEMICAL COMPOSITION %—MAXIMUM

Specification	C	Mn	P	S	Si	Ni	Cr	Mo	Other elements
ASTM A 27-62	.25[2]	.75[2]	.05	.06	.80	.50	.40	.20	Cu .30
	.35[2]	.60[2]	.05	.06	.80	.50	.40	.20	Cu .30
	.25[2]	.75[2]	.05	.06	.80	.50	.40	.20	Cu .30
	.30[2]	.60[2]	.05	.06	.80	.50	.40	.20	Cu .30
	.30[2]	.70[2]	.05	.06	.80	.50	.40	.20	Cu .30
	.35[2]	.70[2]	.05	.06	.80	.50	.40	.20	Cu .30
	.25[2]	1.20[2]	.05	.06	.80	.50	.40	.20	Cu .30
ASTM A 148-60	—[1]	—	.05	.06	—	—	—	—	—
	—[1]	—	.05	.06	—	—	—	—	—
	—[1]	—	.05	.06	—	—	—	—	—
	—[1]	—	.05	.06	—	—	—	—	—
	—[1]	—	.05	.06	—	—	—	—	—
	—[1]	—	.05	.06	—	—	—	—	—
ASTM A 126-63T	.25[11]	.70[11]	.05	.06	.60	.50[5]	.40[5]	.25[5]	Cu .50[5]
	.30	1.00	.05	.06	.60	.50[5]	.40[5]	.25[5]	Cu .50[5]
ASTM A 217-60T	.25	.50–.80	.05	.06	.60	.50[6]	.35[6]	.45–.65	Cu .50[6]
	.20	.50–.80	.05	.06	.60	.70–1.10	.50–.80	.45–.65	Cu .50[6]
	.20	.40–.70	.05	.06	.60	.60–1.00	.50–.90	.90–1.20	Cu .50[6]
	.20	.50–.80	.05	.06	.60	.50[6]	1.00–1.50	.45–1.50	Cu .50[6]
	.18	.40–.70	.05	.06	.60	.50[6]	2.00–2.75	.90–1.20	Cu .50[6]
	.20	.40–.70	.05	.06	.75	.50[6]	4.00–6.50	.45–.65	Cu .50[6]
	.20	.35–.65	.05	.06	1.00	.50[6]	8.00–10.00	.90–1.20	Cu .50[6]

CARBON AND LOW ALLOY CAST STEELS—*(continued)*
PHYSICAL PROPERTIES

SPECIFICATION & HEAT TREATMENT			MECHANICAL PROPERTIES—MINIMUM			
Specification	Class	Heat treatment #	Tensile strength p.s.i.	Yield point p.s.i.	Elong. in 2″ %	Red. of area %	Other tests: bend, impact hardness†

Specification	Class	Heat treatment #	Tensile strength p.s.i.	Yield point p.s.i.	Elong. in 2″ %	Red. of area %	Other tests: bend, impact hardness†
ASTM A 352-60T	LCB	N or NT (or QT)	65,000	35,000	24	35	Impact-Ft.-Lbs.[7] 15@ -50°
	LC1	N or NT (or QT)	65,000	35,000	24	35	15@ -75°
	LC2	N or NT (or QT)	65,000	40,000	24	35	15@ -100°
	LC3	N or NT (or QT)	65,000	40,000	24	35	15@ -150°
ASTM A 356-60T [12]	1	NT	70,000	36,000	20	35	—
	2	NT	65,000	35,000	22	35	—
	3	NT	80,000	50,000	18	35	—
	4	NT	90,000	60,000	16	35	—
	5	NT	70,000	40,000	22	35	—
	6	NT	70,000	45,000	22	35	—
	7	NT	70,000	40,000	22	35	—
	8	NT	80,000	50,000	18	35	—
	9	NT	95,000	60,000	15	35	—
	10	NT	85,000	55,000	20	35	—
ASTM A 389-60T	C23	1850 N, 1250 T (Min) 1 hr/in	70,000	40,000	18	35	—
	C24	1850 N, 1250 T (Min) 12 hr	80,000	50,000	15	35	—
ASTM A 486-63T	70	N or NT or QT	70,000	36,000	22	30	25[14]
	90	N or NT or QT	90,000	60,000	20	40	25[14]
	120	QT	120,000	95,000	14	30	30[14]

STEEL CASTING SPECIFICATIONS

COMPOSITION

CHEMICAL COMPOSITION %—MAXIMUM

Specification	C	Mn	P	S	Si	Ni	Cr	Mo	Other elements
ASTM A 352-60T	.30	1.00	.05	.06	.60	—	—	—	—
	.25	.50–.80	.05	.06	.60	—	—	.45–.65	—
	.25	.50–.80	.05	.05	.60	2.00–3.00	—	—	—
	.15	.50–.80	.05	.05	.60	3.00–4.00	—	—	—
ASTM A 356-60T (12)	.35[2]	.70	.05	.05	.60	—	—	—	V [8]
	.25[2]	.70[2]	.05	.05	.60	—	—	—	V [9]
	.25	.50–.80	.05	.05	.60	—	—	.40–.60	V .15–.25
	.20	.50–.80	.05	.05	.60	—	—	.90–1.20	
	.25[2]	.70[2]	.05	.05	.60	—	—	.90–1.20	
	.20	.50–.80	.05	.05	.60	—	.40–.70	.40–.60	
	.20	.50–.80	.05	.05	.60	—	1.00–1.50	.40–.60	
	.20	.50–.80	.05	.05	.60	—	1.00–1.50	.40–.60	V .15–.25
	.20	.50–.80	.05	.05	.60	—	1.00–1.50	.90–1.20	
	.20	.50–.80	.05	.05	.60	—	1.00–1.50	.90–1.20	V .15–.25
	.20	.50–.80	.05	.05	.60	—	2.00–2.75	.90–1.20	
ASTM A 389-60T	.20	.30–.80	.05	.06	.60	—	1.00–1.50	.45–.65	V .15–.25
	.20	.30–.80	.05	.06	.60	—	.80–1.20	.90–1.20	V .15–.25
ASTM A 486-63T	.35	.90	.05	.06	.80	—	—	—	
	.35	(1)	.05	.06	(1)	(1)	(1)	(1)	(1)
	.35	(1)	.05	.06	(1)	(1)	(1)	(1)	(1)

CARBON AND LOW ALLOY CAST STEELS—*(continued)*
PHYSICAL PROPERTIES

Specification	SPECIFICATION & HEAT TREATMENT		MECHANICAL PROPERTIES—MINIMUM				
	Class	Heat treatment #	Tensile strength p.s.i.	Yield point p.s.i	Elong. in 2" %	Red. of area %	Other tests: bend, impact hardness†
ASTM A 487-63T	1–N	NT	85,000	55,000	22	40	
	2–N	NT	85,000	53,000	22	35	
	3–N	NT	90,000	60,000	20	35	
	4–N	NT	90,000	60,000	20	40	
	5–N	NT	105,000	70,000	18	35	
	6–N	NT	115,000	80,000	18	30	
	7–N[13]	NT	110,000	70,000	18	35	
	8–N	NT	85,000	55,000	20	35	
	9–N	NT	90,000	60,000	20	35	
	10–N	NT	100,000	70,000	18	35	
	1–Q	QT	90,000	65,000	22	45	
	2–Q	QT	90,000	65,000	22	40	
	3–Q	QT	105,000	85,000	17	30	
	4–Q	QT	105,000	85,000	17	35	
	5–Q	QT	105,000	85,000	15	30	
	6–Q	QT	120,000	95,000	12	25	
	7–Q[13]	QT	125,000	100,000	15	30	
	8–Q	QT	105,000	85,000	17	30	
	9–Q	QT	105,000	85,000	16	35	
	10–Q	QT	125,000	100,000	15	35	

STEEL CASTING SPECIFICATIONS

COMPOSITION

CHEMICAL COMPOSITION %—MAXIMUM

Specification	C	Mn	P	S	Si	Ni	Cr	Mo	Other elements
ASTM A 487-63T	.30	1.00	.05	.06	.80	.50[5]	.35	—	V .07–.15, Mo+W .25 Cu .50
	.30	1.00–1.40	.05	.06	.80	.50[16]	.35	.10–.30	
	.35	1.35–1.75	.05	.06	.80	.50[16]	.35	.25–.55	
	.35	1.00	.05	.06	.80	.40–.80	.40–.80	.15–.30[15]	
	.30	1.00–1.40	.05	.06	.80	.40–.80	.40–.80	.15–.25[15]	
	.38	1.30–1.70	.05	.06	.80	.40–.80	.40–.80	.30–.40[15]	
	.20	.60–1.00	.04	.05	.80	.70–1.00	.40–.80	.40–.60[15]	V .03–.10+B .002– .006+Cu .15–.50
	.20	.50–.90	.05	.06	.80	—	2.00–2.75	.90–1.10[15]	
	.33	.60–1.00	.05	.06	.80	.50[16]	.75–1.10	.15–.30	
	.33	.60–1.00	.05	.06	.80	1.40–2.00	.55–.90	.20–.40[15]	
	.30	1.00	.05	.06	.80	.50[5]	.35	—	V .07–.15, Mo+W .25 Cu .50
	.30	1.00–1.40	.05	.06	.80	.50[16]	.35	.10–.30	
	.35	1.35–1.75	.05	.06	.80	.50[16]	.35	.25–.55	
	.35	1.00	.05	.06	.80	.40–.80	.40–.80	.15–.30[15]	
	.30	1.00–1.40	.05	.06	.80	.40–.80	.40–.80	.15–.25[15]	
	.38	1.30–1.70	.05	.06	.80	.40–.80	.40–.80	.30–.40[15]	
	.20	.60–1.00	.04	.05	.80	.70–1.00	.40–.80	.40–.60[15]	V .03–.10+B .002– .006+Cu .15–.50
	.20	.50–.90	.05	.06	.80	—	2.00–2.75	.90–1.10[15]	
	.33	.60–1.00	.05	.06	.80	.50[16]	.75–1.10	.15–.30	
	.33	.60–1.00	.05	.06	.80	1.40–2.00	.55–.90	.20–.40[15]	

CARBON AND LOW ALLOY CAST STEELS—*(continued)*
PHYSICAL PROPERTIES

SPECIFICATION & HEAT TREATMENT

Specification	Class	Heat treatment #
SAE Automotive	0022	A or N or NT
	0025	A or N or NT
	0030	A or N or NT or QT
	0050A	N or NT
	0050B	QT
	080	A or N or NT or QT
	090	NT or NQT
	0105*	NQT
	0120*	NQT
	0150*	NQT
	0175*	NQT
AAR M 201-62	A	Unannealed
	A	A or N
	B	A or N
	C	NT or QT
	D	QT
	E	QT
A.B.S.	1	A or NT
	2	A or NT
	Hull	A or NT

MECHANICAL PROPERTIES—MINIMUM

Tensile strength p.s.i.	Yield point p.s.i	Elong. in 2″ %	Red. of area %	Other tests: bend, impact hardness†
				BHN
—	—	—	—	187 Max.
60,000	30,000	22	30	187 Max.
65,000	35,000	24	35	131–187
85,000	45,000	16	24	170–229
100,000	70,000	10	15	207–255
80,000	50,000	22	35	163–207
90,000	60,000	20	40	187–241
105,000	85,000	17	35	217–248
120,000	95,000	14	30	248–311
150,000	125,000	9	22	311–363
175,000	145,000	6	12	363–415
60,000	30,000	22	30	—
60,000	30,000	26	38	—
70,000	38,000	24	36	—(10)
90,000	60,000	22	45	—
105,000	85,000	17	35	—
120,000	100,000	14	30	—
				Bend—Degrees
60,000	30,000	24	35	120
70,000	36,000	22	30	90
60,000	30,000	24	35	120

STEEL CASTING SPECIFICATIONS 365

COMPOSITION

CHEMICAL COMPOSITION %—MAXIMUM

Specification	C	Mn	P	S	Si	Ni	Cr	Mo	Other elements
SAE Automotive	.12–.22	.50–.90	.05	.06	.60	—	—	—	—
	.25	.75[2]	.05	.06	.80	—	—	—	W .10+Cu .50[3]
	.30[2]	.70[2]	.05	.06	.80	.50[3]	.25[3]	—	—
	.40–.50	.50–.90	.05	.06	.80	—	—	—	—
	.40–.50	.50–.90	.05	.06	.80	—	—	—	—
	—	—	.05	.06	—	—	—	—	—
	—	—	.05	.06	—	—	—	—	—
	—	—	.05	.06	—	—	—	—	—
	—	—	.05	.06	—	—	—	—	—
AAR M 201-62	—	.85	.05	Basic	—	—	—	—	—
	—	.85	.05	.05	—	—	—	—	—
	—	.85	.05	Acid	—	—	—	—	—
	.35	—	.05	.06	—	—	—	—	—
	—	—	.05	—	—	—	—	—	—
	—	—	.05	—	—	—	—	—	—
A.B.S.	—	—	—	—	—	—	—	—	—
	—	—	—	—	—	—	—	—	—

CARBON AND LOW ALLOY CAST STEELS—*(continued)*
PHYSICAL PROPERTIES

SPECIFICATION & HEAT TREATMENT			MECHANICAL PROPERTIES—MINIMUM				
Specification	Class	Heat treatment #	Tensile strength p.s.i.	Yield point p.s.i.	Elong. in 2" %	Red. of area %	Other tests: Bend, impact hardness†
Lloyds	—	A	26–35[4]	—	20	—	120
	—	A	28–35[4]	—	20	—	120
	—	A	28–35[4]	—	TS+E =57	—	180
	—	A	28–35[4]	—	20	30	120
	—	A	28–35[4]	—	20	30	120
	—	A	28–40[4]	—	20	30	120
	—	A	30–45[4]	—	20	30	120

EXPLANATION OF SYMBOLS

( ) Permitted only if agreed upon by purchaser.
\# Heat Treatment: A—Full Annealed; N—Normalised; T—Tempered; Q—Liquid Quenched.
* Hardenability requirements when specified.
† Hardness tests when specified in contract or order.
[ ] Figures in brackets are expected values only. No test required unless specified.
(1) Alloying elements shall be selected by the manufacturer.
(2) For each reduction of .01% C below the max. specified, an increase of .04% Mn above the max. specified will be permitted to a max. of 1.0% Mn (1.40 max. for grade 70–40, A27).
(3) Total max. content of undesirable elements is 1.0%. For each .10% below the specified max. alloy content of 1.0%, an increase of .02% in the Cr plus Mo content and .06% in the Ni and Cu contents above the max. will be permitted.
(4) Tons (2240)/in².
(5) Total max. content of undesirable elements in 1.0%.

COMPOSITION

CHEMICAL COMPOSITION %—MAXIMUM

Specification	C	Mn	P	S	Si	Ni	Cr	Mo	Other elements
Lloyds	—	—	.05	.05	—	—	—	—	
	—	—	.05	.05	—	—	—	—	
	—	—	.05	.05	—	—	—	—	
	.25	.5–.10	.05	.05	.15–.5	—	—	—	
	.25	1.0–1.5	.05	.05	.15–.5	—	—	—	
	.25	.5–1.0	.05	.05	.15–.5	—	—	.4–.7	
	.25	.5–.8	.05	.05	.15–.5	—	.8–1.4	.45–.65	

(6) Restrictions on unspecified alloy elements: W .10% for all grades; total max. content of unspecified elements 1.0% for all grades except WC4 and WC5 where total max. is .60%.
(7) Charpy keyhole notch impact test required at temperature specified by customer.
(8) Vanadium 0.01% minimum may be specified or up to 0.05% permitted.
(9) Vanadium 0.05% minimum may be specified.
(10) Hardenability requirement of Rc=40 max. at 10/16″.
(11) For each reduction of .01% below the specified maximum C content, an increase of .04% Mn above the specified max. will be permitted up to a max. of 1.1%.
(12) The use of aluminium is prohibited in the making of all steels except grade 1.
(13) Proprietary Steel Composition.
(14) Charpy V-Notch at 70°F. Values apply only to sections to 2 inches.
(15) Restriction on unspecified elements Cu 0.50, W 0.10, total 0.60%.
(16) Restriction on unspecified elements Cu 0.50, W 0.10, total 1.00%.

High Alloy Cast Steels
Physical Properties

Specification & Heat Treatment			Mechanical Properties—Minimum			
Specification	Class	Heat treatment #	Tensile strength p.s.i.	Yield point p.s.i.	Elong. in 2" %	Red. of area %	Other tests: bend, impact hardness
ASTM A-128-60	—	Q	—	—	—	—	Bend-Degrees (150)
ASTM A296-63T	CA-15	NT or A	90,000	65,000	18	30	BHN-241 max.
	CA-40	NT or A	100,000	70,000	15	25	—
	CB-30	N or A	65,000	30,000	—	—	BHN-241 max.
	CB-50	N or A	55,000	—	—	—	BHN-241 max.
	CE-30	—*	80,000	40,000	10	—	—
	CF-3	—*	65,000	28,000	35	—	—
	CF-3M	—*	70,000	30,000	30	—	—
	CF-8	—*	65,000	28,000	35	—	—
	CF-8C[5]	—*	70,000	30,000	30	—	—
	CF-8M	—*	70,000	30,000	30	—	—
	CF-16F[2]	—*	70,000	30,000	25	—	—
	CF-20	—*	70,000	30,000	30	—	—
	CG-8M	—*	75,000	35,000	25	—	—
	CG-12	—*	70,000	28,000	35	—	—
	CH-20[3]	—*	70,000[7]	30,000	30	—	—
	CK-20	—*	65,000	28,000	30	—	—
	CN-7M	—*	62,500	25,000	35	—	—

STEEL CASTING SPECIFICATIONS

COMPOSITION

CHEMICAL COMPOSITION %—MAXIMUM

Specification	C	Mn	P	S	Si	Ni	Cr	Mo	Other elements
ASTM A 128-60	1.00–1.40	10.0[1]	.10	.05	—	—	—	—	—
ASTM A 296-63T	.15	1.00	.04	.04	1.50	1.00	11.5–14	.50	[6]
	.20–.40	.40–1.00	.04	.04	1.50	1.00	11.5–14	.50	
	.30	1.00	.04	.04	1.00	2.00	18–21	—	
	.50	1.00	.04	.04	1.00	4.00	26–30	—	
	.30	1.50	.04	.04	2.00	8–11	26–30	—	
	.03	1.50	.04	.04	2.00	8–12	17–21	—	
	.03	1.50	.04	.04	1.50	9–13	17–21	2–3	
	.08	1.50	.04	.04	2.00	8–11	18–21	—	
	.08	1.50	.04	.04	2.00	9–12	18–21	—	Cb[5]
	.08	1.50	.04	.04	2.00	9–12	18–21	2–3	
	.16	1.50	.04[2]	.04[2]	2.00	9–12	18–21	[2]	Se[2]
	.20	1.50	.04	.04	2.00	8–11	18–21	—	
	.08	1.50	.04	.04	1.50	9–13	18–21	3–4	
	.12	1.50	.04	.04	2.00	10–13	20–23	—	
	.20[3]	1.50	.04	.04	2.00	12–15	22–26	—	
	.20	2.00	.04	.04	2.00	19–22	23–27	2–3	
	.07	1.50	.04	.04	1.50	27.5–30.5	19–22	—	Cu 3–4

HIGH ALLOY CAST STEELS
PHYSICAL PROPERTIES

Specification	Class	Heat treatment #	Tensile strength p.s.i.	Yield point p.s.i.	Elong. in 2" %	Red. of area %	Other tests: bend, impact hardness
ASTM A 297-63	HC	—	55,000	—	—	—	—
	HD	—	75,000	35,000	8	—	—
	HE	—	85,000	40,000	9	—	—
	HF	—	70,000	35,000	25	—	—
	HH	—	75,000	35,000	10	—	—
	HI	—	70,000	35,000	10	—	—
	HK	—	65,000	35,000	10	—	—
	HL	—	65,000	35,000	10	—	—
	HN	—	63,000	—	8	—	—
	HT	—	65,000	—	4	—	—
	HU	—	65,000	—	4	—	—
	HW	—	60,000	—	—	—	—
	HX	—	60,000	—	—	—	—
ASTM A 351-63T	CA15	NT	90,000	65,000	18	30	
	CF3	N or Q	70,000	30,000	35	—	
	CF8	N or Q	70,000	30,000	35	—	
	CF3M	N or Q	70,000	30,000	30	—	
	CF8M	N or Q	70,000	30,000	30	—	
	CF8C	N or Q	70,000	30,000	30	—	
	CH8	N or Q	65,000	28,000	30	—	
	CH10	N or Q	70,000	30,000	30	—	
	CH20[9]	N or Q	70,000	30,000	30	—	
	CT35	N or Q	65,000	28,000	15	—	
	CF10MC	N or Q	70,000	30,000	20	—	
	CN7M	N or Q	62,500	25,000	35	—	
	CK45	N or Q	65,000	35,000	10	—	

STEEL CASTING SPECIFICATIONS

COMPOSITION

CHEMICAL COMPOSITION %—MAXIMUM

Specification	C	Mn	P	S	Si	Ni	Cr	Mo	Other elements
ASTM A 297-63	.50	1.00	.04	.04	2.00	4.00	26–30	.50[8]	
	.50	1.50	.04	.04	2.00	4–7	26–30	.50[8]	
	.20–.50	2.00	.04	.04	2.00	8–11	26–30	.50[8]	
	.20–.40	2.00	.04	.04	2.00	8–12	18–23	.50[8]	
	.20–.50	2.00	.04	.04	2.00	11–14	24–28	.50[8]	
	.20–.50	2.00	.04	.04	2.00	14–18	26–30	.50[8]	
	.20–.60	2.00	.04	.04	2.00	18–22	24–28	.50[8]	
	.20–.60	2.00	.04	.04	2.00	18–22	28–32	.50[8]	
	.20–.50	2.00	.04	.04	2.00	23–27	19–23	.50[8]	
	.35–.75	2.00	.04	.04	2.50	33–37	13–17	.50[8]	
	.35–.75	2.00	.04	.04	2.50	37–41	17–21	.50[8]	
	.35–.75	2.00	.04	.04	2.50	58–62	10–14	.50[8]	
	.35–.75	2.00	.04	.04	2.50	64–68	15–19	.50[8]	
ASTM A 351-63T	.15	1.00	.04	.04	1.50	1.00	11.5–14	.50	
	.03	1.50	.04	.04	2.00	8–12	17–21		
	.08	1.50	.04	.04	2.00	8–11	18–21		
	.03	1.50	.04	.04	1.50	9–12	17–21	2–3	
	.08	1.50	.04	.04	1.50	9–12	18–21	2–3	
	.08	1.50	.04	.04	2.00	9–12	18–21		[5]
	.08	1.50	.04	.04	1.50	12–15	22–26		
	.10	1.50	.04	.04	2.00	12–15	22–26		
	.20[9]	2.00	.04	.04	2.00	12–15	22–26		
	.35	1.50	.04	.04	2.50	33–37	13–17	.50	
	.10	1.50	.04	.04	1.50	13–16	15–18	1.75–2.25	[10]
	.07	1.50	.04	.04	1.50	27.5–30.5	19–22	2–3	Cu 3–4
	.20–.45	1.50	.04	.04	1.75	23–27	19–22		

HIGH ALLOY CAST STEELS
PHYSICAL PROPERTIES

SPECIFICATION & HEAT TREATMENT			MECHANICAL PROPERTIES—MINIMUM			
Specification	Class	Heat treatment #	Tensile strength p.s.i.	Yield point p.s.i.	Elong. in 2" %	Red. of Area %	Other tests: bend, impact hardness
ASTM A447-50	I II	— —	80,000[11] — 80,000[11] 20,000[12]	— — —	9 — 4 8	— — —	M. Perm. 1.70 Rupt. 5000 psi M. Perm. 1.05 Rupt. 8000 psi
ASTM A448-50	—	—	65,000	—	4	—	—
MILITARY MIL-S -16993A	1 2	NT NT	90,000 90,000	65,000[15] 65,000	18 18	30 30	— —
MILITARY MIL-S- 867A	I II III	Q Q Q	70,000 70,000 70,000	28,000 30,000 30,000	35 30 30	— — —	— — —

COMPOSITION

CHEMICAL COMPOSITION %—MAXIMUM

Specification	C	Mn	P	S	Si	Ni	Cr	Mo	Other elements
ASTM A447-50	.20–.45 .20–.45	2.50 2.50	.05 .05	.05 .05	1.75 1.75	10–14[13] 10–14[13]	23–28 23–28	— —	N .20 N .20
ASTM A448-50	.35[14]–.75	.50–2.00	.04	.04	.80–2.50	33–37[13]	14–17	—	Ni+Cr 48 min
MILITARY MIL-S-16993A	.15 .15	1.00 1.00	.05 .05	.05 .05	1.50 .50	1.00 .65–1.0	11.5–14.0 11.5–14.0	.50 .50–.70	
MILITARY MIL-S-867A	.08[16] .08 .08	1.50 1.50 1.50	.05 .05 .05	.05 .05 .05	2.00 2.00 2.00	8–11 9–12 9–12	18–21 18–21 18–21	— — 2–3	Cb+Ta 1.1[17]

EXPLANATION OF SYMBOLS

* Heat treatment as agreed upon by manufacturer and purchaser.

(1) Mn—Minimum per cent.

(2) For free machining properties the composition of grade CF-16F may contain suitable combinations of Se, P, and Mo (grade CF-16F) or of S and Mo (grade CF-16Fa) as follows:

Selenium, phosphorus, and molybdenum:	Sulfur and molybdenum:
Selenium, % .20 to .35	Sulfur, % .20 to .40
Phosphorus, max., % .17	Molybdenum, % .40 to .80
Molybdenum, max., % 1.50	

(3) For more severe general corrosive conditions, and when so specified, carbon content shall not exceed .10%. Low carbon grade shall be designated as grade CH-10.

(4) Chemical analysis not normally required for P, S, and Mo, but if present in amounts over those stated may be cause for rejection.

(5) Grade CF-8C shall have a columbium content of not less than eight times the carbon content and not more than 1.0 per cent.

(6) For grade CB-30 a copper content of .90 to 1.20% is optional.

(7) A tensile strength of 65,000 psi min. is permitted when the carbon content is .06% max., or the silicon content is 1.00% max. or both.

(8) Castings having a specified molybdenum range agreed upon by the manufacturer and the purchaser may also be furnished under these specifications.

(9) By agreement, the carbon content of grade CH20 may be restricted to 1.0% maximum. The grade designation shall be CH10.

(10) Grade CF10MC shall have a columbium content of not less than 10 times the carbon content but not over 1.20%.

(11) Properties after aging.

(12) Short-Time High-Temperature Properties at 1600° F. ± 10° F.

(13) Commercial nickel usually carries a small amount of cobalt, and within the usual limits cobalt shall be counted as nickel.

(14) Within carbon range, manufacturer and purchaser shall agree upon nominal carbon content of not less than .45% nor more than .65%. Tolerance on such specified carbon content shall be ±0.10%.

(15) 0.2% offset.

(16) If chromium is over 20% and nickel is over 10%, a maximum carbon content of .12 will be permitted.

(17) Columbium or columbium plus tantalum shall be not less than 10 times the carbon content and not more than 1.10% (tantalum shall not exceed .4 times the sum of the columbium and tantalum content) or titanium content shall be not less than 6 times the carbon content and not more than .75%.

STEEL CASTING SPECIFICATIONS 375

BRITISH SPECIFICATIONS FOR STEEL CASTINGS
SUMMARY OF COMPOSITIONS, TENSILE STRENGTHS AND YIELD STRESSES

Specification	Grade	Type of steel	C	Si	Mn	Ni	Cr	Mo	Cu max.	Nb or Ti	S max.	P max.	Tensile strength (min. unless a range is indicated) tons/in²	kg/mm²	Yield stress minimum tons/in²	kg/mm²
BS 592	A	Carbon steel	0.25 max.	0.60 max.	1.00 max.	0.40 max.*	0.25 max.*	0.15 max.*	0.30*	—	0.060	0.060	28	44	14	22
	B	Carbon steel	0.35 max.	0.60 max.	1.00 max.	—	—	—	—	—	0.060	0.060	32	50	16	25
	C	Carbon steel	0.45 max.	0.60 max.	1.00 max.	—	—	—	—	—	0.060	0.060	35	55	17.5	28
BS 1398		Carbon-molybdenum steel	0.15/0.25	0.20/0.50	0.50/1.00	0.40 max.*	0.25 max.*	0.40/0.70	0.40*	—	0.050	0.050	30	47	16	25
BS 1456	A	1¼% manganese steel	0.18/0.25	0.50 max.	1.20/1.70	—	—	—	—	—	0.050	0.050	35/45	55/71	60% T.S.	60% T.S.
	B		0.25/0.33	0.50 max.	1.20/1.70	—	—	—	—	—	0.050	0.050	45/55	71/87	60% T.S.	60% T.S.
BS 1457		Austenitic manganese steel	1.0/1.35	1.0 max.	11.0 min.	—	—	—	—	—	0.060	0.10	—	—	—	—
BS 1458		45/55 ton alloy steel	—	—	—	—	—	—	—	—	0.050	0.050	45/55	71/87	32	51
BS 1459		55/65 ton alloy steel	—	—	—	—	—	—	—	—	0.050	0.050	55/65	87/102	38	60
BS 1461		3% Chromium-molybdenum steel	0.25 max.	0.75 max.	0.30/0.70	0.40 max.*	2.50/3.50	0.35/0.60	0.40*	—	0.050	0.050	40/50	63/79	24	38
BS 1462		5% Chromium-molybdenum steel	0.20 max.	0.75 max.	0.40/0.70	0.40 max.*	4.00/6.0	0.45/0.65	0.40*	—	0.040	0.040	40	63	27	43
BS 1463		9% Chromium-molybdenum steel	0.20 max.	1.00 max.	0.30/0.70	0.40 max.*	8.00/10.0	0.90/1.20	0.40*	—	0.040	0.040	40	63	27	43
BS 1617	A	High magnetic permeability mild steel	0.15 max.	0.60 max.	0.50 max.	0.40 max.*	0.25 max.*	0.15 max.*	0.40*	—	0.060	0.060	28 max.	44	—	—
	B		0.25 max.	0.60 max.	0.50 max.	0.40 max.*	0.25 max.*	0.15 max.*	0.40*	—	0.060	0.060	26/32	41/50	13	20
BS 1630	A	13% Chromium steel	0.15 max.	1.00 max.	1.00 max.	1.00 max.	11.5/13.5	0.15 max.*	0.40*	—	0.050	0.050	35	55	24	38
	B		0.12/0.20	1.25 max.	1.00 max.	1.00 max.	11.5/13.5	0.15 max.*	0.40*	—	0.050	0.050	40	63	29	46
	C		0.20/0.30	1.00 max.	1.00 max.	1.00 max.	11.5/13.5	0.15 max.*	0.40*	—	0.050	0.050	45	71	30	48

*Residuals

The above extracts from BS 3100 : 1957 are reproduced by permission of the British Standards Institution, 2 Park Street, London, W.1., from whom copies of the complete standard may be purchased.

376   THE FOUNDRYMAN'S HANDBOOK

SUMMARY OF COMPOSITIONS, TENSILE STRENGTHS AND YIELD STRESSES

Specification	Grade	Type of steel	C	Si	Mn	Ni	Cr	Mo	Cu max.	Nb	Ti	S max.	P max.	Tensile strength (min. unless a range is indic.)* tons/in²	kg/mm²	Yield stress minimum tons/in²	kg/mm²
BS 1631	A	Austenitic Chromium-nickel steel	0.12 max.	2.00 max.	2.00 max.	7.00 min.	17.0 min.	—	—	—	—	0.045	0.045	30	47	13.5	21
	B.Nn		0.12 max.	2.00 max.	2.00 max.	8.5 min.	17.0/20.0	—	—	8×C/1.10	—	0.045	0.045	30	47	13.5	21
	B.Ti		0.12 max.	2.00 max.	2.00 max.	7.5 min.	17.0/20.0	—	—	—	4×C/0.70	0.045	0.045	30	47	13.5	21
						Ni+Cr: 25.0 min.											
BS 1632	A	Austenitic Chromium-nickel-molybdenum steel	0.08 max.	1.50 max.	2.00 max.	11.0/14.0	18.0/20.0	3.00/4.00	—	—	—	0.045	0.045	30	47	13.5	21
	B		0.08 max.	1.50 max.	2.00 max.	10.0 min.	16.5/18.5	2.00/3.00	—	—	—	0.045	0.045	30	47	13.5	21
	C. Nb		0.12 max.	1.50 max.	2.00 max.	10.0 min.	16.5/18.5	2.25/2.75	—	8×C/1.10	—	0.045	0.045	30	47	13.5	21
	C. Ti		0.12 max.	1.50 max.	2.00 max.	10.0 min.	16.5/18.5	2.25/2.75	—	—	4×C/0.70	0.045	0.045	30	47	13.5	21
	D		0.08 max.	1.50 max.	2.00 max.	8.00 min.	16.5/18.5	2.00/3.00	—	—	—	0.045	0.045	30	47	13.5	21
BS 1648	A	Heat resisting alloy steel	0.25 max.	2.0 max.	1.0 max.	—	12.0/16.0	—	—	—	—	—	—	—	—	—	—
	B		1.0 max.	2.0 max.	1.0 max.	—	25.0/30.0	—	—	—	—	—	—	—	—	—	—
	C		1.0/2.0	2.0 max.	1.0 max.	6.0 min.	25.0/30.0	—	—	—	—	—	—	—	—	—	—
	D		0.4 max.	2.0 max.	2.0 max.	10.0/14.0	17.0/22.0	—	—	—	—	—	—	—	—	—	—
	E		0.5 max.	2.5 max.	2.0 max.	17.0/22.0	22.0/27.0	—	—	—	—	—	—	—	—	—	—
	F		0.5 max.	3.0 max.	2.0 max.	23.0/28.0	22.0/27.0	—	—	—	—	x	—	—	—	—	—
	G		0.5 max.	3.0 max.	2.0 max.	36.0/46.0	15.0/20.0	—	—	—	—	—	—	—	—	—	—
	H		0.5 max.	3.0 max.	2.0 max.	55.0/65.0	15.0/25.0	—	—	—	—	—	—	—	—	—	—
	K		0.75 max.	3.0 max.	2.0 max.	—	10.0/20.0	—	—	—	—	—	—	—	—	—	—
BS 1760	A	Carbon steel for surface hardening	0.40/0.50	0.60 max.	1.00 max.	0.40 max.*	0.25 max.	0.15 max.*	0.40 max.*	—	—	0.060	0.060	40	63	20	31
	B		0.55/0.65	0.60 max.	1.00 max.	0.40 max.*	0.25 max.	0.15 max.*	0.40 max.*	—	—	0.060	0.060	45	71	22	35
BS 1956	A	1 per cent Chromium steel (abrasion resisting)	0.45/0.55	0.75 max.	0.50/1.00	—	0.80/1.20	—	—	—	—	0.060	0.060	45	71	—	—
	B		0.45/0.55	0.75 max.	0.50/1.00	—	0.80/1.20	—	—	—	—	0.060	0.060	—	—	—	—
	C		0.55/0.65	0.75 max.	0.50/1.00	—	0.80/1.50	0.20/0.40	—	—	—	0.060	0.060	—	—	—	—

*Residuals.

SECTION IX
APPLICATION OF INSULATING AND EXOTHERMIC RISERS TO CASTINGS

INTRODUCTION

The use of feeding aids in both jobbing and mass production foundries is now so well established that it has been decided to devote a complete section to the practical applications. The first part of the Section is given over to the traditional approach with manual methods of calculation, etc., while the remainder deals with the more up-to-date computer based programmes that have been written to refine and automate the techniques. Using these latest methods minimum safety factors and highest possible yields are readily determinable and can be called up and utilised literally at the press of a button.

By far the largest casting demand and therefore the greatest commercial potential for feeding aids lies in the ferrous castings field, primarily in respect of iron castings but steel is also important. These notes are therefore concerned mainly with these applications but other metals and alloys can also be treated similarly provided the differing solidification characteristics and volumes are taken into consideration. The different feed demands in terms of volume feed requirements are suggested in the table on page 392 and these can be included in the appropriate calculations.

Determination of Exothermic Risers and Pads

The complex design of most castings eliminates the possibility of one universal risering formula, and, furthermore, there are variations between foundries involving melting and pouring practice, temperature control, and even the moulding materials used which must be taken into consideration. A method of conventionally risering castings, which take into account the effect of volume/surface area ratio on freezing time, and the different feed demands of various geometric shapes such as cubes, plates, bars, etc., has been developed.

Similar data have been developed for lined risers, and for the practical foundryman much of the mathematical work has been summarised in simple graphical form, thereby permitting the rapid calculation of correctly proportioned risers.

Graphs, such as shown in Figs. 9.1, indicate the minimum diameters of sleeves which are necessary to feed certain shapes of sections. Then the heights of the sleeves are determined by the weight of the section, and the number of risers on relatively uniform sections are determined by the feed distance from each riser. These basic calculations apply equally to risers lined with either exothermic or insulating sleeves. The thickness of exothermic material required for a given diameter of riser is determined by:

$$1/10 \times \text{diameter of riser} + 5 \text{ mm}.$$

However, from a practical standpoint this formula is never accurately followed, and generally certain standard wall thicknesses for a range of riser diameters are used. A typical range is given in Table 1.

TABLE 1

Internal diameter (in)	Wall thickness (in)	Internal diameter (mm)	Wall thickness (mm)
Below 3	$\frac{1}{2}$	Below 75	12
4–6	$\frac{3}{4}$	100–150	20
7–9	1	175–225	25
10–12	$1\frac{1}{4}$	250–305	31
13–15	$1\frac{1}{2}$	330–380	37
16–18	$1\frac{3}{4}$	405–455	43
19–21	2	480–535	50
22–24	$2\frac{1}{4}$	560–610	56
25–27	$2\frac{1}{2}$	635–685	62
28–30	$2\frac{3}{4}$	710–760	68

Riser Volume

The volume of the riser, as in conventional practice, is determined by the shape factor or cooling rate of the casting. Both for economy and most efficient feeding, the height should be not less than one-half the diameter, nor greater than $1\frac{1}{2}$ times the diameter and it is preferable to regard 1:1 as the ratio for which to aim. The steps involved in risering a casting with sleeves can be summarised as:

1. Determine casting or section weight to be fed.
2. Determine geometric type of casting.
3. Determine from feed distance data, end effect, effect of chill use, etc., the numbers and location of the risers (Table 2).
4. Determine feed metal requirement.
5. Determine the diameter of sleeve necessary to give a sound feed, from the shape classification charts (Fig. 9.1). Where there is a choice, use the larger diameter sleeve.

6. Determine the height of the sleeve, knowing the diameter and weight required (Tables 3 and 4).
7. Determine sleeve thickness required from Table 1 (exothermic sleeves only).

TABLE 2. FEEDING DISTANCES OF RISERS

Section design thickness	Feeding distance, D_r in terms of casting thickness, T		Square root of T
	Horizontal feeding	Vertical feeding	
	Square bars ($W=T$)		
1	$4T$	$4T$	$1=1.00$
2	$4T$	$4T$	$2=1.41$
4	$3T$	$3.5T$	$3=1.73$
8	$2.1T$	$2.5T$	$4=2.00$
12	$1.9T$	$2.0T$	$5=2.24$
2 to 12	$6\sqrt{T}$	$6\sqrt{T}$	$6=2.45$
			$7=2.65$
	Semi-plates ($W=2T$)		$8=2.83$
1	$6T$	$6T$	$9=3.00$
2	$6T$	$6T$	$10=3.16$
4	$4.5T$	$4.8T$	$11=3.32$
8	$4.0T$	$4.5T$	$12=3.46$
12	$3.3T$	$3.5T$	
2 to 12	$18\sqrt{T}-16.2$	$16\sqrt{T}-14$	
	Plates ($W=3T$)		
1	$4.5T$	$4.5T$	
2	$4.5T$	$4.5T$	
4	$4.5T$	$4.5T$	
8	$3.5T$	$3.5T$	
12	$2.8T$		
2 to 12	$11.8\sqrt{T}-5.2$		

A Typical Example using British Units

As a practical example, consider a carbon steel casting in the form of a ring with an outside diameter of 84 in, and with a cross-section of $11\frac{1}{2}$ in wide × 10 in deep. Originally this casting was fed by four risers each 18 in. in diameter and 18 in high. Neglecting the running system and the padding involved (in practice about 6 in thick at the riser base blended down the casting side for about 8 in), the total weight of the casting and risers was 12,365 lb with a contact area of about 780 in^2.

The calculations involved and the various considerations taken into account when risering this casting with exothermic sleeves were:

APPLICATION OF RISERS TO CASTINGS 381

1. *Casting weight.* 7180 lb.
2. *Geometric type of casting.* Since the casting is basically a horizontal square bar bent into a curve, risering and feeding range rules applicable to a bar square section may be used. This was considered for calculation purposes as being $10\frac{1}{2}$ in square section.
3. *Feeding distance.* Since the casting is circular, end effects which normally increase feeding distances are absent. According to Steel Founders' Society of America rule the feeding distance from the outside diameter of the exothermic sleeve is 1.9 times the casting thickness on either side, the use of four risers amply meets the feed range requirements.
4. *Feed metal requirements.* A fairly heavy sectioned casting of an average type requires about 25% head metal. In the present case the total riser weight is therefore 1795 lb. Since four risers are required, each riser must therefore weight about 450 lb.
5. *Sleeve diameter.* Figure 6 shows that the minimum riser diameter required to feed a $10\frac{1}{2}$ in square bar is $10\frac{1}{2}$ in. In practice, were this a relatively short bar with substantial end effects giving additional heat extraction, it would be possible to use a 10 in diameter head; the residual head would probably not be flat, but slightly dished but the casting would be sound. However, in the ring casting example there is no end effect, and because such castings are usually run by centre sprues, the metal has to flow some distance before entering the mould cavity; the metal entering the riser is therefore often much colder than is ideally suitable for establishing favourable thermal gradients.

Therefore, experience has taught that a slight increase in riser diameter is preferable in these cases, in order to create directional solidification from the casting to the riser. An 11 in diameter head would therefore be selected as the correct size of head to use in the present case. Since many foundries making their own sleeves have a set of core boxes which gives sleeves in 2 in diameter increments, when they get a 10 in diameter and over, a 12 in diameter head would normally be specified for the present example.

6. *Sleeve height.* Table 3 shows that a 12 in i.d. sleeve must be at least 14 in high to contain 450 lb of steel. In practice, 15 in height is specified to allow for short pouring and hot topping.
7. *Sleeve thickness.* Table 1 shows that the sleeve thickness should be $1\frac{1}{2}$ in.
8. *Further practical considerations.* Following the procedure outlined, it was established that a 12 in diameter, 15 in riser would feed the casting. However, from Table 3 it can be seen that a 14 in diameter, 11 in high riser would also feed the casting, and it was this riser that was used. This riser was chosen since the exothermic material required to make the sleeves was 28 lb less than for the 12 in diameter 15 in high sleeves.

The poured weight when risered conventionally was 12,364 lb, and only 8990 lb when risered with exothermic sleeves. Therefore a weight saving of 3384 lb was realised which was then available for pouring other castings; alternatively, with the same melting capacity, four of these castings could be poured in place of three.

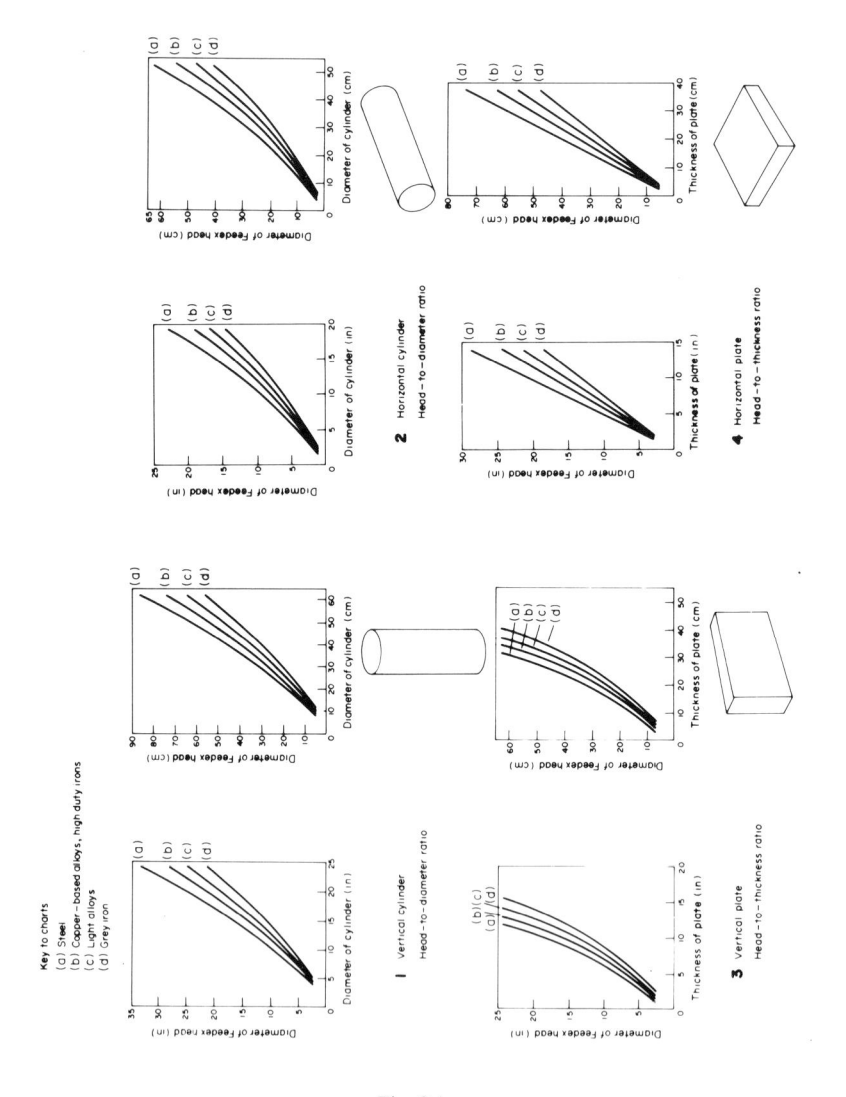

Fig. 9.1.

APPLICATION OF RISERS TO CASTINGS 383

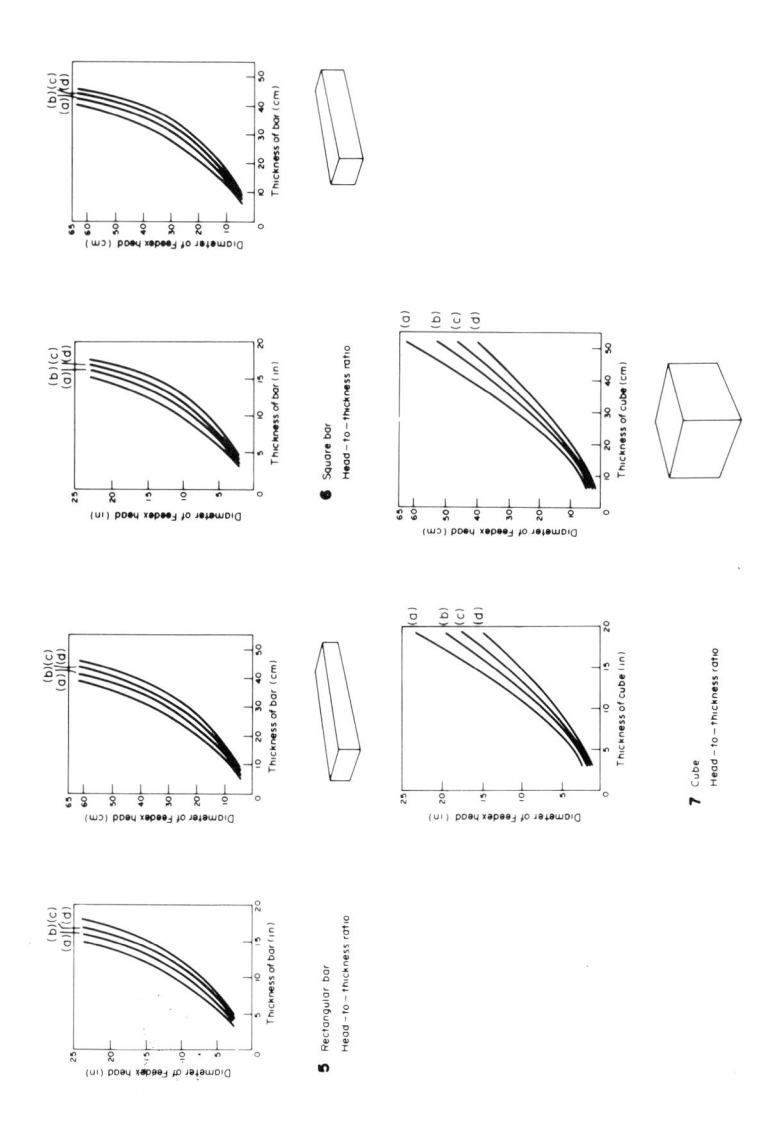

Fig. 9.1 (cont.)

IMPORTANT NOTES FOR CAST IRONS—BREAKER CORES

The long freezing ranges of grey irons and, in some cases, malleable, nodular, and high duty irons, are further prolonged by the insulating properties of riser sleeves. This can lead to the development of a coarse grain structure immediately under the head.

Coarsening of the grain structure can be avoided by the use of breaker or knock-off cores, thus restricting feeding necks. Breaker cores must have high dry strength in order to withstand ferrostatic pressures, and must have a low rate of gas evolution on heating.

A core sand base incorporated in a riser shape is by far the most convenient and consistent method of employing breaker cores. The sketch shows a typical riser sleeve of this design.

It will be noted that the sleeve material is blended into the breaker core and the corresponding increase in thickness ensures that no premature freezing off at the neck can occur.

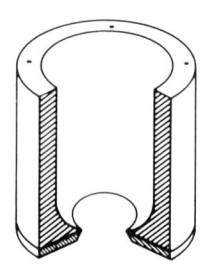

Fig. 9.2

The following table gives the dimensions of breaker cores commonly in use:

TABLE 3

Diameter of head		Feeding orifice (circular or square)		Thickness of core	
(in)	(mm)	(in)	(mm)	(in)	(mm)
2.5	63	1.0	25	0.3	8
3.0	76	1.2	30	0.3	8
3.5	88	1.4	36	0.4	10
4.0	102	1.6	41	0.4	10
4.5	114	1.8	46	0.5	12
5.0	127	2.0	51	0.5	12
5.5	139	2.2	56	0.6	15
6.0	152	2.4	61	0.6	15
6.5	164	2.6	66	0.7	18
7.0	178	2.8	71	0.7	18
7.5	190	3.0	76	0.8	20
8.0	203	3.2	81	0.8	20
8.5	215	3.4	86	0.9	23
9.0	229	3.6	91	0.9	23
9.5	241	3.8	97	1.0	25
10.0	254	4.0	102	1.0	25

TABLE 4
RISER WEIGHT PER 6 INCH HEIGHT

I.D. (in)	Volume (in³)	Weight steel (pounds)	Weight iron (pounds)	Weight brass (pounds)	Weight aluminium (pounds)
1	5	1	1	1	½
1½	11	3	3	3	1
2	19	5	5	6	2
2½	29	8	8	9	3
3	42	12	11	13	4
3½	58	16	15	18	5
4	75	21	20	23	7
4½	96	27	25	30	9
5	118	33	31	37	11
5½	143	40	37	44	13
6	170	48	44	53	16
6½	199	56	52	62	18
7	231	65	60	72	21
7½	265	75	69	82	24
8	302	86	79	94	28
8½	340	96	88	105	31
9	382	108	99	118	35
9½	425	120	111	131	39
10	471	133	123	146	43
10½	520	147	135	161	48
11	570	161	148	177	52
11½	623	176	162	193	57
12	679	192	177	210	63
13	796	225	206	247	73
14	924	261	240	286	85
15	1060	300	276	329	98
16	1206	341	314	374	111
17	1362	385	354	422	125
18	1527	432	397	473	140
19	1701	481	442	527	156
20	1885	533	490	584	173
22	2281	646	593	707	210
24	2715	768	706	842	250
26	3186	902	828	988	293

APPLICATION OF RISERS TO CASTINGS

TABLE 5
RISER WEIGHT PER 150 MM HEIGHT

I.D.	Volume	Weight steel	Weight iron	Weight brass	Weight aluminium
(mm)	(dm³)	(kg)	(kg)	(kg)	(kg)
30	.10	.83	.76	.85	.27
40	.18	1.47	1.35	1.52	.48
50	.29	2.31	2.12	2.39	.75
60	.42	3.33	3.05	3.43	1.08
70	.57	4.53	4.15	4.67	1.47
80	.75	5.92	5.42	6.10	1.93
90	.95	7.49	6.86	7.72	2.44
100	1.17	9.25	8.48	9.54	3.01
110	1.42	11.20	10.26	11.55	3.65
120	1.69	13.33	12.21	13.74	4.34
130	1.99	15.64	14.33	16.12	5.09
140	2.30	18.14	16.62	18.70	5.91
150	2.65	20.83	19.08	21.47	6.78
160	3.01	23.70	21.71	24.43	7.72
170	3.40	26.76	24.51	27.58	8.71
180	3.81	30.00	27.48	30.91	9.77
190	4.25	33.42	30.62	34.44	10.88
200	4.71	37.04	33.93	38.17	12.06
220	5.70	44.82	41.06	46.19	14.60
240	6.78	53.34	48.86	54.97	17.37
260	7.96	62.60	57.34	64.51	20.39
280	9.23	72.60	66.50	74.82	23.64
300	10.60	83.34	76.34	85.89	27.14
320	12.06	94.83	86.86	97.72	30.88
340	13.62	107.06	98.07	110.33	34.87
360	15.27	120.02	109.94	123.68	39.09
380	17.01	133.73	122.50	137.81	43.55
400	18.85	148.17	135.73	152.70	48.26
450	23.85	187.53	171.78	193.25	61.07
500	29.45	231.52	212.08	238.59	75.40
550	35.64	280.14	256.62	288.70	91.24
600	42.41	333.39	305.40	343.57	108.58
650	49.78	391.27	358.42	403.22	127.43
700	57.73	453.78	415.68	467.64	147.79

HOW TO MAKE FEEDEX EXOTHERMIC SLEEVES

Mixing

FEEDEX exothermic material is mixed with water, either by hand, in a sand mill, or in a mixer until the consistency of a mixed core sand is obtained. Note that the green bond is somewhat lower than that of an average oil-sand mixture.
Water addition must be strictly controlled and varies with each type of FEEDEX as shown in the table.

Type of FEEDEX	Optimum % Water	Pints Water per 25 lb	Litres Water per kg	Milling Time sec	Optimum Drying Temperature
3	4	$\frac{4}{5}$	0.04	60–180	180–200°C. (350–390°F.)
4	5	1	0.05	60–180	180–220°C. (350–390°F.)
50	10	2	0.10	60–290	180–350°C. (350–650°F.)
93	10	2	0.10	180	180–200°C. (350–390°F.)

Note: Once water has been added FEEDEX should be moulded and dried within 2 or 3 hours.

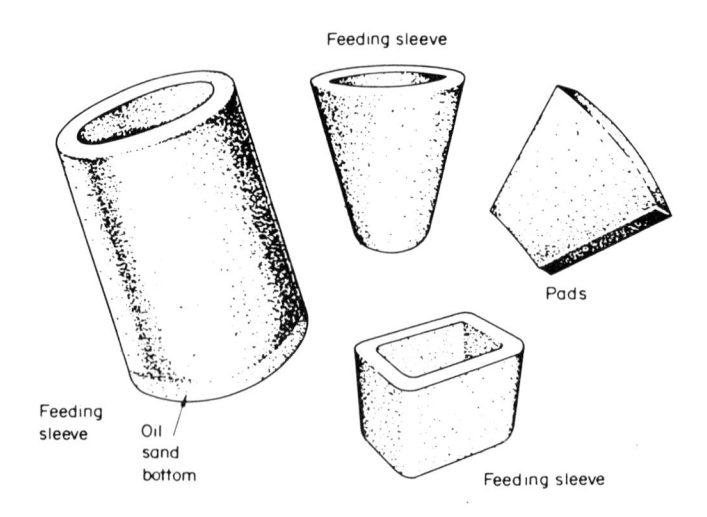

Fig. 9.3

APPLICATION OF RISERS TO CASTINGS 389

Moulding

The mixed FEEDEX is moulded into sleeves or required shapes as shown on page 388. A layer of core sand 13 mm thick is incorporated at the end of the sleeves which will be in contact with the casting, otherwise the sleeve may "burn on" to the casting.

Venting

Venting is necessary to exhaust the gases of combustion and allow free access of air to complete the exothermic reaction. Sleeves should be vented to within 10 mm of the bottom and it is also important to vent the backing sand. This is clearly illustrated in the drawings shown on this page.

Drying

Drying temperature of FEEDEX exothermic materials 3 and 4 should not exceed 200°C (390°F) otherwise there is a risk of premature ignition. Drying time depends largely on the local conditions, a fair guide is to dry for one hour per 25 mm wall thickness. FEEDEX 50 may be dried at between 180°C (360°F) and 350°C (660°F). Due to the increased water content of this material the drying time must be increased by 50%.

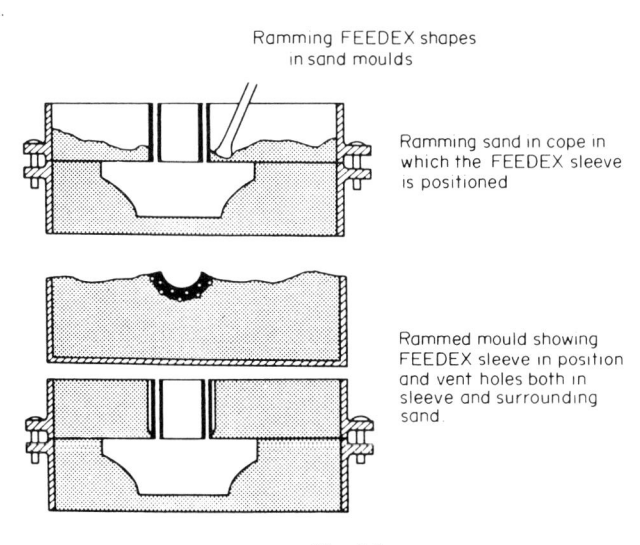

Ramming FEEDEX shapes
in sand moulds

Ramming sand in cope in
which the FEEDEX sleeve
is positioned

Rammed mould showing
FEEDEX sleeve in position
and vent holes both in
sleeve and surrounding
sand.

Fig. 9.4.

PRACTICAL GUIDE TO THE DETERMINATION OF FEEDER SLEEVE DIMENSIONS

The following is a very general guide and deals only with open top feeders. See also "The application of Mouldable Exothermic Materials to Steel Castings" page 395.

The same calculations can and should be employed for determining feeder sleeve dimensions when using the alternative fully insulating (KALMIN) or self-heating refractory sleeves (KALMINEX). These newer materials are more effective for some applications than a fully exothermic sleeve and are far more resilient and able to withstand rough use. Because of their extremely high short- and long-term insulation value loss of heat from the open top of a riser assumes increased importance and all sleeved heads should be covered with a suitable hot topping or anti-piping compound immediately on completion of pouring.

STAGE 1

To determine the sleeve dimensions required for a particular section of casting which is to be fed, the type of alloy and ruling section of the casting are the major considerations.

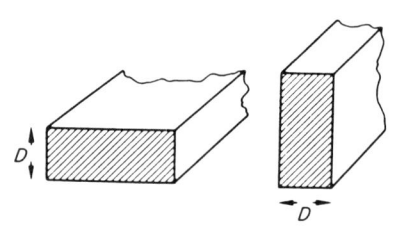

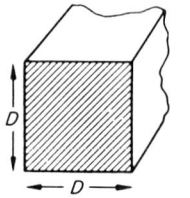

D = ruling section representing part of casting to be fed. This dimension determines FEEDEX sleeve diameter. (See table below.)

Fig. 9.5.

Alloy	Internal sleeve diameter
Steel	D or greater than D*
High Duty Irons	$\tfrac{2}{3}$D to D*
Malleable Iron	$\tfrac{2}{3}$D to D*
Ductile Iron	$\tfrac{2}{3}$D to D*
Grey Iron	$\tfrac{1}{2}$D to $\tfrac{2}{3}$D*
Copper Base Alloys	$\tfrac{2}{3}$D to D*
Light Alloys	$\tfrac{1}{2}$D to D*

*When the section to be fed is deep or has sections attached to it, the internal sleeve diameter should be greater than D or approaching the upper limit of the recommendations given above.

Note on Sleeves for Grey Iron

FEEDEX sleeves with an integral breaker core are preferred for grey iron castings. This minimises the formation of a coarse structure beneath the riser—a disadvantage which may result from the use of straightforward cylindrical sleeves. Separate oil sand breaker cores, used with a cylindrical sleeve can be used if desired. To obtain correct breaker core dimensions or given size of sleeve calculate as follows:—

Breaker core orifice diameter—divide internal sleeve diameter by 2.5.
Breaker core thickness—divide internal sleeve diameter by 10.

STAGE 2

Next determine sleeve height. This should ideally be from 1–1.5 times the internal sleeve diameter. FEEDEX sleeve height is very often influenced by depth of moulding boxes, heights of runner bushes, etc. In practice, therefore, FEEDEX sleeve heights are often from 0.8–2.0 times their diameter.

STAGE 3

From the formula $3.14 \times r^2 \times h$, where r is sleeve radius and h sleeve height, calculate the internal volume of the sleeve. The resulting volume should approximate to the percentage quantity of liquid metal required to adequately feed that cast section to which the sleeve is attached.

Volume of Feed Metal Required for Various Alloys

Steel	15% to 25%
High Duty Irons	12% to 17%
Malleable Iron	12% to 17%
Ductile Iron	12% to 17%
Grey Iron	Approx. 6%
Copper Base Alloys	12% to 17%
Light Alloys	10% to 15%

STAGE 4

Selecting the Right Type of Riser Sleeve Material

FEEDEX exothermic materials for sleeve making are available in several grades so that ignition sensitivity and burning rate are suited to all types of alloy.

The following table gives principal recommendations taking into consideration sleeve diameters. Other types of FEEDEX material are available for special applications.

TABLE 3

Alloy	FEEDEX 3 (mm)	FEEDEX 4 (mm)	FEEDEX 50 (mm)	FEEDEX 93 (mm)
Steel	50 to 150	Up to 50	Over 150	50 to 150
Iron (all types)	50 to 200	Up to 50	Over 200	50 to 200
Copper base alloys	50 to 200	Up to 50	Over 200	50 to 200
Light alloys	Over 100	Up to 100	—	—

FEEDEX 3 and 93 are frequently interchangeable. Steel foundries generally require a clean strip on heads so FEEDEX 3 is preferred.

Determining Sleeve Thickness

Refer to the following table for determining sleeve thickness requirements.

APPLICATION OF RISERS TO CASTINGS 393

English		Metric	
Internal Diameter (in)	Wall Thickness (in)	Internal Diameter (mm)	Wall Thickness (mm)
Below 3	$\frac{1}{2}$	75 or less	12
4–6	$\frac{3}{4}$	100–150	18
7–9	1	175–225	25
10–12	$1\frac{1}{4}$	225–305	31
13–15	$1\frac{1}{2}$	330–380	37
16–18	$1\frac{3}{4}$	405–455	43
19–21	2	480–535	50
22–24	$2\frac{1}{4}$	560–610	56
25–27	$2\frac{1}{2}$	635–685	62
28–30	$2\frac{3}{4}$	710–760	68

Use the table below to determine the type of insulating material required for a non-exothermic application.

Alloy	Riser diameter up to 125 mm	Riser diameter above 125 mm
Aluminium	KALMIN 33	KALMIN 33
Copper base	KALMIN 44	KALMIN 44
Cast iron	KALMIN 4170	KALMINEX 30
Steel	KALMIN 4170	KALMINEX 30

NEW TECHNOLOGY IN FEEDING AIDS

Insulators

Whilst other methods of increasing feeding yields, by using exothermic powders formed into sleeves around the feeder head have proven highly successful, they do have their drawbacks. Fume evolution or a tendency to promote mould/metal reaction are among the disadvantages. To overcome these problems, insulating materials of a fibrous nature have been developed. These have the inherent advantage of being super lightweight compared to their forerunners and are highly effective and extremely resilient.

Kalmin 4170. White, or slightly off white, plain cylindrical sleeves and domes. Developed as a general riser insulating material, i.e. for all metals, KALMIN 4170 is particularly cost effective on the higher melting point alloys such as iron and steel.

The extremely low thermal capacity means that this material is a highly efficient insulator and does not chill the metal on first contact.

Feeders of up to 8 in diameter can be insulated with KALMIN 4170.

Kalmin 33. Yellow/white, low density, insulating riser sleeves designed for light alloy applications. KALMIN 33 may be thought of as a specialised application of KALMIN 4170, i.e. for alloys such as the aluminium (LM) series. These sleeves are available for lining risers of up to 6 in diameter.

Kalmin 44. Dark brown, low density, cylindrical sleeves for insulating copper-based alloys. Again, this may be considered a specialised KALMIN 4170 type material—a recipe applicable to the "yellow" metals. Units are available to line risers of up to 10 in diameter.

Exothermic Sleeves

The concept of fibrous, lightweight insulating sleeves was extended further to embrace the exothermic qualities of the earlier FEEDEX powders and retain good insulating characteristics. Thus, low density, but robust sleeves, with low fume evolution and high permeability can produce intense heat output without undue initial chill to the metal in the head.

Kalminex 2000. Grey flecked exothermic/insulating feeder sleeves for iron and steel castings. Application may also be extended to copper and nickel alloys subject to suitability trials with regard to sleeve sensitivity. Efficiency of KALMINEX 2000 sleeves is such that a modulus extension factor (MEF) of 1.65 may be used in standard feeder size calculations.

Cylindrical and domed units up to 7 in diameter are available in this material.

Kalminex 30. Dark brown exothermic and highly insulating feeding sleeves in cylinders and domes up to 20 in diameter. KALMINEX 30 may be applied to iron, steel and non-ferrous castings covering feeder head sizes beyond the range of the above-mentioned feeding aids.

Overall efficiency is slightly less than KALMINEX 2000, MEF value being an average of 1.3 since the sensitivity of the exothermic reaction is designed to develop heat slowly over an extended period.

Insert Sleeves

The latest and most exciting feeding aids technology lies with insert sleeves. As the name implies, these sleeves, accurately formed internally and

APPLICATION OF RISERS TO CASTINGS 395

externally, are designed to be inserted into a pre-formed cavity like a core. The advantage is that mass production moulding systems can now utilise highly effective exothermic/insulating materials whereas previously there was no access or application.

Kalminex 2000S. A grey, low density fibrous material having highly exothermic characteristics. Application of these units is to iron and steel castings, where the sleeve is inserted into the mould just prior to closing.

Kalminex 2000HS. Dark brown, low density material with exactly the same applications as KALMINEX 2000S, the difference being that this material is suitable for top feeding Ductile iron castings (there is a danger of fade underneath top feeders when using KALMINEX 2000S with S.G. iron). Also, KALMINEX 2000HS is of a harder constitution, able to withstand high moulding pressures without mould springback.

Both insert sleeve materials are available in units suitable for risers up to approximately 6 in diameter. They are highly efficient and extremely cost effective, allowing efficient utilisation of box space.

The above outline of Foseco feeding aids products is comprehensive but by no means exhaustive. Also available is a product called KALBORD designed to insulate large feeder heads in any metal. A range of FERRUX and FEEDOL anti-piping compounds will insulate open topped feeders with a degree of exothermicity as required, whilst various grades of FEEDEX, (see other sections), can be obtained in powder or sleeve form and thus satisfy many applications, e.g. feeders for materials like aluminium, iron and copper-base alloys.

APPLICATION OF FEEDING AIDS. STEEL CASTINGS

Steps Involved in Calculating Feederhead Sizes

Basically, the size of the feeder head supplying liquid metal to a steel casting section needs to be based on two criteria:

1. Volume demand.
2. Thermal requirements.

The volume requirements depend upon the shrinkage characteristics of the alloy and can thus be found from knowing the weight (or volume) of the casting section and liquid and liquid-solid shrinkage factor of the metal. (Alloy shrinkage factors usually include shrinkage in the liquid state from an averaged pouring temperature and volume reduction in the "mushy" state).

Thermal requirements are found to be related directly to the modulus of the casting section, (i.e. volume/surface area). It has been said that to be thermally satisfactory, a feeder head must have a modulus 20% larger than that of the casting it is feeding.

Initially, then, it is necessary to find the weight of the casting section being fed and its modulus. However, before this is done, feeding distances must be checked to ensure that the feeding range to be covered by the feederhead does not exceed the feeding distance for the casting section thickness, (see table of Feeding Distances).

In summary, the steps are:

1. Determine the number of feeders required and their placement by consulting feeding distance charts. (N.B. Chills will increase total feed distance).
2. For each casting section determine the weight. (Hint—break section down into simpler geometric shapes).
3. For each casting section determine the modulus.
4. Use the weight and the modulus to obtain the correct feeder size.

Graphs are available from Foseco which correlate these two variables and provide a quick method of arriving at feeder size in sand, insulating or exothermic material. Some of these graphs are included in this handbook.

If these graphs are not used and feeder size calculation is carried out "long-hand", then the procedure becomes much more complex, i.e. volume and thermal requirements must be determined and re-determined individually until a feeder head size is obtained which satisfies both criteria.

This would involve:

(a) Working out a feeder size volumetrically (N.B. A cylindrical feeder of height to diameter ratio 1:1 is only approximately 15% efficient!).
(b) Ensuring that this feeder satisfies the casting thermally, i.e. its modulus is 20% larger than that of the casting. (Adjustments must be made if insulating or exothermic materials are used by way of MEF or ASAF (Apparent Surface Alteration Factor) constants relating to these materials).
(c) If step (b) fails, re-iteration of the process is necessary using a larger feeder if the original feeder modulus is undersize.

The Foseco Calculator

Before the more accurate nomograms came into being, the Foseco calculator provided the Methods Engineer in the Steel Foundry with a quick guide to the necessary feeder size for the job. Designed to be carried in

APPLICATION OF RISERS TO CASTINGS 397

the pocket, this plastic calculator is based on the "inscribed sphere method", and relates this dimension to the weight of the casting section via the slide rule principle.

Thus the Methods Engineer finds the diameter of largest sphere which will fit into the casting section in question—he does not need to work out the modulus. Obviously, taking the inscribed sphere without other factors (such as a second dimension and the number of non-cooling faces) rather than the modulus will be less accurate than the "modulus method" but is much easier and quicker to determine.

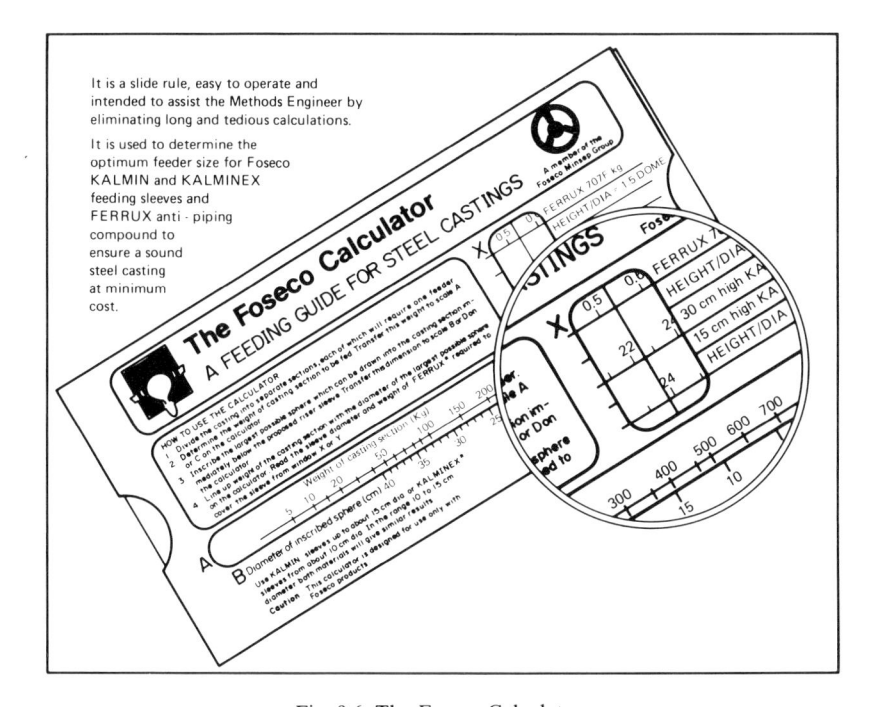

Fig. 9.6. The Foseco Calculator

The Foseco Feeding Graphs

Product application graphs are produced by Foseco (F.S.) Limited, Tamworth (U.K.) to assist in selecting the most efficient and cost effective feeding aid for any particular steel casting. These graphs or nomograms take on a simple, easy-to-use format and cover feeder head sizes from 15 to 1500 mm in diameter and utilising exothermic and insulating materials.

398 THE FOUNDRYMAN'S HANDBOOK

Direct application of a particular product will generally involve reading off two scales, one for the weight of the casting section, the other for the modulus of the section. These scales are joined with a straight edge, with the resultant line cutting a third, intermediate axis. This third scale will reveal the size of sleeve to be used.

Other graphs allow comparison between various products and their respective sleeve sizes. These graphs also assume the nomogram format.

The following four graphs illustrate the above.

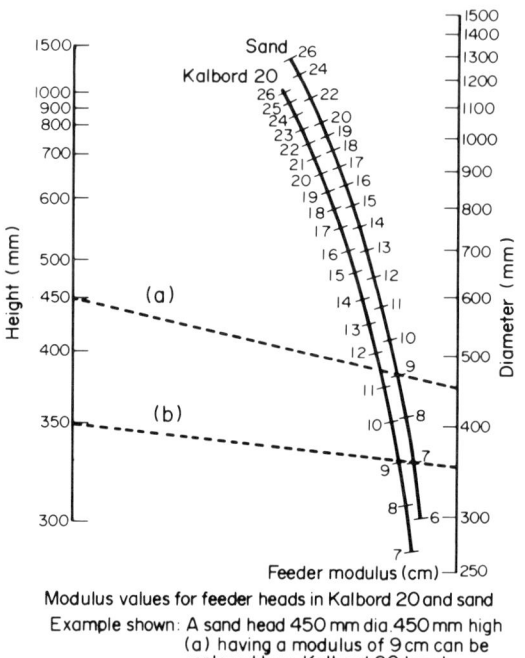

Modulus values for feeder heads in Kalbord 20 and sand
Example shown: A sand head 450 mm dia. 450 mm high
(a) having a modulus of 9 cm can be
replaced by a Kalbord 20 head
350 mm dia. 350 cm high (b) which
also has a modulus of 9 cm

N.B. The modulus values allow for the bottom surface of the feeder head as a non-cooling surface

Fig. 9.7. Modulus Values for Feeder Heads in KALBORD 20 and Sand.

APPLICATION OF RISERS TO CASTINGS

Conversion for sand feeders, insulating and exothermic sleeves

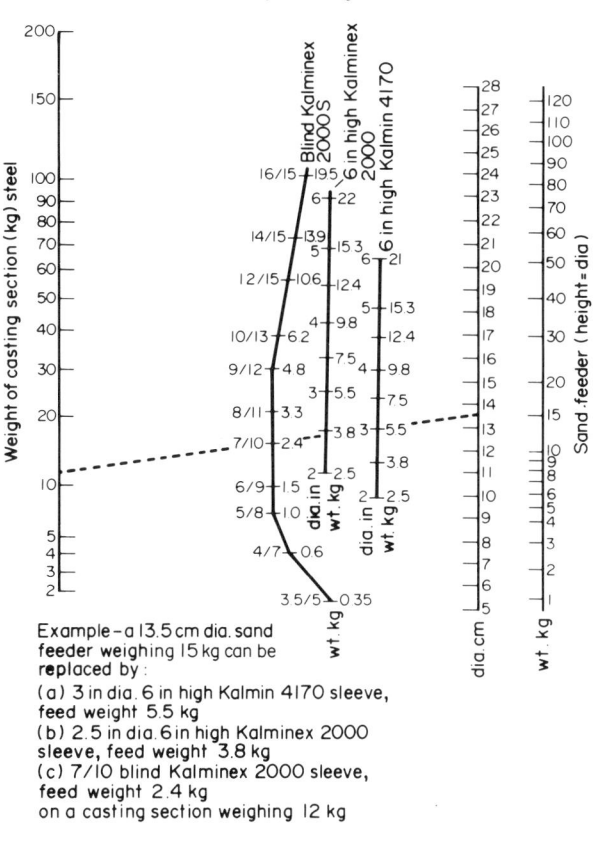

Example – a 13.5 cm dia. sand feeder weighing 15 kg can be replaced by:
(a) 3 in dia. 6 in high Kalmin 4170 sleeve, feed weight 5.5 kg
(b) 2.5 in dia. 6 in high Kalminex 2000 sleeve, feed weight 3.8 kg
(c) 7/10 blind Kalminex 2000 sleeve, feed weight 2.4 kg
on a casting section weighing 12 kg

Fig. 9.8.

400 THE FOUNDRYMAN'S HANDBOOK

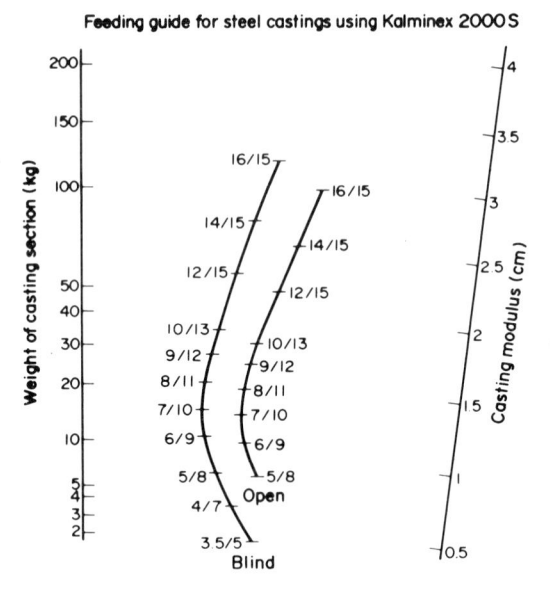

Fig. 9.9.

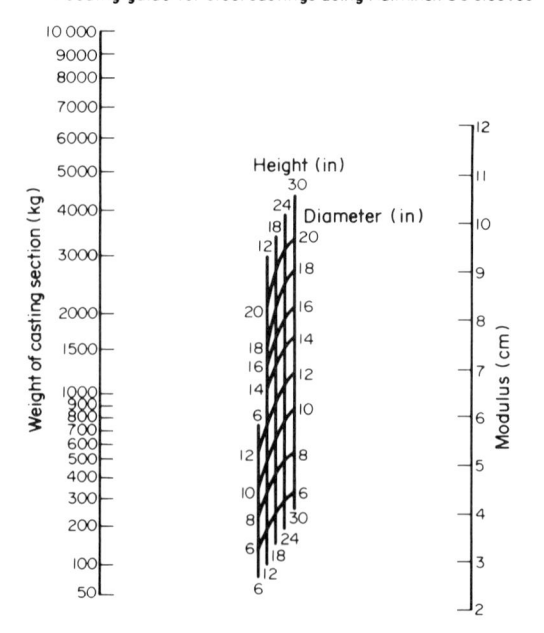

Fig. 9.10.

Feedercalc

Originally developed as an aid to methoding services provided by Foseco for their customers, Feedercalc is a piece of computer software now available to many foundries (iron and steel), assisting in many functions other than feeding.

The simple-to-use program is completely menu-driven and virtually self explanatory in that all the user needs to do is answer easy questions asked by computer. Responses to the questions are evaluated by the program followed by a display of Feeding options open to the operator for that particular casting. The Methods Engineer then selects his preference or else asks the computer to highlight the most cost effective feeding material.

As well as the Feeder size subroutine, the program is equipped with Weight Estimation, Cost Analysis of Feeding System, Side Neck calculation and Feeding Distance routines accessible from a Main Menu as shown below.

Feedercalc Program Menu

1. *Feeder Size:* This routine forms the core of the entire program, allowing the operator to select an economical feeder for a given casting section based on a series of inputs.
2. *Feeding Distance:* Feeder placement and number of feeders per casting are often determined via feeding distance calculations. The section of FEEDERCALC which does this allows the user to examine casting sections with and without end-effect and in the presence and absence of chills.
3. *Cost Analysis:* Feeding cost determinations may be carried out with option 3 for a single feeder or for a comparison of two feeders, e.g. cost of sand feeder versus cost of sleeved feeder.
4. *Weight Calculations:* Intended as a rapid means to deriving the weight of the casting section to be fed, this section contains in-built formulae for volumes of simple geometric shapes. Breaking the casting down into similar shapes saves a vast amount of time on the "calculator" method!
5. *Side Neck Calculation:* This sub-program is ancillary to the feeder size program and is designed to assist with the determination of the feeder/casting contact.

402 THE FOUNDRYMAN'S HANDBOOK

6. *Print Customer Details:* A heading for hardcopy print-outs may be obtained using this option. The heading prints customer name and casting details.

Following is an example of a steel casting method obtained from using the Feedercalc Computer Program. Although the pattern of thinking and judgement are not taken away from the duties of the Engineer, the example shows the simplification, speed, efficiency and consistency brought about by using such a powerful tool.

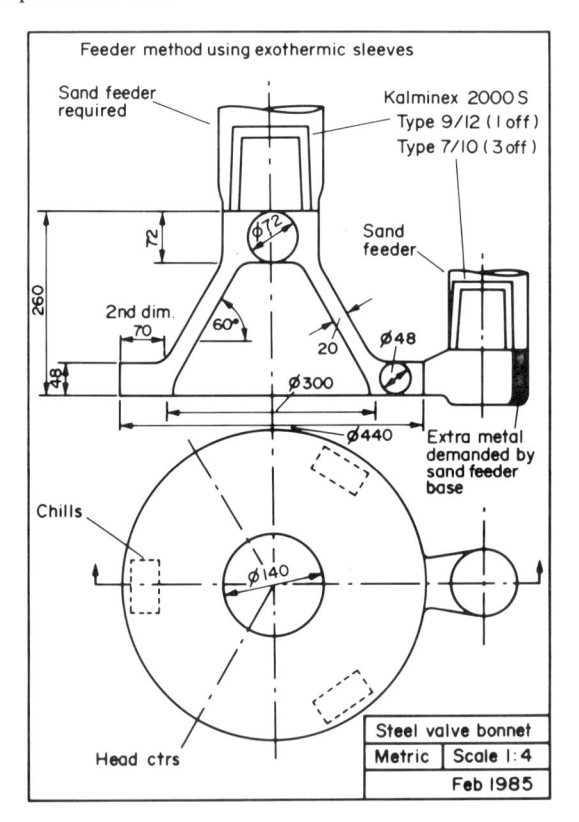

Fig. 9.11.

APPLICATION OF RISERS TO CASTINGS 403

WEIGHT ESTIMATION

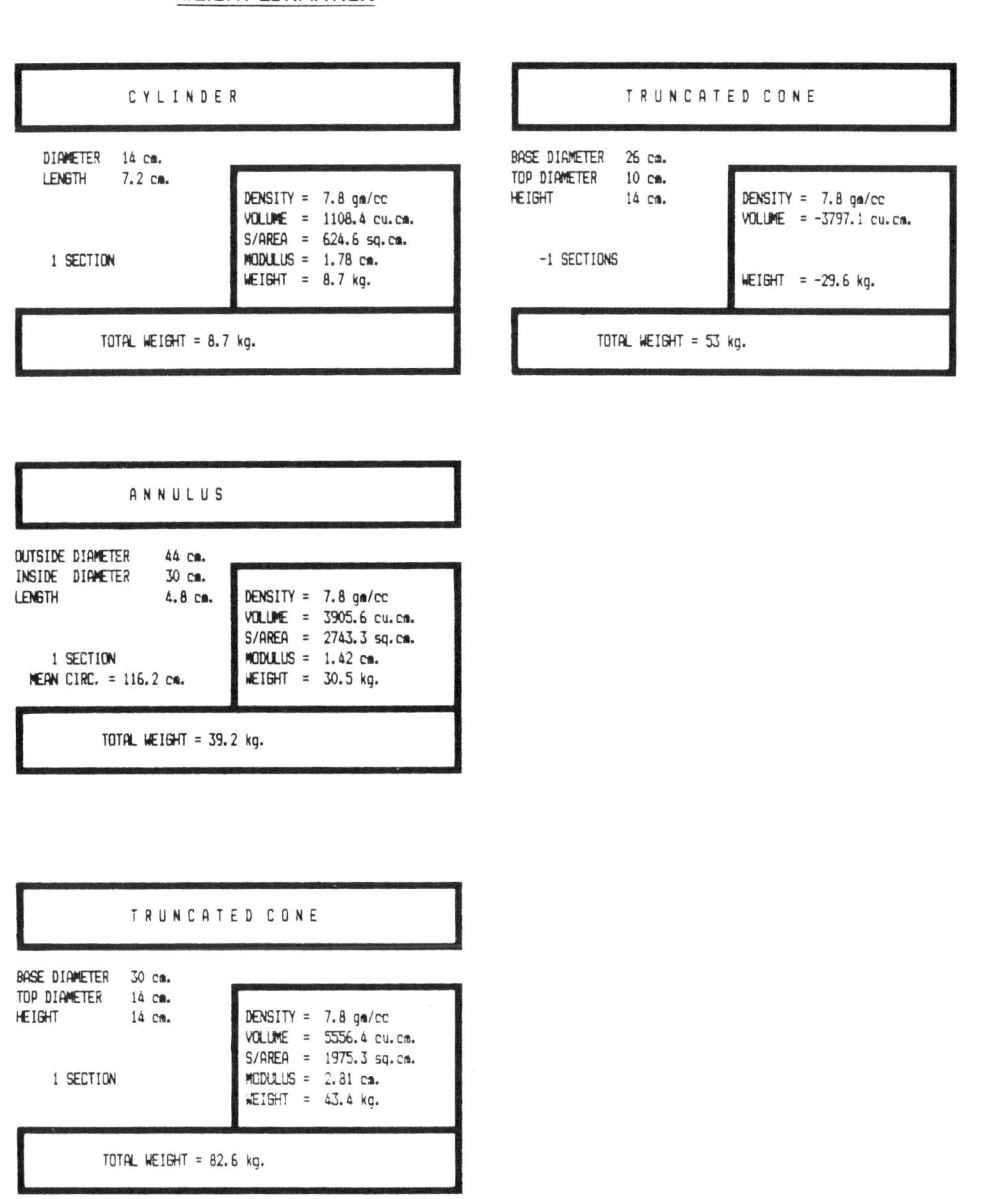

```
CYLINDER

DIAMETER   14 cm.
LENGTH     7.2 cm.
                    DENSITY =  7.8 gm/cc
                    VOLUME  =  1108.4 cu.cm.
                    S/AREA  =  624.6 sq.cm.
1 SECTION           MODULUS =  1.78 cm.
                    WEIGHT  =  8.7 kg.

TOTAL WEIGHT = 8.7 kg.
```

```
TRUNCATED CONE

BASE DIAMETER  26 cm.
TOP DIAMETER   10 cm.
HEIGHT         14 cm.
                     DENSITY =  7.8 gm/cc
                     VOLUME  = -3797.1 cu.cm.
-1 SECTIONS
                     WEIGHT  = -29.6 kg.

TOTAL WEIGHT = 53 kg.
```

```
ANNULUS

OUTSIDE DIAMETER  44 cm.
INSIDE  DIAMETER  30 cm.
LENGTH            4.8 cm.
                          DENSITY =  7.8 gm/cc
                          VOLUME  =  3905.6 cu.cm.
                          S/AREA  =  2743.3 sq.cm.
1 SECTION                 MODULUS =  1.42 cm.
MEAN CIRC. = 116.2 cm.    WEIGHT  =  30.5 kg.

TOTAL WEIGHT = 39.2 kg.
```

```
TRUNCATED CONE

BASE DIAMETER  30 cm.
TOP DIAMETER   14 cm.
HEIGHT         14 cm.
                     DENSITY =  7.8 gm/cc
                     VOLUME  =  5556.4 cu.cm.
                     S/AREA  =  1975.3 sq.cm.
1 SECTION            MODULUS =  2.81 cm.
                     WEIGHT  =  43.4 kg.

TOTAL WEIGHT = 82.6 kg.
```

Fig. 9.12.

404   THE FOUNDRYMAN'S HANDBOOK

FEEDING DISTANCE/NO. FEEDERS CALCULATION

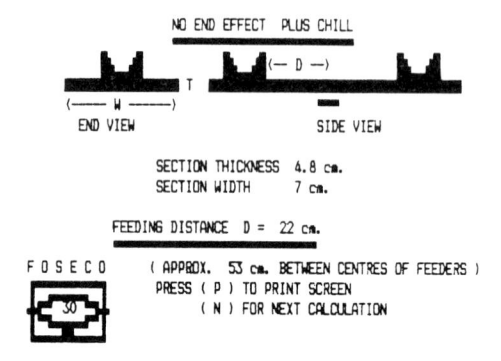

```
                NO END EFFECT PLUS CHILL
   ╦╦         ╦╦(— D —)        ╦╦
   ┃┃   T     ┃┃                ┃┃
(——— W ———)                    ━━
  END VIEW              SIDE VIEW

            SECTION THICKNESS  4.8 cm.
            SECTION WIDTH      7 cm.

            FEEDING DISTANCE  D = 22 cm.

 F O S E C O   ( APPROX.  53 cm. BETWEEN CENTRES OF FEEDERS )
               PRESS ( P ) TO PRINT SCREEN
               ( N ) FOR NEXT CALCULATION
```

Fig. 9.13.

Mean Circumference—116 cm (around flange)

$$\therefore \text{Number of Feeders required} = \frac{116 \text{ cm}}{53 \text{ cm}} = 2.2, \text{ i.e. } 3 \text{ Feeders.}$$

CASTING SECTION WEIGHTS FED BY FEEDERS

Bottom Feeders

The three bottom feeders supply liquid metal to the flange plus half of the main body (truncated hollow cone):

i.e. Each head feeds $\frac{1}{3}$ (flange wt + $\frac{1}{2}$ body wt)

$$= \tfrac{1}{3} (30.5 + \tfrac{1}{2} (13.8))$$
$$= 12.5 \text{ kg.}$$

Top Feeder

This riser feeds the valve bonnet head plus half of the body:
i.e. $8.7 + \tfrac{1}{2}$ (13.8) kg
$$= 15.6 \text{ kg.}$$

APPLICATION OF RISERS TO CASTINGS

BOTTOM FEEDER SIZES

```
WEIGHT OF CASTING SECTION  12.5 kg.           F O S E C O
INSCRIBED SPHERE DIA. 4.8 cm.      20 %SAFETY
SHRINKAGE = 6 %    2.4 N.C.F.      SIDE FEEDER    30
7.8 gm/cc    INGATE 2 DIM = 7 cm.  BINDER 4
```
Using Exothermic Fibre Sleeves

```
KALMINEX 'S' BLIND SFP 7/10K KC3160    Weight 2.4 kg.

            FEEDING YIELD =  71 % ( 75 %)

SAFETY MARGIN = 37 %
APPROX. WT. OF FEEDER BASE =  2.8 kg.
MIN NECK MODULUS = 1.9 cm.
```

```
WEIGHT OF CASTING SECTION  12.5 kg.           F O S E C O
INSCRIBED SPHERE DIA. 4.8 cm.      20 %SAFETY
SHRINKAGE = 6 %    2.4 N.C.F.      SIDE FEEDER    30
7.8 gm/cc    INGATE 2 DIM = 7 cm.  BINDER 4
```
Using Sand Feeder.

```
SAND FEEDERS    10.8 cm.dia.  10.8 cm.high  Weight 8 kg.
  YIELD = 50 %
                10 cm.dia.    15 cm.high    Weight 9 kg.
  YIELD = 46 %
                9.4 cm.dia.   18.8 cm.high  Weight 10 kg.
  YIELD = 44 %
          ADD .06 kg. FERRUX 707F
APPROX. WT. OF FEEDER BASE =  5.5 kg.
MIN NECK MODULUS = 1.9 cm.
```

Fig. 9.14.

NECK DIMENSIONS FOR SIDE FEEDERS

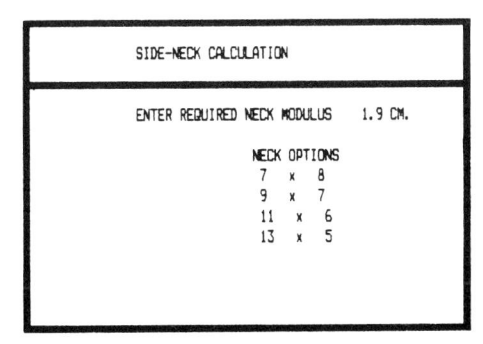

```
              SIDE-NECK CALCULATION

      ENTER REQUIRED NECK MODULUS    1.9 CM.

                      NECK OPTIONS
                       7  x   8
                       9  x   7
                      11  x   6
                      13  x   5
```

Fig. 9.15.

THE FOUNDRYMAN'S HANDBOOK

TOP FEEDER DIMENSIONS

```
WEIGHT OF CASTING SECTION  15.6 kg.(+ 6.4 kg.)   F O S E C O
INSCRIBED SPHERE DIA. 7.2 cm.      20 %SAFETY
SHRINKAGE = 6 %    0 N.C.F.        TOP FEEDER  [  30  ]
7.8 gm/cc    INGATE 4 DIM = 14 cm.  BINDER 4
```

Exothermic Fibre Sleeves.

```
KALMINEX 'S' BLIND  SFP 9/12K KC3596    Weight 4.8 kg.

++++++++++++++++++++++++++++++++++++++++++++++++++++++++
            FEEDING YIELD = 76 % ( 81 %)
++++++++++++++++++++++++++++++++++++++++++++++++++++++++
SAFETY MARGIN = 40 %
```

```
WEIGHT OF CASTING SECTION  15.6 kg.(+ 6.4 kg.)   F O S E C O
INSCRIBED SPHERE DIA. 7.2 cm.      20 %SAFETY
SHRINKAGE = 6 %    0 N.C.F.        TOP FEEDER  [  30  ]
7.8 gm/cc    INGATE 4 DIM = 14 cm.  BINDER 4
```

Sand Feeder

```
SAND FEEDERS   14.6 cm.dia.   14.6 cm.high  Weight 19 kg.
YIELD = 45 %
               13.7 cm.dia.   20.5 cm.high  Weight 24 kg.
YIELD = 40 %
               12.9 cm.dia.   25.8 cm.high  Weight 26 kg.
YIELD = 37 %
       ADD .17 kg. FERRUX 707F
```

Fig. 9.16.

APPLICATION OF RISERS TO CASTINGS 407

COST ANALYSIS SHOWING SAVINGS WITH EXOTHERMIC SLEEVES

PRICE OF SCRAP STEEL PER TONNE	82	
COST OF FURNACE POWER/TONNE MELTED	45	
OTHER FURNACE COSTS/TONNE MELTED	43	Costing Rates
ALLOY COSTS/TONNE MELTED	14	
LADLE COSTS/TONNE MELTED	18	
FOUNDRY & CUSTOMER SCRAP RATE (%)	5 %	
MELTING LOSSES (%)	9 %	
LOSSES FOR BURNING & GRINDING (%)	8 %	

METAL & MELTING COSTS PER CASTING

	SAND FEEDERS	SLEEVES	SAVING	
WEIGHT OF CASTING	53 KG.	53 KG.		
WEIGHT OF FEEDERS	59.5 KG.	20.4 KS.	39.1 KG.	
WT OF RUNNERS ETC	10 KG.	10 KG.		
YIELD	43.2 %	63.5 %		Cost Comparisons
COST OF FURNACE POWER	6.93	4.71	2.22	
OTHER FURNACE COSTS	6.62	4.5	2.12	
ALLOY COSTS	2.15	1.46	.69	
COST OF MELTING LOSSES	1.13	.77	.36	
COST OF BURNING LOSSES	.45	.19	.26	
LADLE COSTS	2.77	1.88	.89	
SLEEVE COSTS		3.1	-3.1	
TOTAL COSTS	20.05	16.61	3.44	
COST / FINISHED TONNE	378.3	313.39	64.91	

Fig. 9.17.

N.B.

It should be borne in mind that the Feedercalc software allows the Engineer to calculate Feeder sizes by either the "Modulus" or the "Inscribed Sphere" method. Each is as accurate as the other provided the correct information is entered via the keyboard. Worthy of note, however, is the fact that the latter method is by far the simpler and quicker to use and is more tolerant to errors of the information input!

For further information on Feedercalc, the types of hardware for which it is designed, details of licensing, etc. kindly contact Foseco (F.S.) Ltd., Metallurgical Division, Tamworth, Staffordshire B78 3TL.

APPLICATION OF FEEDING AIDS. IRON CASTINGS

Collectively the term is used to describe Grey, Malleable, Ductile (S.G.) and Compacted Graphite (C.G.) varieties of cast iron. From a feed calculation aspect only the S.G. (and to a lesser extent the C.G.) grades pose real problems due to its erratic and irregular feed characteristics.

An S.G. Casting may be:

(a) fed conventionally from top risers.
(b) fed from the running system.
(c) not fed at all.

Some of the different factors outlined in the steel casting section will apply to ductile iron in variable degree, with weight (or volume) being the most important.

Most cast irons will solidify with a net expansion due to the formation of graphite (as flake or spheroid) during the freezing process. The extent is related directly to the Carbon Equivalent Value. Therefore provided the mould material is unyielding, potential shrinkage sites will be filled by the expanding metal mass and, assuming the net effect is a balance or slight expansion, no feed metal is required. This, however, is an idealised situation and from a practical point of view, the pouring temperature, casting mass and position and number of ingates can all affect the calculation.

The only safe and consistent way to ensure sound ductile castings is to adopt conventional feeding practices that can accommodate variations in practice. The most effective method is based on KALMINEX 2000HS sleeves using a "weight" calculation as being the most influential variable. A graph has been developed using these parameters and is shown below. A FEEDERCALC Computer Program for ductile iron is also available and may be used in a similar manner to that previously described for steel castings. An example is shown.

APPLICATION OF RISERS TO CASTINGS 409

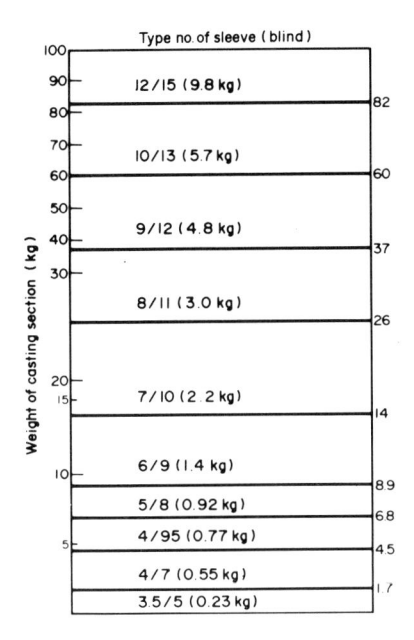

Fig. 9.18.

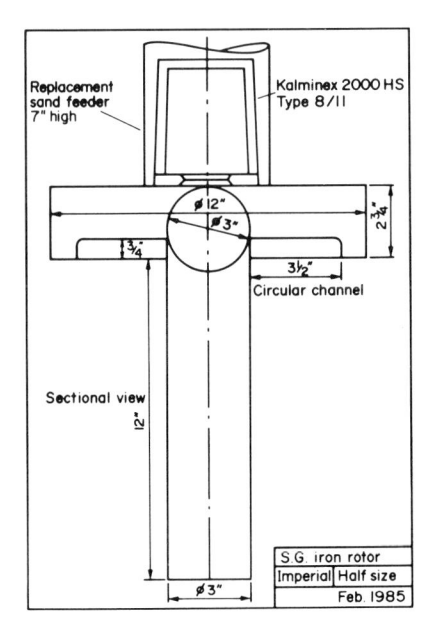

Fig. 9.19.

410   THE FOUNDRYMAN'S HANDBOOK

WEIGHT ESTIMATION

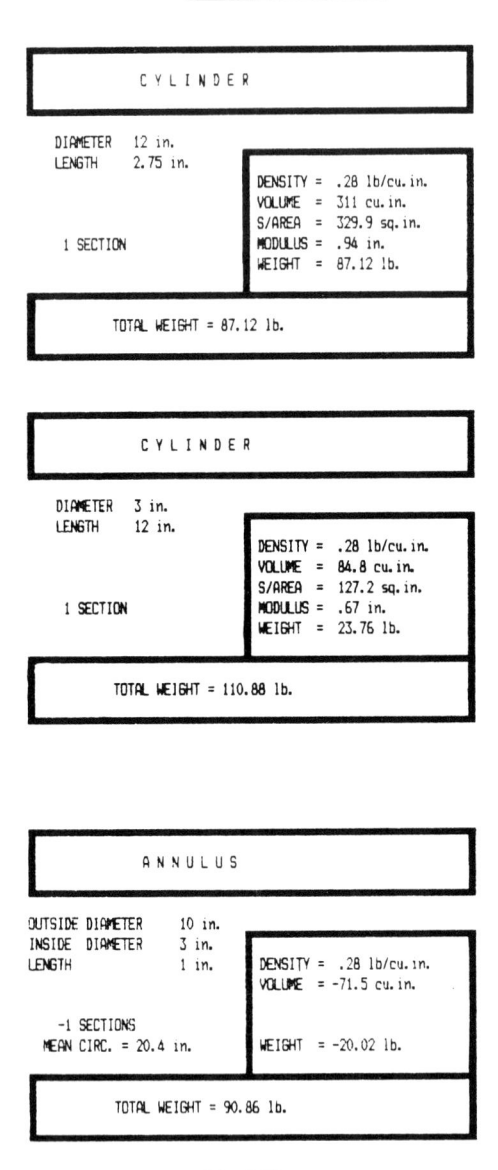

```
┌─────────────────────────────────────────────────┐
│   ┌─────────────────┐                           │
│   │   C Y L I N D E R │                         │
│   └─────────────────┘                           │
│                                                 │
│   DIAMETER   12 in.                             │
│   LENGTH     2.75 in.                           │
│                      ┌──────────────────────────┤
│                      │ DENSITY = .28 lb/cu.in.  │
│                      │ VOLUME  = 311 cu.in.     │
│                      │ S/AREA  = 329.9 sq.in.   │
│   1 SECTION          │ MODULUS = .94 in.        │
│                      │ WEIGHT  = 87.12 lb.      │
│                                                 │
│                                                 │
│        TOTAL WEIGHT = 87.12 lb.                 │
└─────────────────────────────────────────────────┘


┌─────────────────────────────────────────────────┐
│   ┌─────────────────┐                           │
│   │   C Y L I N D E R │                         │
│   └─────────────────┘                           │
│                                                 │
│   DIAMETER   3 in.                              │
│   LENGTH     12 in.                             │
│                      ┌──────────────────────────┤
│                      │ DENSITY = .28 lb/cu.in.  │
│                      │ VOLUME  = 84.8 cu.in.    │
│                      │ S/AREA  = 127.2 sq.in.   │
│   1 SECTION          │ MODULUS = .67 in.        │
│                      │ WEIGHT  = 23.76 lb.      │
│                                                 │
│        TOTAL WEIGHT = 110.88 lb.                │
└─────────────────────────────────────────────────┘


┌─────────────────────────────────────────────────┐
│   ┌─────────────────┐                           │
│   │   A N N U L U S   │                         │
│   └─────────────────┘                           │
│                                                 │
│   OUTSIDE DIAMETER    10 in.                    │
│   INSIDE  DIAMETER    3 in.                     │
│   LENGTH              1 in.  ┌──────────────────┤
│                              │ DENSITY = .28 lb/cu.in. │
│                              │ VOLUME  = -71.5 cu.in.  │
│                                                 │
│   -1 SECTIONS                                   │
│   MEAN CIRC. = 20.4 in.      │ WEIGHT = -20.02 lb. │
│                                                 │
│        TOTAL WEIGHT = 90.86 lb.                 │
└─────────────────────────────────────────────────┘
```

Fig. 9.20.

APPLICATION OF RISERS TO CASTINGS 411

FEEDER SIZE CALCULATION

```
WEIGHT OF CASTING SECTION  91 lb.              F O S E C O
INSCRIBED SPHERE DIA. 3 in.         20 %SAFETY
TEMP = 1390   C.E. = 4.4    BLIND   TOP FEEDER
.28 lb/cu.in.  INGATE 3             BINDER  6
```
Exothermic Fibre Sleeve

```
KALMINEX 'S' BLIND  SFP 8/11K KC3164    Weight 7.2 lb.

               FEEDING YIELD =  93 %

SAFETY MARGIN = 29 %
```

```
WEIGHT OF CASTING SECTION  91 lb.              F O S E C O
INSCRIBED SPHERE DIA. 3 in.         20 %SAFETY
TEMP = 1390   C.E. = 4.4    BLIND   TOP FEEDER
.28 lb/cu.in.  INGATE 3             BINDER  6
```

```
SAND FEEDERS  4.7 in.dia.   7 in.high    Weight 34 lb.
YIELD = 73 %
              4.4 in.dia.   8.8 in.high  Weight 38 lb.
YIELD = 71 %
```
Sand Feeder

Fig. 9.21.

COST ANALYSIS

```
PRICE OF SCRAP STEEL PER TONNE            82
COST OF FURNACE POWER/TONNE MELTED        45
OTHER FURNACE COSTS/TONNE MELTED          43      Costing Rates
    ALLOY COSTS/TONNE MELTED              14
    LADLE COSTS/TONNE MELTED              18
FOUNDRY & CUSTOMER SCRAP RATE ( % )        5 %
            MELTING LOSSES ( % )           .9 %
LOSSES FOR BURNING & GRINDING ( % )        8 %
```

```
                    METAL & MELTING COSTS PER CASTING
                                                        Practice Comparison
                    SAND FEEDERS   SLEEVES    SAVING
WEIGHT OF CASTING    41.4 KG.      41.4 KG.
WEIGHT OF FEEDERS    15.5 KG.       3.3 KG.   12.2 KG.
WT OF RUNNERS ETC     5 KG.         5 KG.
            YIELD    66.8 %        83.2 %

COST OF FURNACE POWER    3.5       2.81       .69
OTHER FURNACE COSTS      3.34      2.68       .66
        ALLOY COSTS      1.08       .87       .21
COST OF MELTING LOSSES    .57       .46       .11
COST OF BURNING LOSSES    .13       .05       .08
        LADLE COSTS      1.4       1.12       .28
        SLEEVE COSTS                .89      -.89

        TOTAL COSTS    10.01       8.88      1.13
COST / FINISHED TONNE  242.02    214.49     27.53
```

Fig. 9.22.

Head Size Calculation For Ductile Iron—Summary

Since ductile iron is fed by the "volume control" method, it is essential to know the weight or volume of the casting section to be fed. A fast guide to feeder size is then available from the above-mentioned nomogram.

If the Feedercalc program is utilised other factors need to be known. These are:

1. Weight of casting section.
2. Inscribed sphere of section.
3. Carbon, Silicon and Phosphorus contents.
4. Pouring temperature.
5. Mould medium.
6. Ingate Position.
7. Top or side feeder.
8. Blind or Open Feeder.

APPLICATION OF RISERS TO CASTINGS 413

The computer will make minor adjustments for variables other than the major one, weight.

Many methods are used to calculate feeder sizes for S.G. iron throughout the U.K., and indeed, throughout the world. However, many tend to be "hit or miss" affairs, the most successful ones being those adopting a more conventional approach. Foseco's aids to methoding ductile iron are based empirically on data relating directly to Foseco feeding aid products and, as such, generates a highly efficient system.

The application of the various exothermic, insulating and anti-piping compounds (A.P.C.s) for the individual metal groups and related to the diameter of the riser is given diagrammatically in the following four diagrams.

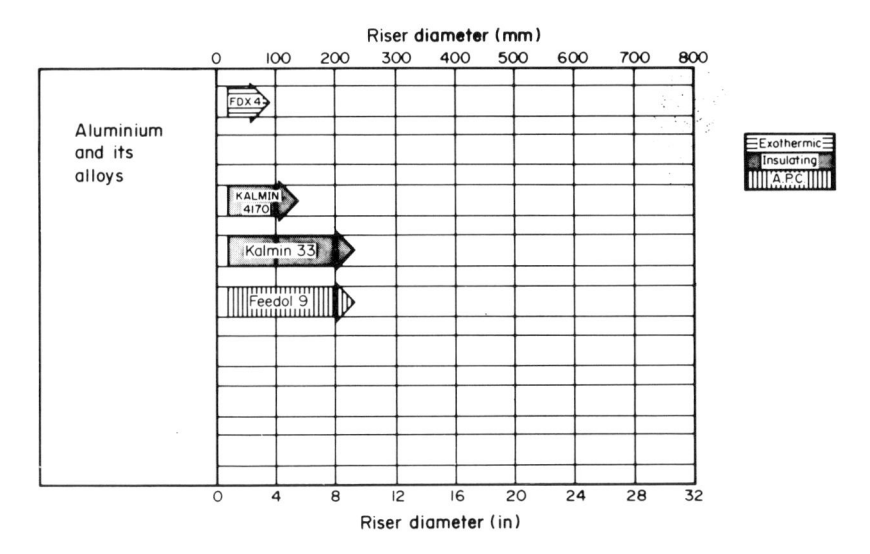

Fig. 9.23.

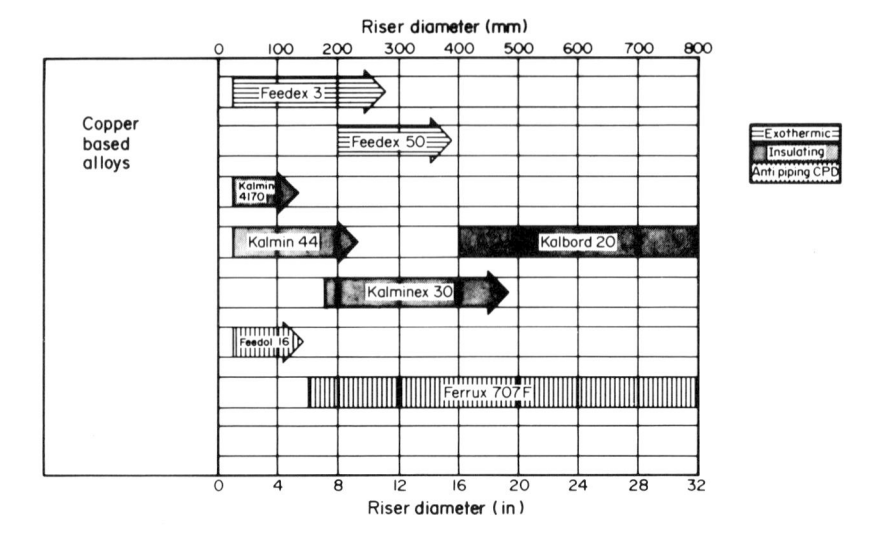

Fig. 9.24.

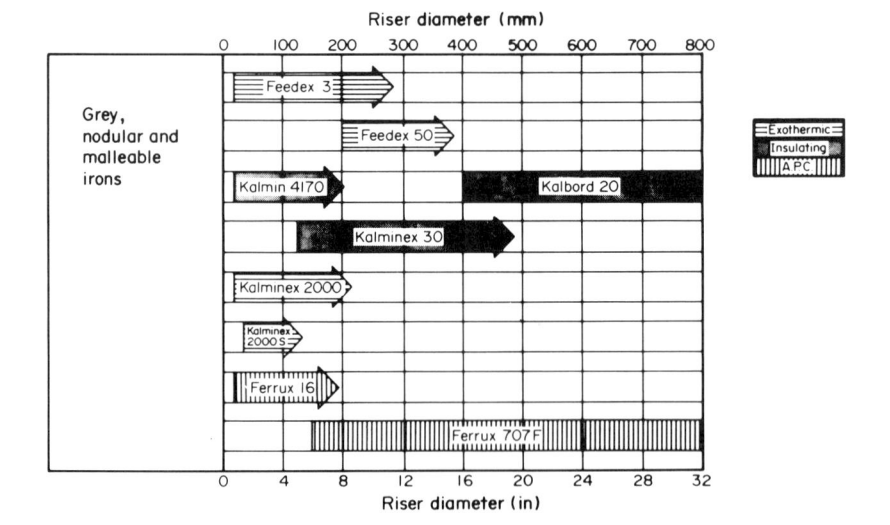

Fig. 9.25.

APPLICATION OF RISERS TO CASTINGS

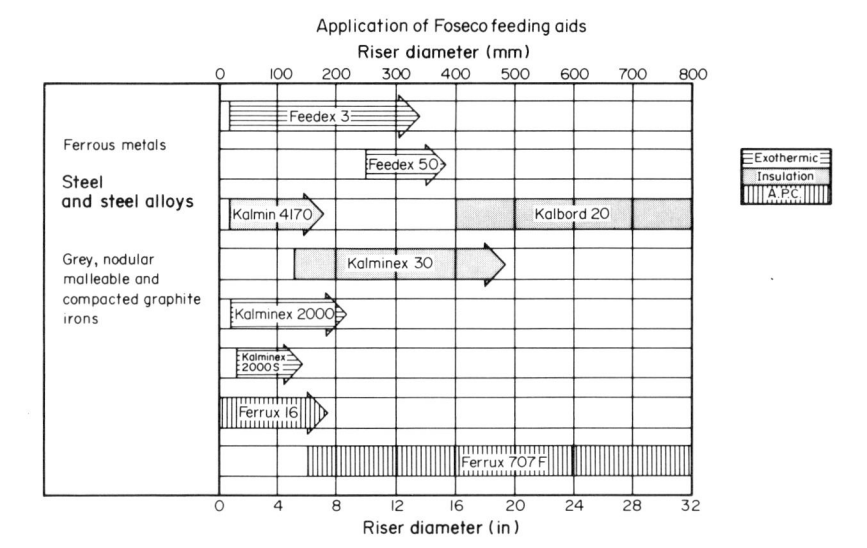

Fig. 9.26.

SECTION X
PRINCIPAL FOSECO PRODUCTS

The following sections list the main generic product names of the various current FOSECO (F.S.) Metallurgical Divison, United Kingdom materials for foundry use.

In practice and in almost every instance the generic name will be followed by a number, for example FEEDEX 3, COVERAL 11, etc. identifying the particular grade of that product for a specialised application. Further information on any product, including Health and Safety Precautions, can be obtained by reference to individual product literature available on request.

At the end of the section there is an indication of other Information Sheets, Wall Charts, etc. that are also available on request and which could prove helpful.

PRODUCTS FOR ALUMINIUM AND MAGNESIUM ALLOYS

ADAL Additive: Bagged or tabletted alloying additives: a range of metals is available.

COVERAL Flux: Covering, protecting and modifying fluxes for all aluminium alloys.

DEGASER Hydrogen Remover: Tabletted degassing agents for the removal of hydrogen from aluminium melts. Some varieties combine the action of degassing and simultaneous grain refinement.

DYCASTAL Addition: A powdered addition to molten aluminium that provides controlled gassing of the metal off-setting shrinkage unsoundness in gravity diecastings.

FEEDEX Exothermic Material: Mouldable or preformed exothermic shapes for application to risers to increase the feed potential thus permitting appreciable reduction in size.

FEEDOL Anti-Piping Compound: Similar to FEEDEX but a loose powder for application to riser surfaces to prevent heat loss to atmosphere.

INERTEX Powder:	A dusting compound for the protection of exposed metal surfaces during the melting and casting of magnesium alloys.
INSURAL Shapes:	Highly refractory non-wettable preforms for the handling and transport of molten aluminium. Mastic and adhesive is also available.
KALMIN 33 Sleeves etc.:	Fibrous highly insulating sleeves and shapes for lining risers, etc.
LOMAG Magnesium Remover:	For the removal of small unwanted amounts of magnesium from molten aluminium alloys.
MAGREX Flux:	Covering and cleaning fluxes for magnesium alloys. Provide effective protection during melting and efficient oxide removal.
NAMETAL Sodium:	Foil wrapped metallic sodium pieces for modification of the eutectic aluminium/silicon alloys.
NAVAC Sodium:	Vacuum processed sodium sealed in rigid pure aluminium containers. Free from oil, oxides and other gas producing materials.
NUCLEANT Grain Refiner:	Tabletted grain refining agents for aluminium and magnesium alloys including the hyper-eutectic aluminium/silicon series.
SIVEX Filters:	White rigid cellular ceramic filters for all aluminium and magnesium alloys. Sizes for all applications.
TILITE Grain Refiner:	Tabletted self sinking nucleating agent for large melts of wrought and cast aluminium alloys.

PRODUCTS FOR COPPER AND NICKEL BASE ALLOYS

ALBRAL Flux:	For removing oxides and cleaning alloys containing significant amounts of aluminium or silicon, e.g. aluminium bronze, high tensile and diecasting brass, etc.

CHROMBRAL Flux:	Provides a similar function as ALBRAL to a specific group of alloys containing chromium in alloying amounts.
CUPREX Flux:	A range of block and powder fluxes for alloys containing below 5% of zinc and requiring oxidising melting conditions to minimise hydrogen pick-up. For gunmetals, phosphor bronze, etc.
CUPRIT Flux:	As for CUPREX but provide reducing conditions for the brasses and alloys containing above about 5% of zinc.
DEOXIDISING TUBES:	A range of tubed deoxidants in various sizes to simplify controlled and accurate additions to different melt sizes.
ELIMINAL Flux:	For the efficient removal of contaminating amounts of aluminium and silicon thus improving fluidity and casting properties. Will also remove some zinc and tin if present.
FEEDEX 3 Exothermic:	A mouldable or pre-formed exothermic compound for lining riser cavities to improve feed potential. Very many patterns are available and significant economies can be shown.
FEEDOL Anti-Piping Compound:	An exothermic hot topping compound for application to riser surfaces to prevent heat loss to atmosphere.
INGOTOL Dressing:	Chill and ingot dressings to improve surface appearance of the billet, ingot, etc.
KALMIN 44 Sleeves:	Highly insulating sleeves for application to risers. Perform a similar function to FEEDEX but are insulating only in action.
LOGAS 50 Tablets:	A perforated tabletted degasser for plunging into copper and nickel base melts to remove dissolved hydrogen.

PLUMBRAL Flux:	Covering and cleaning fluxes for copper alloys containing a high lead content. Helps promote a uniform lead distribution.
RECUPEX Flux:	A range of fluxes for the recovery of fine or dirty scrap and swarf in reverberatory and crucible furnaces. High flux fluidity improves cleansing ability.
REGENEX Flux:	A powdered degassing medium for small copper base melts. Acts similarly to LOGAS 50. Useful for dirty charges.
SIVEX Filters:	Rigid cellular ceramic filters for in-mould filtration of metal as it enters the casting cavity.
SLAX Coagulant:	Slag coagulant and controller to give positive control of slag in furnace or ladle.

PRODUCTS FOR FERROUS METALS

BRIX Cupola Flux:	Supplementary cupola fluxing agents to increase carbon, control sulphur and ensure cleaner working and hotter metal.
CUPOLLOY Briquettes:	Silicon, manganese, chromium, molybdenum and phosphorus ferro-alloys available in fluxed briquette form for making alloying additions to cast irons.
FERAD Addition:	Controlled bismuth and boron additions for malleable iron. A rapid dissolving 4:1 ratio in tablet form for ladle addition.
FERROGEN Fluxes:	Fluxes and synthetics slags for cast irons to scavenge and clean molten metal and ladle/furnace walls.
FERROTUBES Deoxidants:	Degassing and deoxidising materials for flake irons in tube form.

INOCULIN, INOTAB and INOPAK Inoculants:	For the efficient ladle (INOCULIN) or in-mould (INOTAB OR INOPAK) inoculation of all cast irons.
KOMPAK Additive:	A specialised ladle treatment for the production of compacted graphite irons by a method not involving titanium and magnesium.
LADELLOY Additive:	A series of fluxed and graded additions to adjust the composition of cast irons in the ladle.
MSI 90 Inoculin Dispenser:	An automated inoculant dispenser for ladle lip or sprue/runner bush inoculation of cast irons: used in conjunction with INOCULIN 90.
NODULANT Nodularising Additive:	Granular and briquetted additives for the in-ladle production of spheroidal graphite (ductile) iron.
SEDEX Ceramic Filters:	Cellular foam filters for the efficient in-mould filtration of all types of cast iron.
SLAX Coagulant:	A range of slag coagulants for use in the ladle to collect and coalesce liquid slags before metal is poured.
SODA ASH Blocks:	Cupola or air furnace flux addition to adjust and control basicity of the slag and hence slag activity.
STELOGEN Deoxidant:	A deoxidant and killing agent for small steel melts to remove gas before casting.
STELORIT Fluxes:	Covering fluxes for plain carbon and alloy steels to ensure correct melting conditions for small melts.
TELLURIT Mould Paints:	Tellurium—containing paints for localised application on sand moulds or chills to produce an area of white iron in the casting.
TELLURIUM Tubes:	Metallic tellurium in tube packs for ladle addition to grey irons to ensure fully chilled (white iron) structures in the casting.

PRODUCTS FOR THE WHITE METALS

ZINCREX Flux:	Fluxes for cleaning and refining zinc and zinc base alloys, Mazak, Kirksite, Zamak, etc.
PLUMBREX Flux:	For scavenging and drossing off melts of lead and lead alloys, type and Babbitt metals, etc.
STANNEX Flux:	Specifically designed for the tin and tin base field to clean and remove non-metallics etc. into the dross.

MISCELLANEOUS PRODUCTS

Binders and Moulding Materials

AMBERSIL Release Agents:	Aerosol packed silicone release agents and breaking in compounds for foundry use.
ASKURE Catalysts:	A range of acids and acid mixtures for use as setting agents in the cold setting resin processes.
BENTOKOL Sand Additive:	Combined volatile/clay additives providing a one-shot addition to sands for cast iron work.
CARSET Setting Agents:	Ester-based setting agents for use with the self-setting silicate-based sand bonding processes.
CARSIL Binders:	A range of silicate based binders for the Carbon Dioxide Process.
CORFIX Adhesives:	Glues of all types for core assembly, shell moulding, stack moulding, etc.
CORSEAL Sealants:	Joint line fillers and core repair material to prevent internal flash and maintain casting contours.
DOW CORNING Silicone Release Agents:	Bulk silicone lubricating and release agents for foundry purposes.
ECOLOTEC, FUROTEC, FENOTEC and KOOLKAT Resin Binders:	A range of resin binders incorporating PF/UF/FA variations and mixtures for self-setting and cold box application.

GASBINDA Silicate Binders:	A range of straight and additioned silicates for the Carbon Dioxide and ester-setting processes.
HERCULITE Foundry Plaster:	An additioned refractory grade of foundry plaster for the manufacture of moulds, etc. for aluminium castings.
LUSCIN Sand Additive:	A carbonaceous green sand additive for the promotion of lustrous carbon when iron is cast into the moulds.
LUTRON Sand and Binder:	An oil-bonded waterless ready-to-use art sand for high definition aluminium and copper base castings. A binder is also available for mixing with own fine sand.
MIXAD Sand Additives:	A range of sand additives for preventing expansion defects, promoting optimum properties of clay bonds and providing an alternative to starch in steel sands.
PATTREX Plasters:	Hard pattern or stone plasters for the production of odd-sides, patterns, etc.
SEPARIT, SEPAROL and SEPRATEK:	Parting agents to ensure clean accurate lifts of green sand from the pattern. In powder, liquid and emulsion form.
SILISET Setting Agents:	Ranges of ester-based setting agents for the self-setting silicate process to cover most required setting speeds.
SOLOSIL Binder:	A complex additioned silicate for very high quality work requiring easy mixing, long shelf life and exceptional breakdown after casting.
STRIPCOTE Paint and Parting Agent:	One variety consists of a black plastic strippable protective paint for outside storage of patterns, dies etc. The other is an aluminised parting agent for difficult resin and silicate lifts.
TAK Sealing Compound:	Block and ready for use strips and reels of cope/drag sealing compound to prevent flash, run-outs, etc.

TERRADUST Casters Flour:	A brass casters flour to improve parting and surface finish of the casting.
VELOSET Setting Agent:	Gives the fastest possible set for the self-setting silicate-bonded process.

Coatings

ANISCOL Cleaner:	An acidic material for attacking and softening silicate-bonded gravity die coatings.
CHILCOTE Dressing:	Dressings for external metal chills to reduce adhesion to the casting and prevent blowing.
DYCOTE Dressing:	A very wide range of specialised dressings for gravity, low and high pressure diecastings.
FIRIT Dressing:	Silicate-bonded refractory dressings for tools, ladles, etc. to prevent attack by molten metal.
FRACTON Dressing:	Highly refractory dressings for brickwork, crucibles, launders, pig moulds etc. to promote clean parting of skulls, etc.
HARDCOTE Dressing:	Non-refractory dressings to harden the faces and edges of sand moulds to prevent drying out and crumbling.
HOLCOTE Dressing:	A series of semi-thixotropic water-based dressings in ready for use form for severe applications.
INGOTOL Dressing:	Applied to static metal moulds to improve the surfaces of copper-based billet, strip, etc. See also under copper-base section.
ISOMOL Dressings:	Spirit-based high solids content ready to use dressings for severe operating situations.
LUBIX Dressings:	Graphite filled greases principally for use on plungers of high pressure diecasting machines.
MOLCO AND MOLDCOTE Dressings:	Similar spirit-based mould dressings in paste or ready for use forms for average casting sections.

TELLURIT Dressing: A specialised coating—see under Products for Ferrous Metals.
TERRACOTE AND TERRAPAINT Dressings: Powders and pastes for mixing with water to produce conventional foundry coatings.
THINNERS Solvents: A range of various thinners and resin/solvent mixtures for use as thinners for spirit-based dressings.
TRIBONOL Coating: A specialised powder coating applied electrostatically.
ZEROTHERM Dressing: Non-flammable, spirit-based, rapid air-dry coatings.

Feeding Aids

FABREX Insulator: Insulating sleeves and shapes mainly for use in connection with continuous casting.
FEEDEX Exothermic: Highly exothermic mouldable powders or pre-formed shapes to improve feeding efficiency of lined risers.
FEEDOL and FERRUX Anti-Piping Compounds: Applied to the surfaces of non-ferrous (FEEDOL) and ferrous (FERRUX) risers to retain heat and improve feed.
KALBORD Insulator: Segmented board of highly insulating refractory for lining the riser cavities (above 300 mm dia.) of very large castings.
KALMEX and THERMEXO Powders: Strongly exothermic powders for application to risers and which produce molten steel at high temperature.
KALMIN Sleeves and Shapes: Pouring cups, botting cones, sleeves, slabs, etc. in fibrous insulating materials or self heating insulators.
KALMINEX Sleeves: Sleeves, domes and shapes in very accurately preformed shapes etc. to assist feeding of mass produced castings. They are strongly exothermic.
RADEX Ladle Cover: Highly insulating, expanding and strong exothermic covers to prevent heat loss from large ladles of metal.

PRINCIPAL FOSECO PRODUCTS 427

Molten Metal Transfer

INSURAL Shapes, etc:	A specialised pre-formed refractory material for aluminium transfer systems. See also under Products for Aluminium.
KALSEAL Refractory:	An air-drying high performance refractory grout and patching material originally developed for use with KALTEK (q.v.)
KALTEK Slabs and Ladle Liners:	One-piece or constructed disposable ladle liners for steel and cast iron applications ensuring clean ladles each time with temperature conservation.

In addition to the above products there are some items of equipment available as follows:

MSI 90 Inoculin Dispenser:	See under Ferrous Metal Treatment.
FLUX INJECTOR:	A machine for metering and sub-surface injection of a wide range of fluxes for the aluminium alloys.
LIQUIMIXER:	Two sizes of air operated paddle mixers for the accurate preparation and suspension of dressings, etc.
FEEDERCALC:	Computerised feeding calculation for ferrous and non-ferrous castings available as a complete programme subject to license.
FOSECO CALCULATOR:	Manual slide rule method of calculating required feeder sizes for castings.
FLUX GUN:	An air operated hopper fed flux gun for placing flux accurately on the metal surface and on the walls of large reverberatory furnaces. Very useful for furnace cleaning.

FOSECO PUBLICATIONS

Several of the leaflets and other publications below are available also in Bulgarian, Czech, French, German, Hungarian, Italian, Japanese, Portuguese, Rumanian, Russian, Spanish and Swedish and may be obtained from the FOSECO representatives in these countries.

LEAFLETS

Leaflets describing the principal FOSECO products and giving full details of the various grades of each, with detailed instructions for their use, are available, free on request. Useful tables, technical data, and in some instances case histories based on practical experience, are also included.

INFORMATION SHEETS

A series of Information Sheets, giving authoritative and up-to-date information and technical data on a wide variety of foundry and allied subjects, is available free, on request.

Information
sheet No.   FERROUS

48	Nodular Graphite Cast Iron
49	Malleable Cast Iron
52	Inoculated High Duty Grey Cast Iron
54	High Duty Cast Iron
62	Silicate Bonded Sand
63	The Production of Aluminium Matchplates in Herculite Plaster FG
64	Rimming Steel
66	Cupola Operation Data
67	The Application of Feedex Exothermic Material to Castings

Information
Sheet No.   NON-FERROUS

9	Aluminium Bronze
14	High Tensile Brass Alloys
35	Notes on High Lead Content Bronzes
43	High Conductivity Copper
44	Commercial Copper
46	Copper Nickel Alloys
53	Degassing with Lithium
56	Copper-Zinc Alloys for Castings and Die-casting
57	Copper-Tin Alloys—Bronzes and Gunmetals
58	Phosphor Bronze Alloys
62	Silicate Bonded Sand
63	The Production of Aluminium Matchplates in Herculite Plaster MC
67	The Application of Feedex Exothermic Material to Castings
68	Zinc, Tin and Lead

LIGHT ALLOYS

16	Aluminium-Silicon Alloys (11–13% Si)—LM6 and LM20 (Recommendations apply also to LM2 and LM9)

17	Aluminium-Silicon Alloys (3:5-11.0% Si)—LM25 (Recommendations apply also to LM8 (obsolete), 16, 18 and to Aerospace alloys DTD716A, 722A, 727A and 735A)
33	Aluminium-Magnesium Alloy—LM10
39	Aluminium-Silicon-Copper Alloy—LM4 (Recommendations apply also to LM22)
41	Aluminium-Magnesium Alloy—LM5
51	Aluminium-Copper Alloy (4-5% Cu)—LM11 (obsolete) (Recommendations apply also to BS L91, L92, DTD 361B)
55	Aluminium-Copper-Silicon-Zinc Alloy—LM21
60	Frontier 40E Alloy
61	Aluminium Piston Alloys (Alloys covered include LM12, LM13, LM14 (obsolete), LM15 (obsolete), LM26, LM28, LM29 and LM30: also Aerospace alloys L35 and L52)
62	Silicate Bonded Sand
63	Herculite Metal Casting Plaster
67	The Application of Feedex Exothermic Material to Castings

CHARTS

Three Scrap Diagnosis Charts (Ferrous, Non-Ferrous and Light Alloy) listing common casting defects, their appearance, cause, remedy and Foseco products recommended.

A coloured wall chart "The Cupola", shows in very full detail the layout and working principle of the cupola. It is particularly useful for instructing pupils, trainees and foundry personnel generally.

Data charts are available for Iron, Copper Base and Aluminium respectively, giving information on physical properties, melting recommendations and other details for the principal British casting alloys and their equivalents. A further chart describes mould coatings for all types of castings.

INDEX

Acid resisting cast irons, 321
Air consumption of various pneumatic tools, 50
Alloying elements—Influence—on cast iron structure, 265
Alloying—Order of, 200
Alloys—Casting, shrinkage and contraction of, 48–49
Alloys and metals—Corrodibility of, 12–17
Alpax (see Aluminium—Silicon alloys)
Alternative Moulding systems, 109
  V Process, 109
  F Process, 109
  Full Mould, 109
  Fluid Sand, 109
  Randupson, 109
  Replicast, 109
Aluminium alloys for castings—Code colours, 139
Aluminium alloys—Grain refinement, 192
Aluminium alloys—Melting, fluxing and degassing procedures—
  BS 1490 LM4, 139–180
  BS 1490 LM5
  BS 1490 LM6
  BS 1490 LM10
  BS 1490 LM13
  BS 1490 LM16
  BS 1490 LM21
  BS 1490 LM22
  BS 1490 LM24
  BS 1490 LM25
  BS 1490 LM27
  Frontier 40E
Aluminium alloys—Gravity die-casting, 149, 166, 180, 340
Aluminium alloys Hypereutectic, refinement of, 152
  compositions, etc., 157
Aluminium alloys—Pressure die-casting, 322–328, 348
Aluminium Bronze—Gravity die-casting, 344
Aluminium Bronze—Melting, fluxing and degassing, 251
Aluminium Casting alloys—LM series, 126
  composition, etc., 127
  aerospace series, 136
Aluminium casting alloys—Specifications and properties, 128–131
Aluminium—Effect on structure of cast iron, 265
Aluminium—Removal from brass and bronze alloys, 227
Aluminium scrap, swarf, etc.—Reclamation, 190

Aluminium-silicon alloys—Melting, modification, etc., 141
Aluminium alloys—Casting characteristics, 127
Aluminium alloys—Related specifications, 132–133
Areas and circumferences of circles, 42–44
Art castings—Sand mixtures, 74
Atmospheres—Oxidising and reducing, 210
Atomic weight of metals, 10–11
Austenitic cast iron, 317

Blackheart malleable cast iron, 287
Boiling point of metals, 10–11
Botting mixtures for cupola tap hole, 59–60
Botting practice, 59–60
Brass—High tensile, 250
Brass—Pressure die-casting, 351
Brass—Gravity die-casting, 225, 345
Brasses—Copper-zinc alloys, 223–229
  Alpha-beta alloys
  Die-castings
  Effects of added elements
  Mechanical properties—sand cast
  Melting technique
  Mould dressing and sand additions
  Moulding practice
  Removal of aluminium
  Removal of silicon
  Running, gating and feeding
  Test Bars
Brazing metal—Melting and fluxing procedure, 247
Breakdown agents, CO_2 Process, 94
Breaker core dimensions, 384
British Standard Specifications
  Aluminium Alloys, 139–180
  Cast Irons, 274
  Copper base alloys, 202–206
  Grey cast iron, 274, 318
  Malleable cast iron, 286
  Nodular graphite cast iron (S.G. iron), 295
  Phosphor Bronze, 202
  Silicon Bronze, 249
  Steel, 356
Bronzes and Gunmetals—copper tin alloys, 230–237
  Gravity die-casting
  Melting practice
  Metal—mould reaction
  Pinholes
Bronze stick casting—Solid, cored and continuous, 238

431

INDEX

Calcium boride, 218
Carbon-dioxide process, 87 et seq.
Carbon equivalent of cast iron, 280
Carbon—influence on cast iron, 259
Cast alloys—Contraction and shrinkage of, 48
Casting defects—cause and cure, 310
Casting weight from pattern weight, 47
Cast iron—Classification of graphite size and shape for flake and S.G. irons, 281
Cast iron—Influence of alloying elements on structure, 265
Cast iron—Influence of normal constituents, 259-262
Cast iron—Malleable, 286
Cast iron—Nodular graphite—British Standard Specification, 295
Cast iron rolls—Thickness of chills for, 53
Cast iron—Relationship between CE and Tensile strength, 280
Cast iron for resisting corrosion, 321
Cast iron for resisting heat, 315
Cast iron for resisting wear, 326
Cast iron—Typical compositions for various castings, 272
Cast iron scrap—Size and weight for cupolas, 257
Ceramic foam filters, 62-66
Chills for chilled rolls, 53
Chromium in cast iron—Effect on structure, heat, wear and acid resistance, 317, 322, 326
Circles—Table of areas and circumferences of, 42-44
Coal dust in iron foundry sand, 71
Cold Box systems, 84
Coefficient of linear expansion for metals, 11
Coke—Composition for cupola, 258
Commercial copper castings, 215
Compacted Graphite (C.G.) cast iron, 305
Composition of cast iron for different castings, 272
Control tests for sand, 72
Conversion—Brinell hardness to Vickers, Firth, Rockwell, etc., 51
Centimetres to inches, 18
Cubic centimetres to cubic inches, 21
Density to Baume and Twaddell, 39
Factors, 8-9
Gallons U.S. to Imperial, 30
Grammes to ounces (avoirdupois), 22
Kilograms to pounds, 23
Kilograms to long tons, cwts., etc., 27
Kilograms to short tons, cwts., etc., 28
Kilos/sq cm to lb/sq inch, 34
Kilos/sq mm to long tons/sq inch, 32

Kilos/sq mm to short tons/sq inch, 33
Litres to Imperial gallons, 29
Litres to U.S. gallons, 30-31
Long tons to short tons, 24
Metres to feet, 19
Metric tonnes to long tons, 25
Metric tonnes to short tons, 26
Square centimetres to square inches, 20
Temperature Centigrade to Fahrenheit, 36-38
Continuous casting—Bronze, 239
Converter—Lining mixture for acid steel, 78
Copper base alloys—Density and thermal expansion, 210
Copper base alloys—Melting and fluxing procedure, 210 et seq., 244
Copper castings—Moulding procedure, 221
Copper—Effect of impurities and trace elements on electrical conductivity, 215
Copper—Effect of structure on cast iron, 245
Copper—HC and Commercial—Melting and fluxing procedure, 215-221
Copper—nickel alloys—Melting and fluxing procedure, 251-252
Core coatings, 114
Proprietary coatings, 123
Corrodibility of some common metals and alloys, 12-17
Corrosion resisting cast irons, 321
Cupola—Botting clay and practice, 59-61
Cupola charge calculations, 267-268
Cupola charge materials—Composition of, 256
Cupola—Melting gains and losses, 258
Cupola—Operation data, 263-264

Decimal and metric equivalents of fractions of inches, 40-41
Defects in castings—cause and cure, 310
Density of metals, 18-19
Deoxidants for copper-base alloys, 253
Die-casting—Gravity, 340 (see also under appropriate alloy)
Die-casting—Pressure, 348 (see also under appropriate alloy)
Die steels—Brinell hardness, 349
Die steels—Composition of, 341, 349

Electrical conductivity of copper—Effect of impurities and elements, 215
Electrical resistivity of metals, 11
Elektron (see under Magnesium base alloys)
Equivalents—Metric and Decimal of Fraction of 1 inch, 40-41

INDEX

Expanded Polystyrene Systems—Replicast processes, 108
Exothermic feeder sleeves, 388

Feedercalc computerised heading systems, 401
Feeder sleeves, 379, 390
Filtration of molten metal, 62
  filters, 63–66
Fineness number for sand, 86
Flux injection (Al. alloys), 195
Flux washing (Al. alloys), F.I.L.D. Process, 193
Foseco products—List of principal, 418
Foseco publications, 428–429
Fractions of inches with decimal and metric equivalents, 40–41
Frontier 40E—Aluminium alloy, 176
Furan systems, 82
Fusible alloys, 46

Gating terms and methods, 55–56
German silver (see nickel silver)
Grain refinement—Aluminium alloys, 192
Grain refinement—Magnesium alloys, 180
Graphite—size and shape classification in cast irons, 281
Gravity die-casting, 340 (see also under appropriate alloy)
Guide to Feeder sleeve dimensions, 390
Gunmetal castings—Pickling, 243
Gunmetal—Gravity die-casting, 346
Gunmetal—Tin bronze—Melting and fluxing procedure, 212, 230, 247
Gunmetal castings—Physical properties, 208, 210

Hardness conversion chart, Brinell, Vickers, Firth Scleroscope and Rockwell, 51
Hardness of Die steels, 349
Heat resisting cast irons, 315
High Conductivity Copper, 215
  Alloys, 216
  Moulding, 221
High tensile brass (manganese bronze) melting and fluxing procedure, 250
Hot Box systems, 85
Hydrometer conversion table, 39
Hyper-eutectic aluminium-silicon alloys, 152

Inhibiting elements—S.G. and C.G., 298
Inhibitors, sand, 164, 170, 184
Inoculated grey cast iron, 329

Insulating feeder sleeves, 393
International System of Units, 2–7
Iron castings—Chemical composition for various, 272

Ladle lining—Cold lining Systems, 58–59
  Crane ladles, 57
  Steel, 78
Ladles—Method for finding weight of metal contained in, 53–54
Latent heat of fusion of metals, 10
Lead alloys—Gravity die-casting, 346
Lead alloys—Pressure die-casting, 352
Lead bronze—Melting, fluxing and degassing, 213
Limestone, 258
Lining for bottom poured steel ladles, 78
Lining mixtures for pouring ladles and hand shanks, 57
Lining mixtures—Side blown converter for acid steel, 78
Lithium, 217, 345

Magnesium base alloys—Gravity die-casting, 186, 347
Magnesium base alloys—Melting, fluxing and grain refining, 180–190
Magnesium base alloys—Pressure die-casting, 189, 350
Magnesium—Removal from aluminium alloys, 190–191
Malleable cast iron—British Standard Specifications, 286
Malleable cast iron, 286
  Chemical composition
  Mechanical properties
Manganese—Influence on cast iron, 259, 261, 265
Material—Weight in lb per cubic foot, 45
Mazak (see under Zinc base alloy)
Melting gains and losses in the cupola, 258
Metals and alloys—Relative corrodibility of, 12–17
Metal-mould reaction, 163, 170
Melting point of metals, 10–11
Metals—Tables of physical properties, 10–11
Methods of gating, 55–56
Metric equivalent with decimals and fractions of inches, 40–41
Modification of aluminium-silicon alloys, 141, 151
Modulus, 396
Molybdenum—Effect on structure of cast iron, 265
Monel (see under Copper-nickel alloys)
Mould and core coatings, 114

434 INDEX

Nickel brass—Nickel bronze—Nickel silver (see under Copper-nickel alloys)
Ni-hard—Composition and properties, 327
Ni-Resist—Chemical composition, 317, 322
Nodular iron (S.G.) British Standard Specification, 295
Non-Ferrous casting alloys—British Standard and allied specifications, 202-209
Non-Ferrous metals and alloys—Melting and fluxing techniques, 212 et seq.
Aluminium bronze
Brass alloys—Die cast
Brass alloys—Sand cast
Brasses, Commercial
Colour coding
Copper, High conductivity
Copper—nickel alloys, nickel bronze, nickel silver, nickel brass
Gunmetal (tin bronze) alloys
Manganese bronze (high tensile brass)
Phosphor bronze
Silicon bronze

Oil sand mixture for cores, 77

Patching for cupola lining, 312
Patternmakers contraction for casting alloys, 49
Pattern weight to weight of casting, 47
Phosphor bronze—Melting and fluxing procedure, 229
Phosphorus—Influence on cast iron, 239, 242
Physical properties of metals, 10-11
Pickling of brass and gunmetal castings, 243
Pickling of iron castings for sand removal, 314
Pig iron—Typical compositions of, 256
Pneumatic tools—Air consumption, 50
Pouring ladles and shanks—Lining mixtures for, 57
Pouring temperature (see melting and fluxing procedure for metal or alloy concerned)
Pressure Die-casting, 348
Properties of metals, 10-11

Ramming mixture and patching for cupola lining, 338
Reclamation of aluminium, 190
Reclamation of copper base scrap, swarf, etc., 242
Reclamation, sand, 106
Reducing and oxidising atmosphere, 210

Refractory Linings, 78
Replicast expanded polystyrene systems, 108
Resin bonded sand
Furan, 82
Cold Box, 84
Hot Box, 85
Risers—Application of sleeves to castings, 378
Risers—Weight of metal contained in, 385-387
Rolls—Cast iron, 53

Sand—Additives, 71
Sand—Coal dust in, 71
Sand—Control tests, 72-73
Sand—for heavy steel castings, 76
Sand facing—Green for iron castings, 70
Sand mixture for art castings, waterless sand, 74
Sand mixtures for copper base alloys, 74-76
Sand mixture for oil sand cores, 77
Sand mixtures—Moulding and core for steel castings, 76
Sand mixtures—Synthetic for iron castings, 71
Sand test data for castings in different alloys, 70
Sand fineness number and calculation, 86
Sand Resin bonded, 82 et seq.
Silicate bonded, 87 et seq.
Sand sieve grading, British Standard Sieves, 72
Scrap aluminium—Reclamation as ingot metal, 190
Scrap diagnosis—Cause and cure, 310
Seger cones—Bend temperature, 52
Shrinkage and contraction of casting alloys, 48-49
SI—International System of Units, 2-7
Sieves—Comparison of standard sieves, 79-81
Silicate sand—Self setting and CO_2 Process, 87 et seq.
Silicon bronze—Melting and fluxing procedure, 249
Silicon—Influence on cast iron, 315, 323
Silicon—Removal from brass alloys, 228
Silumin (see under Aluminium alloys)
Slush casting, 344
Sodium for modification of aluminium-silicon alloys, 143
Solders and fusible alloys, 46
Specific heat of metals, 10
Spheroidal graphite cast iron, 295, 308, 319
Steam reaction—Copper, 217
Copper base alloys, 231

INDEX

Steel castings—Moulding and core sand mixtures, 76
Steel castings—American and British Specifications, 356
Steel castings—application of Feedings aids, 395
Steel scrap—Size and weight for cupola, 258
Stress values—Kilos/sq mm to tons/sq inch, 35
Sulphur control in cupola melting, 269
Sulphur—influence on cast iron, 269

Taphole or botting clay mixture for cupola, 59
Tellurium chill inducing agent for cast iron, 328
Temperature conversion tables, 36–38
Test Bar Patterns, 159, 160, 175, 229, 240, 241
Thermal conductivity of metals, 11
Tin—Gravity die-casting, 347
Tin—Pressure die-casting, 352
Titanium—Effect on structure of cast iron, 266

Transverse rupture stress for cast iron, 277

Vanadium-effect on structure of cast iron, 266

Water-less Sand—LUTRON, 74
Wear resisting cast irons, 326
Wedge test for cast iron, 334
Weight of casting from weight of pattern, 47
Weight of molten metal in pouring ladles, 53–54
Weight of materials in lb per cu ft, 45
Weight per cu cm of different metals and alloys, 54
Whiteheart malleable cast iron—British Standard Specifications, etc., 287

Zinc base alloys—Pressure die-casting, 351
Zirconium—Effect on structure of cast iron, 266